FRACTIONAL KINETICS IN SOLIDS

Anomalous Charge Transport in Semiconductors, Dielectrics and Nanosystems

FRACTIONAL KINETICS IN SOLIDS

Anomalous Charge Transport in Semiconductors, Dielectrics and Nanosystems

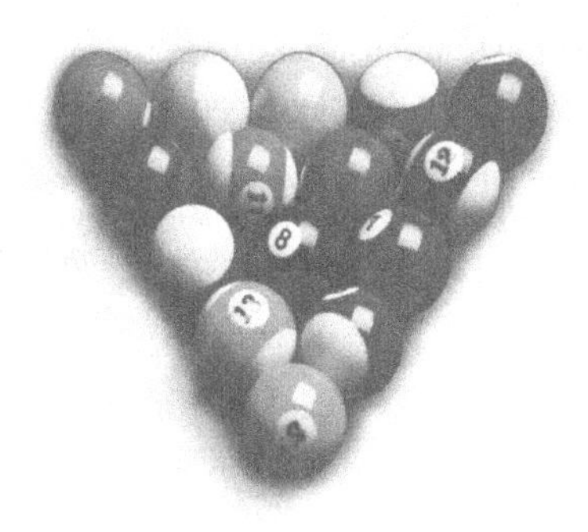

Vladimir Uchaikin
Renat Sibatov

Ulyanovsk State University, Russia

World Scientific

NEW JERSEY • LONDON • SINGAPORE • BEIJING • SHANGHAI • HONG KONG • TAIPEI • CHENNAI

Published by

World Scientific Publishing Co. Pte. Ltd.
5 Toh Tuck Link, Singapore 596224
USA office: 27 Warren Street, Suite 401-402, Hackensack, NJ 07601
UK office: 57 Shelton Street, Covent Garden, London WC2H 9HE

British Library Cataloguing-in-Publication Data
A catalogue record for this book is available from the British Library.

FRACTIONAL KINETICS IN SOLIDS
Anomalous Charge Transport in Semiconductors, Dielectrics and Nanosystems

ISBN 978-981-4355-42-1

Printed in Singapore.

Contents

Preface

There exists a few excellent books introducing in fractional calculus and demonstrating its applications in various fields of science. Nowadays, monographs of other kind begin to be published, which is devoted to interpretation of some fixed field of science by using fractional calculus as a convenient and efficient tool. So, the book [Mainardi (2010)] is completely devoted to viscoelastic media. Similarly, our book is aimed at electronic processes in solids and uses fractional calculus as means of investigations.

At the present time, there is accumulated a large amount of information on charge carriers transport (relaxation) processes in disordered solids – dielectrics, semiconductors, quantum dots and their systems. In these materials, various transport modes are observed. Many of them having ordered (regular, homogeneous) structures reveal the normal mode characterized by Gaussian statistics and described by the standard diffusion equation, others do not. Among anomalous (non-Gaussian) transport regimes, the dispersive transport stands out especially [Scher & Montroll (1975); Zvyagin (1984); Madan & Shaw (1988); Bässler (1993); Tyutnev et al. (2005); West (2006); etc.]. This type of transport is observed in various disordered materials, such as amorphous semiconductors, porous solids, polycrystalline films, liquid-crystalline materials, polymers, etc. The main features of dispersive transport were ascertained from time-of-flight measurements of drift mobility of charge carriers in amorphous semiconductors. The transient current curves $I(t)$ for dispersive transport manifest the so-called *universality*, that is, similarity in the $\log I - \log t$ coordinates. These curves essentially differ from the "normal" step-wise curves corresponding to the Gaussian transport. They decay as very stretched function with two power law sections: $I(t) \propto t^{-1+\alpha}$ for $t < t_T$, and $I(t) \propto t^{-1-\alpha}$ for $t > t_T$. The parameter $\alpha \in (0, 1)$ called the dispersion parameter turns out to be the order of fractional time-derivative in new equations describing such systems.

To describe special characteristics of the dispersive transport, some authors (see [Arkhipov et al. (1983); Nikitenko (2006)]) operate with time-dependent diffusion coefficient and mobility. The fractional approach described in this book is an alternative description of the kinetics. These approaches are not equivalent and gives comparative analysis of their capabilities in description of processes governed by dispersive transport in structures based on disordered semiconductors is of great interest.

It is important to note that dispersion parameter α, extracted from the time-of-flight experiments for some materials (i.e. porous semiconductors) is weakly dependent on temperature (see, eg, [Blom & Vissenberg (1998); Rao et al. (2002)]). Many authors attribute this fact to the topological disorder, but not to the energetic one. The Arkhipov-Rudenko diffusion equation for the multiple trapping mechanism, adapted for phonon-assisted hopping in Nikitenko's papers, is not applicable in the case of a weak temperature dependence of α and the strong structural disorder. The equation does not take into account the percolation nature of the trajectories of hopping conductivity and their sharp distinctions from the Brownian ones. Relationship between fractional equations and subdiffusion models of percolation allows us to hope for progress in this direction in frameworks of the fractional differential approach.

It is well known, that most disordered dielectrics (polymers, biomolecules, colloids, porous materials, doped ferroelectric crystals, etc.) also manifest anomalous kinetics which are expressed in the non-Debye law of relaxation. This fact has been recognized over a hundred years ago by Curie and von Schweidler [Curie (1889); von Schweidler (1907)] and continues to attract attention of researchers throughout the century. The Jonscher analysis [Jonscher (1986)] of numerous experimental data led him to the conclusion that an explanation for the deviation from the pure Debye law should be sought not in terms of relaxation times distribution, but rather as the manifestation of some common law named the *universal relaxation law*. Nevertheless, many others believe that there is indeed a physical cause for a distribution of relaxation times, and occasionally suggestions have been made relating to certain particular cases, e.g. those concerning the occurrence of protonic resonance made by Kliem and Arlt (1987). Apparently, von Borgnis (1938) was the first who used fractional derivatives (but without using this term) in electrodynamics of solid dielectrics.[1] Beginning since sixteens [Nigmatulin & Vyaselev (1964); Nigmatullin & Belavin (1964)], the fractional calculus has been more and more widely embedded into description of the non-Debye relaxation processes [Nigmatullin (1986); Westerlund (1991); Nigmatullin & Ryabov (1987); Mainardi & Gorenflo (2000); Hilfer (2000a); Novikov & Privalko (2002); Coffey et al. (2002); Aydiner (2005); Novikov et al. (2005); Uchaikin & Uchaikin (2007); Tarasov (2008a)].

The fractional calculus allows to describe anomalous relaxation in disordered solids and normal relaxation in simple systems in the framework of a unified approach. Fractional calculus is well developed (see monographs [Oldham & Spanier (1974); Samko et al. (1987); Podlubny (1999); Nakhushev (2003); Pskhu (2005); Kilbas et al. (2006); Uchaikin (2008); Das (2008); Mainardi (2010); Tarasov (2010); Herrmann (2011)]) and has many physical applications.

[1] Two years earlier, fractional derivatives were used by Gemant (1936) in continuum mechanics of viscoelastic materials.

Disorder due to molecular chaos and the lack of a perfect local order motivates going beyond the classical theory based on the Central Limit Theorem and referring to its generalized counterpart developed by P. Lévy. It is known that a consistent statistical analysis of the dynamic equations by reducing the number of variables in the Liouville equation with the use of the Zwanzig-Mori projection operators, or similar methods (method of Bogoliubov, Zubarev's statistical operator, the formalism of Green's functions, the Lee recurrence relations and so on) shows the effects of nonlocality and memory (hereditary) as a result of the presence of hidden variables. The combination of such a generic statistics property as *self-similarity* with integral representations of spase- and time-nonlocality of dynamic systems that forms that ground on which the fractional differential kinetic theory rises (see for detail, [Uchaikin (2012)]).

Inverse-power waiting time distributions we are talking about produce random processes characterized by presence of a special property: intermittency. In particular, this means that long random gaps in the pulses flow possessing the property will be observed at any scales: passing to larger scales does not make the observed flow homogeneous. This can be referred to as a time disorder. A spatial disorder, often expressed in terms of fractal concepts, play also an important role, especially in mesoscopic systems. Randomly distributed heterogeneity and quantum interference effects lead to fluctuations in the conductance g in the ensemble of mesoscopic samples. The distribution of the conductance is usually studied in three different modes: metal (when the localization length ξ is much larger than the characteristic size of a system L), dielectric ($\xi \ll L$) and intermediate ($\xi \sim L$). The most detailed theoretical results were obtained for one-dimensional and quasi-1D systems in the two approaches to the description of the coherent conductance and localization in disordered wires: field theory and transfer matrix theory. Frahm (1995) and Brouwer (1996) have shown their equivalence in the case of a large number of transverse modes of $N \gg 1$ for all symmetry classes. In the transfer matrix theory, electron transport is seen as a scattering problem. The conductance at zero temperature is associated with the quantum-mechanical transmission matrix of $\mathbb{T}$ by the Landauer formula. Dorokhov, Mello, Pereyra and Kumar [Dorokhov (1982); Mello et al. (1988)] derived an equation, known as DMPK-equation for the distribution of the $\mathbb{T}$-matrix eigenvalues in the weak localization regime. Muttalib and Klauder [Muttalib & Klauder (1999)] proposed a generalized DMPK-equation for the description of electron transport in highly disordered systems. They used the assumption that the small variations of the scale parameters of the system (e.g., length) lead to the small variations of the transfer parameters. However, this assumption does not always hold for the strong localization. For example, it is known that the small fluctuations of the potential barrier parameters for quantum tunneling can lead to very large fluctuations of its tunneling transparency. The effect is also manifested at mesoscopic scales, for example, in the dispersive transport in disordered semiconductors. If one assumes that the fluctuations of the transport parameters are self-similar, then it is possible to avoid the expansion over a small parameter and to generalize the DMPK method to the case of the strong localization and higher dimensions within the fractional differential approach. It should also be noted, that equations of the Fokker-Planck type in both cases (weak and strong localization) were obtained for a regular distribution of inhomogeneities. However, experiments [Kohno & Yoshida (2004); Kohno (2008); Hegger et al. (1996)] and numerical simulations [Leadbeater et al. (1998); Amanatidis et al. (2012); Fernández-Marin et al. (2012)] show that the disorder in mesoscopic systems can be of a fractal (self-similar) type.

Interest in non-Gaussian transport theory has recently revived in connection with anomalous relaxation-diffusion processes observed in nanoscale systems: nanoporous silicon, glasses doped by quantum dots, quasi-1D systems, and arrays of colloidal quantum dots. These systems are very promising for applications in spintronics and quantum computing. They can also be useful for studying the fundamental concepts of physics of disordered solids: localization, nonlinear effects associated with long-range Coulomb correlations, occupancy of traps and Coulomb blockade. Due to the preparation method of colloidal nanocrystals, the energy disorder is always presented in these systems, which is confirmed by experiments on fluorescence blinking of single quantum dots (CdSe, CdS, CdSe/ZnS, CdTe, InP, etc). As shown in some recent papers [Novikov et al. (2005); Sibatov (2011b)], the Lévy statistics plays a crucial role in the interpretation of experiments with charge transfer in QD arrays.

In many samples of colloidal QD arrays, the power-law decay $t^{-\nu}$ $(0 < \nu < 1)$ of current response to a step voltage is observed. The exponent ν depends on nanocrystal size and temperature. Novikov et al. (2005) argue that the observed current is the conduction current, rather than the bias current, since the integral of the current (the charge) diverges. Nonexponential relaxation of the current can be explained by the time dependence of the system. Ginger and Greenham (2000) proposed that the flux of charge decreases due to charge injection from the suppression of contact. This suppression is due to the effect of the Coulomb blockade by electrons trapped in the nanocrystals. In Ref. [Morgan et al. (2002)], the power-law decay of $I(t)$ is explained by the representation of non-equilibrium population of electrons distributed over the OD array in the form of the Coulomb glass.

Whatever it was, these phenomena exhibit the properties postulated by the universal relaxation law. Combined with some other physical proposition, this law can form a basis for phenomenological approach to the processes considered in this book. Let us list the main motivations for this approach:

(1) the fractional kinetics belongs to the domain of influence of the universal relaxation law;
(2) fractional kinetic equations are connected with the known models of random processes and limit theorems of the probability theory;
(3) one can create the unified formalism describing the normal and anomalous kinetics;

(4) it is possible to take into account energetic and topological types of disorder in common.

Various systems evolute in a similar way. We observe here the situation similar to that taking place in the Central Limit Theorem, conclusion of which almost does not depend on details of "microscopic" distributions except their mean values and variances. However, the form of these distributions and the law of their spreading differ from those given by this theorem. The long heavy tails observed in these processes make leaving the scope of classical statistics (the Central Limit Theorem, Gaussian and Poisson random processes, the classical kinetic and diffusion schemes) and refer to its generalized counterpart including the Generalized Limit Theorem, Lévy walk and fractional Poisson process, fractional versions of kinetic and diffusion equations. The totality of these concepts together with correspondent equations and methods of their solution developed in the fractional calculus forms what we call the *fractional kinetics* and consider as an essential part of the anomalous kinetic theory (see [Uchaikin & Zolotarev (1999); Chandre et al. (2008); Luo & Afraimovich (2010); Klages et al. (2008); Baleanu et al. (2012)]).

In the theory of charge transport in semiconductors, the fractional derivatives were used for the first time in [Arkhipov et al. (1983); Babenko (1986); Tiedje (1984)]. Babenko used them to find time dependence of concentration at the boundary of p-n junction during normal transport at a given current density by means of factorization of the normal diffusion operator. It should be noted that the integro-differential operator with power-law kernel was used in the theory of dispersive transport in Ref. [Arkhipov et al. (1983)] in 1983; the authors expressed the relationship between concentrations of free and localized carriers through the fractional integral (non-explicitly). In other papers (see for example [Arkhipov & Rudenko (1982); Emelyanova & Arkhipov (1998)]), they chose to use a different approximate relation between concentrations of localized and free carriers, which they called "the master equation of dispersive transport". This relation is believed to hold for any density of localized states and permits expressing results through elementary functions in the case of an exponential density. The Arkhipov-Rudenko master equation leads to the diffusion equation with time-dependent diffusion coefficient and mobility [Arkhipov et al. (1983)].

Based on kinetic trapping-emission equations written in [Noolandi (1977)], Tiedje (1984) derived a transport equation neglecting diffusion in terms of Laplace integral transformation. The inverse Laplace transform of this equation represents a fractional equation [Sibatov & Uchaikin (2007)].

Barkai (2001) made use of the Fokker-Planck fractional differential equation proposed by Metzler, Barkai, and Klafter (1999) to account for transient photocurrent relaxation in amorphous semiconductors. He showed the agreement between selected results obtained by the fractional differential approach and predicted by the Scher-Montroll model (1975). Barkai (2001) justified introduction of the fractional Fokker-Planck equation as follows: "Transport in ordered media is often modeled using diffusion equation, this approach being the simplest and most widely used. Dispersive Scher-Montroll transit time type experiments, observed in a large number of disordered systems, can be described phenomenologically using the fractional Fokker-Planck equation. This is only one example of physical phenomena in which a different type of calculus, i.e., noninteger calculus, plays a central role." Another fractional diffusion equation was proposed by Hilfer (2000b) who simply replaced of the first time derivative by the fractional one. Bisquert (2003, 2005) used it to describe the multiple trapping and to explain relaxation of luminescence in semiconductors with exponential density of localized states.

Power-law decay of photoluminescence in amorphous semiconductors was described in Refs. [Seki et al. (2003, 2006)] based on the generalized random walk model with recombination by tunnel radiative transitions. Recombination was limited by dispersive diffusion of the carriers. In the framework of this model, Seki et al. (2006) compiled a fractional differential equation for the first passage time distribution density. The recombination rate was found by using the integral Laplace transform of this equation.

As shown in Refs [Uchaikin (1999); Uchaikin & Sibatov (2007)] the main asymptotic terms of solutions to random walk equations in the Scher-Montroll model are solutions of fractional differential equations, the Green functions of which are expressed through fractionally stable densities. Solution of the equation proposed in Bisquert's paper [Bisquert (2003)] was found in terms of stable densities [Uchaikin & Sibatov (2005)].

The latter authors have shown that fractional differential kinetics is a consequence of dispersive transport self-similarity [Uchaikin & Sibatov (2007)]. The Fokker-Planck fractional differential equation applied by Barkai (2001), the equation for delocalized carriers used in [Bisquert (2003)], and the generalized equation of Ref. [Sibatov & Uchaikin (2007)] are related via the expression obtained by Arkhipov, Popova, and Rudenko [Arkhipov et al. (1983)].

Listing many problems connected to charge kinetics in disordered solids, we do not commit ourselves to solve all of them in this book. Our aim is more modest: we want to show how the fractional calculus works in this field believing that its dissemination will have a significant impact on the further development of the electrodynamics of disordered media.

Chapter 1 provides an overview of statistical grounds of the fractional kinetic theory. Interplay between the Lévy statistics, fractional calculus and anomalous transport is demonstrated. The fractional stable statistics is introduced. The memory concept is discussed in connection with fractional Poisson process. Fractional generalizations of the branching Furry process is introduced. Lévy flights on a one-dimensional stochastic solid fractal are considered. Such features of particles walk are caused by correlations in path lengths arising due to quenched disorder of solid fractal medium.

Chapter 2 is devoted to theoretical description of the dispersive charge carrier transport in disordered semiconductors. A relationship between universality of transient current curves, self-similarity of dispersive transport, and kinetic equations with fractional derivatives is established. Fractional differential equations provide a unified mathematical framework for describing normal and dispersive transport. In frames of this approach, it is possible to take into account energetic and topological types of disorder in common.

In **Chapter 3**, we describe transient processes in structures based on disordered semiconductors under conditions of dispersive transport. Transient current decays in time-of-flight experiments are calculated for different situations. Transport in non-homogeneous samples is described. We calculate transient characteristics in a diode and frequency characteristics of its conductance and diffusion capacitance.

Chapter 4 is devoted to processes of anomalous kinetics in nanoscale systems. Statistics of blinking quantum dot fluorescence, transport in colloidal quantum dot arrays, and conductance through fractal quasi-one-dimensional wire are considered. Power-law blinking statistics is analyzed on the base of the bifractional renewal process. Results for photon counting statistics are presented. A new modification of the Scher-Montroll model is proposed for the charge transport in colloidal quantum dot array. It takes into account Coulomb blockade forbidding multiple occupancy of nanocrystals and influence of energetic disorder of interdot space. The model explains power law current transients and the presence of memory effects. Considering scattering problem in mesoscopic systems, we derive a fractional analogue of the DMPK-equation for the joint probability distribution of the transmission eigenvalues in case of fractal disorder characterized by a heavy-tailed distribution of distances between scatterers. Solutions of this equation lead to a new class of universal conductance distributions related to the Lévy stable statistics. Analytical results agree with Monte-Carlo simulations.

In **Chapter 5**, we consider the fractional differential approach describing anomalous dielectric relaxation, discuss interrelation between different time and frequency domain responses and corresponding fractional relaxation equations and describe some memory phenomena in disordered dielectrics. The phenomenon of "memory regeneration" is revealed, described by fractional relaxation equations, corroborated experimentally and explained physically.

In concluding **Chapter 6**, the concept of intermediate asymptotics is introduced on the base of truncated Lévy flight. This leads to the correspondence principle connecting the classical statistics with the fractional stable statistics of mesoscopic systems.

It is a great pleasure for us to thank all those who has supported us in writing this book. We would like to thank the Administration of the Ulyanovsk State University (Russia) for good conditions for fruitful work. We are grateful to Ambrozevich S. A., Bulyarskii S. V., Galiyarova N. M., Grushko N. S., Meilanov R. P.,

Nigmatullin R. R., Nikitenko V. R., Rekhviashvili S. Sh., Sandev T., Timashev S. F., Virchenko Yu. P. for fruitful discussions. We are grateful to V. V. Saenko and V. V. Shulezhko for their help in preparation of the manuscript.

We also wish to thank the Russian Foundation for Basic Research for financial support (grant 10-01-00618, grant 12-01-97031).

The book is addressed to students and post-graduate students, engineers and physicists, specialists in the theory of solids, in mathematical modeling and numerical simulations of complex physical processes, to all who wish to make themselves more familiar with application of fractional differentiation method to the description of charge kinetic phenomena in disordered solids: conductors, semiconductors, dielectric quantum dots and their arrays.

Chapter 1

Statistical grounds

The main motivations for the theory of anomalous kinetics based on transport equations with fractional-order derivatives are these facts: (1) the fractional kinetics belongs to the domain of influence of the universal relaxation law; (2) fractional kinetic equations are connected with the known models of random processes and limit theorems of the probability theory; (3) one can create the unified formalism describing the normal and anomalous kinetics; (4) it is possible to take into account energetic and topological types of disorder in common. But before we begin to consider specific physical situations, we will study in more detail the interplay between the Lévy statistics, fractional calculus and anomalous transport.

1.1 Lévy stable statistics

1.1.1 *Generalized limit theorems*

The classical statistical mechanics predicts the behavior of matter in macroscopic scales from the knowledge of the nature of molecules comprising the system as well as intermolecular interactions. As a rule, these predictions concern with the thermodynamic limit ($N \to \infty$, $V \to \infty$, $N/V = \text{const}$) leading directly to the *Central Limit Theorem* (CLT). The latter states, that the sum $\Sigma_n = \sum_{j=1}^{n} X_j$ of independent identically distributed random variables X_j *having a finite variance* σ_1^2 is distributed closely to the Gauss law if the number n of terms is large enough.

Writing letter G for standard Gaussian random variable (with zero mean and unit variance) and symbol $\overset{d}{\sim}$ for asymptotical equivalence, we get (for centered terms):

$$\frac{\Sigma_n - a_n}{\sigma_n} \overset{d}{\sim} G, \quad n \to \infty, \tag{1.1}$$

$$a_n = a_1 n, \quad \sigma_n = \sigma_1 n^{1/2}.$$

The Gaussian distribution possesses a special property: sum of two independent random variables distributed according to this law is again distributed according to it (except parameters which change their values following the rules of the probability theory). Such distributions and corresponding random variables are called Lévy-stable distributions and Lévy-stable variables. Therefore, if we exactly double the size n of a microscopic system by combining it with an exact replica of itself, the extensive parameters a (the mean value) and σ^2 will be twice as much but the form of the distribution law remains the same. Repeating this operation, we shall obtain a consequence of self-similar systems.

All statistical mechanics of equilibrium state (in its classical version) can be derived from the only one this relation as it is done by Khintchin (1943) (see also [Kubo (1965)] and [Lavenda (1991)]). Dividing a homogeneous system into n equal parts, Khintchin considered X_j as random energies of the subsystems and Σ_n as the total energy of the system belonging to the canonical Gibbs ensemble. As follows from abovesaid, the resulting distribution and all following from it do not depend on details of the subsystems. Thus, the results of statistical mechanics have a very general character, are expressed in a simple analytic and handy form and applicable to systems of various nature.

French mathematician P. Lévy extended the Central Limit Theorem to the case of an infinite variance, when

$$\mathrm{Prob}(X_j > x) \sim a_+ x^{-\alpha}, \quad x \to \infty,$$

$$\mathrm{Prob}(X_j < x) \sim a_- |x|^{-\alpha}, \quad x \to -\infty,$$

where

$$a_+ \geq 0, \quad a_- \geq 0, \quad a_+ + a_- \equiv c > 0$$

and

$$0 < \alpha < 2.$$

Under these conditions[1] the *Generalized Limit Theorem* (GLT) takes place, which says that

$$\frac{\Sigma_n - a_n}{b_n} \overset{d}{\sim} S^{(\alpha,\beta)}, \quad n \to \infty,$$

where

$$a_n = n a_1, \quad a_1 = \begin{cases} 0, & \alpha < 1, \\ \langle X_j \rangle, & \alpha > 1, \end{cases}$$

$$b_n = n^{1/\alpha} b_1, \quad b_1 = \{\pi c / [2\Gamma(\alpha)\sin(\alpha\pi/2)]\}^{1/\alpha}, \quad \beta = \frac{a_+ + a_-}{c}$$

and $S^{(\alpha,\beta)}$ is an *α-stable random variable.* Its pdf $g^{(\alpha,\beta)}(x)$ is not expressible through elementary functions in general case, only characteristic functions can be represented:

$$\tilde{g}^{(\alpha,\beta)}(k) = \left\langle e^{ikS^{(\alpha,\beta)}} \right\rangle = \exp\{-|k|^\alpha [1 - i\beta \tan(\alpha\pi/2)\,\mathrm{sign}(k)]\}, \quad -\infty < k < \infty.$$

[1] They may be weakened by replacing coefficients $a_\pm$ with the *slowly varying* (in Karamata's sense) functions $h_\pm(t) > 0$, $t > 0$, that for any $k > 0$ satisfy the limit condition $\lim[h(kt)/h(t)] \to$ const $< \infty$ as $t \to \infty$, for example, $\ln t$ (see, for detail, [Feller (1967)]).

This is the *standard characteristic function in the form A*. There exist a few other forms [Uchaikin & Zolotarev (1999)], the most popular from them is form C (observe the difference in designations):

$$\widetilde{g}(k;\alpha,\theta)=\left\langle e^{ikS(\alpha,\theta)}\right\rangle=\exp\left\{-|k|^{\alpha}\exp[-i\alpha\theta(\pi/2)\,\mathrm{sign}(k)]\right\},$$

where α is the same exponent as in form A, and the new skewness parameter $\theta\in[-\theta_\alpha,\theta_\alpha]$, $\theta_\alpha=\min\{1,2/\alpha-1\}$, is linked to β via relation

$$\beta=\tan(\theta\alpha\pi/2)/\tan(\alpha\pi/2).$$

For corresponding random variables, the following relation takes place:

$$S(\alpha,\theta)\stackrel{d}{=}[\cos(\theta\alpha\pi/2)]^{1/\alpha}S^{(\alpha,\beta)}.$$

Notice, that $\cos(\theta\alpha\pi/2)\geq 0$ for any $\theta\in[-\theta_\alpha,\theta_\alpha]$, and $\alpha\in(0,2]$.

1.1.2 *Two subclasses of stable distributions*

Two subclasses are most popular in applications. First of them is formed by symmetrical stable distributions whose characteristic function is

$$\widetilde{g}(k;\alpha,0)\equiv\widetilde{g}(k;\alpha)=e^{-|k|^{\alpha}},\quad -\infty<k<\infty,\quad 0<\alpha\leq 2.$$

Observe, that

$$g(x;2)=\frac{1}{2\sqrt{\pi}}e^{-x^2/4},\quad -\infty<x<\infty,$$

is the Gaussian, and

$$g(x;1)=\frac{1}{\pi(1+x^2)},\quad -\infty<x<\infty,$$

is the Cauchy distribution.

The second subclass consists of one-sided stable distributions (subordinators):

$$g_+(x;\alpha)\equiv g(x;\alpha,1),\quad S_+(\alpha)\equiv S(\alpha,1),\quad 0<\alpha<1.$$

It is more convenient to describe them by means of the Laplace characteristic function. In particular, for a subordinator distributed on the positive semiaxis,

$$\hat{g}(\lambda;\alpha,1)\equiv\hat{g}_+(\lambda;\alpha)=\left\langle e^{-\lambda S(\alpha,\theta)}\right\rangle=e^{-\lambda^{\alpha}},\quad 0<\alpha\leq 1.$$

The example is given by the Lévy-Smirnov distribution:

$$g_+(x;1/2)=\frac{1}{2\sqrt{\pi x^3}}\exp\left(-\frac{1}{4x}\right),\quad x>0.$$

Note, that when $\alpha>1$ the mean value of X exists, that is integral $\int_{-\infty}^{\infty}xp(x)dx=a$ converges and the observed sum Σ_n has the mean value proportional to n. The variance of all stable random variables with $\alpha<2$ is infinite, therefore we must use another measure of scattering, say, the half-width of pdf on some fixed level. Let us denote this measure by $\Delta\Sigma_n$. Thus, for $\alpha \in (1,2)$

$$\langle\Sigma_n\rangle \propto n, \quad \Delta\Sigma_n \propto n^{1/\alpha},$$

and relative fluctuations

$$\frac{\Delta\Sigma_n}{\langle\Sigma_n\rangle} \propto \frac{1}{n^{1-1/\alpha}},$$

and the system remains extensive in one parameter (Σ_n) but fluctuations becomes larger, the very law of their dependence on size n changes from normal $n^{-1/2}$ to $n^{-(1-1/\alpha)}$. When α approaches to 1, relative fluctuations tends to be independent of n and really becomes constant, when α crosses the value 1. Note that it should be again redefined: the average diverges now, and we must use some fixed point on the distribution instead, for example, the position of maximum (all Lévy-stable densities are unimodal). Denoting this point by $\hat{\Sigma}_n$ and taking into account that

$$\hat{\Sigma}_n \propto n^{1/\alpha}, \quad \Delta\Sigma_n \propto n^{1/\alpha}$$

we see that for $\alpha < 1$ indeed

$$\frac{\Delta\Sigma_n}{\hat{\Sigma}_n} \propto \text{const}.$$

Thus, interpreting n as the size of a physical system, we can say that we meet here the model possessing some hallmarks of a mesoscopic system: nonextensivity and nondissapearing fluctuations. The crucial parameter is the characteristic exponent α, determining the low of falling of tails in original distributions. The pdfs of symmetric and one-sided stable variables are presented in Fig. 1.1.

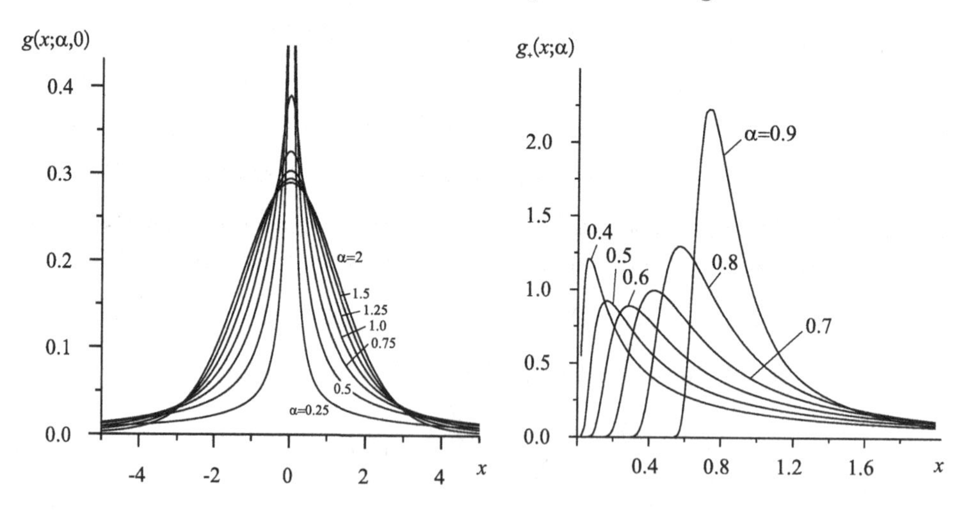

Figure 1.1 Symmetrical (left panel) and one-sided (right panel) Lévy stable densities.

1.1.3 *Fractional stable distributions*

The next generalization of the limit theorem is related to the number n of random terms: now it will be considered as a random variable N defined by a random Markov chain as follows. Let $\Theta_j,\ j = 1,2,3,\dots,$ be independent identically distributed positive random variables such that $\mathrm{Prob}(\Theta > t) \sim b_+ t^{-\omega}$, $0 < \omega < 1$, $t \to \infty$ and $T_n = \sum_{j=1}^{n} \Theta_j$. Then $N(t)$ is determined via relations

$$T_{N(t)} < t < T_{N(t)+1}.$$

As proved in [Korolev & Uchaikin (2000)], under this condition

$$\frac{\Sigma_{N(t)} - a_{N(t)}}{c_1 t^{\omega/\alpha}} \overset{d}{\sim} F(\alpha,\omega,\theta), \qquad t \to \infty,$$

where $F(\alpha,\omega,\theta)$ denotes the random variable distributed according to the *fractional stable distribution*

$$p_{F(\alpha,\omega,\theta)} \equiv q(x;\alpha,\omega,\theta) = \int_0^\infty g(xy^{\omega/\alpha};\alpha,\theta) g_+(y;\omega) y^{\omega/\alpha} dy.$$

Representing it in the form

$$q(x;\alpha,\omega,\theta) = \int_0^\infty dz \int_0^\infty \delta(xy^{\omega/\alpha} - z)\ g(z;\alpha,\theta) g_+(y;\omega) y^{\omega/\alpha} dy$$

$$= \int_0^\infty dz \int_0^\infty \delta(x - y^{-\omega/\alpha} z)\ g(z;\alpha,\theta) g_+(y;\omega) dy$$

and comparing the result with the formula

$$q(x;\alpha,\omega,\theta) = \langle \delta(x - F(\alpha,\omega,\theta)\rangle,$$

one can express fractional stable random variable $F(\alpha,\beta,\theta)$ through two independent stable variables in the following way:

$$F(\alpha,\omega,\theta) \overset{d}{=} S(\alpha,\theta)/[S_+(\omega)]^{\omega/\alpha}. \tag{1.2}$$

The reason of the term "fractional stable" is that any random variable with distribution belonging to this class is equal in distribution to the ratio of two independent random variables. Notice that one can use the mixed (AC) representation in which

$$q^{(\alpha,\omega,\beta)}(x) = \int_0^\infty g^{(\alpha,\beta)}(xy^{\omega/\alpha}) g_+(y;\omega) y^{\omega/\alpha} dy$$

is pdf of the random variable

$$F^{(\alpha,\omega,\beta)} \overset{d}{=} S^{(\alpha,\beta)}/[S_+(\omega)]^{\omega/\alpha}.$$

Notice, that for $\omega = 1$ the fractional stable pdfs coincide with corresponding stable pdfs, and for $\alpha = 2, \omega = 1$ they recover normal distribution with variance equal 2:

$$q(x;2,1,0) = \frac{1}{2\sqrt{\pi}} e^{-x^2/4}, \quad -\infty < x < \infty.$$

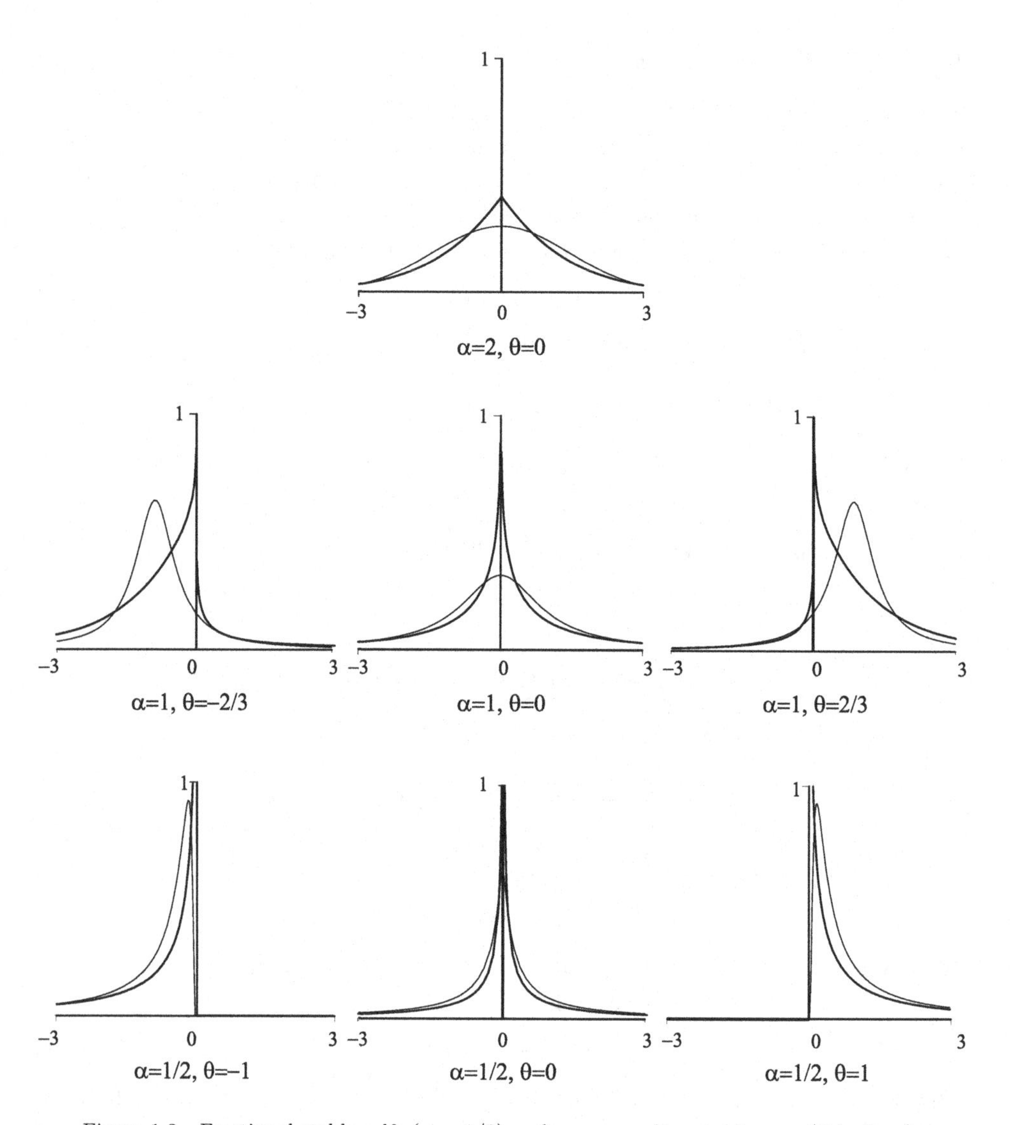

Figure 1.2 Fractional stable pdfs ($\omega = 1/2$) and corresponding stable ones (thin lines).

Let us stress that the fractional stable laws form a rich class of distributions including strictly stable distributions as a subclass. We can meet among them Gaussian $q(x; 2, 1, 0)$ and Cauchy distribution $q(x; 1, 1, 0)$, symmetric $q(x; \alpha, \omega, 0)$ and asymmetric $q(x; \alpha, \omega, \theta \neq 0)$ distributions, two-sided $q(x; \alpha, \omega, \theta \neq \pm 1)$ and one-sided $q(x; \alpha, \omega, \theta = \pm 1)$ distributions (see Fig. 1.2 and 1.3). Even δ-function finds a place among them:

$$q(x; 1, 1, 1) = \delta(x - 1), \quad q(x; 1, 1, -1) = \delta(x + 1).$$

First, this kind of distributions appeared in [Kotulski (1995)] where the one-dimensional Montroll-Weiss problem (1965) was solving. Later, these distributions

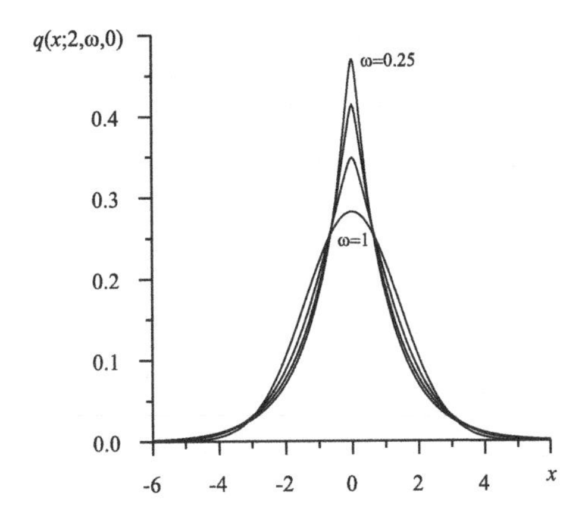

Figure 1.3 Fractional stable densities $q(x; 2, \omega, 0)$ for different ω (0.25, 0.5, 0.75, 1.0).

were named *fractional stable distributions* (FSD) [Kolokoltsov et al. (2001)], analytically investigated [Uchaikin (2002b); Bening et al. (2004)], generalized to multidimensional case [Uchaikin (2002d, 2003c)], numerically calculated [Uchaikin (2002d)] and applied to solving various problems [Weron et al. (2005); Lagutin & Uchaikin (2009); Xia Yuan et al. (2009); Uchaikin & Sibatov (2009b)].

1.1.4 *Self-similar processes: Brownian motion and Lévy motion*

Considering the random function $X(t) = \sum_{N(t)}$ and interpreting it as a trajectory of some particle moving along the x-axis, we deal with a special kind of motion – *stochastic motion*. At first glance, there is no any common in this motion and dynamical motion, when a particle trajectory has a form of smooth differentiable curve. The Brownian trajectory has a form of extraordinarily fractured and highly irregular line, which is continuous but nowhere differentiable. Nevertheless, some common feature in the two motions is observed: they are *self-similar*, at least at small scales.

Self-similarity (scaling) is a special kind of symmetry, when change in scale of some variables of the dynamical system can be compensated by change in scale of others.

A random process $\{X(t)\}$ is called *self-similar with index $H > 0$ in the strict sense* if for any $a > 0$ and any $n \geq 1$, $t_1, t_2, \dots, t_n$ the joint distribution of the random variables $X(at_1), \dots, X(at_n)$ is identical with the joint distribution of $a^H X(t_1), \dots, a^H X(t_n)$:

$$(X(at_1), \dots, X(at_n)) \stackrel{d}{=} (a^H X(t_1), \dots, a^H X(t_n)).$$

For self-similarity of a Markov process, it is enough self-similarity of the one-dimensional distribution:

$$X(at) \stackrel{d}{=} a^H X(t).$$

or equivalently

$$X(t) \stackrel{d}{=} t^H X(1),$$

In terms of pdf $p(x,t)$, the one-dimensional H-self-similar Markov process is characterized by the relation

$$p(x,t) = t^{-H} p(xt^{-H}, 1)$$

following from the chain of evident equalities

$$\int_{-\infty}^{x} p(x', t)dx' = \text{Prob}(X(t) < x)$$

$$= \text{Prob}(t^H X(1) < x) = \text{Prob}(X(1) < xt^{-H}) = \int_{-\infty}^{xt^{-H}} p(x', 1)dx'$$

and after differentiating it with respect to x.

Formally, the Newtonian motion of a free particle with a constant velocity v can be described as well in terms of δ-pdf $p(x,t) = \delta(x - vt)$ and we have

$$p\,(x,t) = \delta(x - vt) = t^{-1}\delta(x/t - v) = t^{-H} p(xt^{-H}, 1), \ \ H = 1$$

as before.

For Brownian motion with the diffusion coefficient K

$$X(t) \stackrel{d}{=} \sqrt{2Kt}\, G$$

and

$$X(at) \stackrel{d}{=} \sqrt{2Kat}\, G = a^H X(t), \quad H = 1/2.$$

In terms of densities we have

$$p(x,t) = \frac{1}{2\sqrt{Kt}} e^{-x^2/4Kt} = t^{-1/2} \frac{1}{2\sqrt{K}} e^{-(xt^{-1/2})^2/4K} = t^{-H} p(xt^{-H}, 1), \quad H = 1/2.$$

Really, there exists a whole family of self-similar processes, the Newtonian and Brownian motions are only individual members of them. Nevertheless, it is quite reasonable to admit that namely self-similarity of Brownian (Gaussian) motion provided the great progress of classical statistics.

Recall, that the Brownian motion (Bm-process) can be defined as a homogeneous in time and space Markov process with a finite variance. It possesses the self-similarity property.

Let us define L-motion (Lm-process) as a homogeneous in time and space Markov process with a self-similar distribution

$$p(x,t) = t^{-1/\alpha} g^{(\alpha)}(xt^{-1/\alpha}),$$

where $\alpha = 1/H$, and $g^{(\alpha)}(x)$ is still unknown pdf. Passing from Bm to Lm is realized via replacing the variance finiteness demand by the distribution self-similarity, therefore, it can not make the Lm class to be more narrow than the Bm class. To find out new terms of the family, we consider random positions of the process at two times t and $t+\tau$. Under the above conditions,

$$X(t+\tau) = X(t) + X(\tau)$$

where $X(t)$ and $X(\tau)$ are independent and the sum pdf is expressed through the convolution of summands pdfs:

$$p(x, t+\tau) = p(x,t) * p(x,\tau) \equiv \int_{-\infty}^{\infty} p(x-x', t) p(x', \tau) dx'. \tag{1.3}$$

This is the Chapman-Kolmogorov equation expressing the defining property of Markov processes.

Passing to characteristic functions, which reflects the self-similarity property by formula

$$\tilde{p}(k,t) = \tilde{g}^{(\alpha)}\left(kt^{1/\alpha}\right),$$

we arrive at the functional equation

$$\tilde{g}^{(\alpha)}\left(k(t+\tau)^{1/\alpha}\right) = \tilde{g}^{(\alpha)}\left(kt^{1/\alpha}\right)\tilde{g}^{(\alpha)}\left(k\tau^{1/\alpha}\right),$$

determining the family of Lévy-stable (more exactly, strictly stable) laws. We saw their solutions above, so that for the process we obtain:

$$\widetilde{p^{(\alpha,\beta)}}(k,t) = \exp\{-t|k|^\alpha[1 - i\beta\tan(\alpha\pi/2)\operatorname{sign}(k)\}, \tag{1.4}$$

or

$$\widetilde{p}(k,t;\alpha,\theta) = \exp\left\{-t|k|^\alpha \exp[-i\theta(\alpha\pi/2)\operatorname{sign}(k)]\right\}. \tag{1.5}$$

Below we consider differential equations for them.

1.1.5 *Space-fractional equations*

As it is readily seen from (1.4) and (1.5), the characteristic functions satisfy the evolution equations

$$\frac{\widetilde{p^{(\alpha,\beta)}}(k,t)}{\partial t} = -|k|^\alpha[1 - i\beta\tan(\alpha\pi/2)\operatorname{sign}(k)]\,\widetilde{p^{(\alpha,\beta)}}(k,t), \tag{1.6}$$

and

$$\frac{\partial\widetilde{p}(k,t;\alpha,\theta)}{\partial t} = -|k|^\alpha \exp[-i\theta(\alpha\pi/2)\operatorname{sign}(k)]\,\widetilde{p}(k,t;\alpha,\theta) \tag{1.7}$$

with the initial condition

$$\widetilde{p^{(\alpha,\beta)}}(k,0) = \widetilde{p}(k,0;\alpha,\theta) = 1.$$

For $\alpha = 2$ we recognize here the Fourier image of the ordinary diffusion equation:

$$\frac{\partial\widetilde{p}(k,t;2,0)}{\partial t} = -|k|^2\widetilde{p}(k,t;2,0).$$

When $\alpha < 2$ the multiplier in a right-hand side of (1.6) can be rewritten as

$$|k|^\alpha[1 - i\beta \mathrm{tg}(\alpha\pi/2)\mathrm{sign}\ k] = A(-ik)^\alpha + B(ik)^\alpha.$$

Recall that $(-ik)^\alpha$ and $(ik)^\alpha$ are Fourier images of the Riemann-Liouville fractional operators

$$_{-\infty}\mathsf{D}_x^\alpha f(x,t) = \frac{1}{\Gamma(1-\{\alpha\})}\frac{\partial^{[\alpha]+1}}{\partial x^{[\alpha]+1}}\int\limits_{-\infty}^{x}\frac{f(\xi,t)d\xi}{(x-\xi)^{\{\alpha\}}}$$

and

$$_x\mathsf{D}_\infty^\alpha f(x,t) = \frac{1}{\Gamma(1-\{\alpha\})}\frac{\partial^{[\alpha]+1}}{\partial x^{[\alpha]+1}}\int\limits_{x}^{\infty}\frac{f(\xi,t)d\xi}{(\xi-x)^{\{\alpha\}}},$$

where $\{\alpha\}$ is a fractional part of α and $[\alpha]$ is its integer part

$$\alpha = [\alpha] + \{\alpha\}, \qquad 0 < \{\alpha\} < 1.$$

Taking into account formula

$$(\pm ik)^\alpha = |k|^\alpha \exp[\pm i(\alpha\pi/2)\ \mathrm{sign}(k)],$$

we perform next trivial transformation:

$$A(-ik)^\alpha + B(ik)^\alpha =$$

$$= |k|^\alpha\{A\exp[-i(\alpha\pi/2)\ \mathrm{sign}(k)]A(-ik)^\alpha + B\exp[i(\alpha\pi/2)\ \mathrm{sign}(k)]\} =$$

$$= |k|^\alpha[(A+B)\cos(\alpha\pi/2) - i(A-B)\sin(\alpha\pi/2)\ \mathrm{sign}(k)] =$$

$$= (A+B)\cos(\alpha\pi/2)|k|^\alpha[1 - i(A-B)/(A+B)\tan(\alpha\pi/2)\ \mathrm{sign}(k)].$$

From equality

$$|k|^\alpha[1 - i\beta\mathrm{tg}(\alpha\pi/2)\ \mathrm{sign}(k)] =$$

$$= (A+B)\cos(\alpha\pi/2)|k|^\alpha[1 - i(A-B)/(A+B)\tan(\alpha\pi/2)\ \mathrm{sign}(k)]$$

we find:

$$A = \frac{1+\beta}{2\cos(\alpha\pi/2)}, \quad B = \frac{1-\beta}{2\cos(\alpha\pi/2)}.$$

Consequently, the equation for $p^{(\alpha,\beta)}(x,t)$ can be written in the following form

$$\frac{\partial p^{(\alpha,\beta)}(x,t)}{\partial t} = \triangle^{(\alpha/2,\beta)}p^{(\alpha,\beta)}(x,t) \tag{1.8}$$

with the initial condition

$$p(x,0) = \delta(x)$$

and one-dimensional fractional Laplacian

$$\triangle^{(\alpha/2,\beta)} = -\frac{1}{\cos(\alpha\pi/2)}\left[\frac{1+\beta}{2}\ {}_{-\infty}\mathsf{D}_x^\alpha + \frac{1-\beta}{2}\ {}_x\mathsf{D}_\infty^\alpha\right]$$

including the Riemann-Liouville fractional operators ${}_{-\infty}\mathsf{D}_x^\alpha$ and ${}_x\mathsf{D}_\infty^\alpha$.

As follows from these expressions, solution of the equation

$$\frac{\partial f^{(\alpha,\beta)}(x,t)}{\partial t} = K\triangle^{(\alpha/2,\beta)} f^{(\alpha,\beta)}(x,t) + \delta(t)\delta(x) \tag{1.9}$$

is represented as

$$f^{(\alpha,\beta)}(x,t) = (Kt)^{-1/\alpha} g^{(\alpha,\beta)}((Kt)^{-1/\alpha}). \tag{1.10}$$

There exists another representation of the equation in terms of fractional Feller operator [?]. We'll just notice here that in the symmetric case when $\beta = 0$

$$\frac{(-ik)^\alpha + (ik)^\alpha}{2\cos(\alpha\pi/2)} = |k|^\alpha = (k^2)^{\alpha/2} = \mathcal{F}(-\triangle)^{\alpha/2}$$

we obtain symmetric diffusion fractional equation

$$\frac{\partial p^{(\alpha,0)}(x,t)}{\partial t} = -(-\triangle)^{\alpha/2} p^{(\alpha,0)}(x,t), \tag{1.11}$$

which reduces to the ordinary diffusion equation when $\alpha \to 2$.

For the case $\alpha = 1$, form A is not applicable because of denominator $\cos(\alpha\pi/2)$ becoming zero and we should consider another form, form C. In order to transform the equation (1.7) for the characteristic function to the equation for corresponding pdf $p(x,t;\alpha,\theta)$ we rewrite (1.7) in the form

$$|k|^{-\alpha(1-\theta)}\mathcal{F}\frac{\partial p(x,t;\alpha,\theta)}{\partial t} = |k|^{\alpha\theta}\exp[-i(\theta\alpha\pi/2)]\ \mathrm{sign}(k)\mathcal{F}p(x,t;\alpha,\theta).$$

After inverting this relation we have

$$(-\triangle)^{\alpha(\theta-1)/2}\frac{\partial p(x,t;\alpha,\theta)}{\partial t} = -\ {}_{-\infty}\mathsf{D}_x^{\alpha\theta} p(x,t;\alpha,\theta)$$

or

$$\frac{\partial p(x,t;\alpha,\theta)}{\partial t} = -(-\triangle)^{\alpha(1-\theta)/2}\ {}_{-\infty}\mathsf{D}_x^{\alpha\theta} p(x,t;\alpha,\theta). \tag{1.12}$$

In the symmetrical case ($\theta = 0$)

$$\frac{\partial p(x,t;\alpha,0)}{\partial t} = -(-\triangle)^{\alpha/2} p(x,t;\alpha,0)$$

In the extremely asymmetrical case ($\theta = 1$), $X(t) > 0$ if $\alpha < 1$, and (1.12) takes the form

$$\frac{\partial p(x,t;\alpha,1)}{\partial t} = -\ {}_0\mathsf{D}_x^\alpha p(x,t;\alpha,1). \tag{1.13}$$

Performing the Laplace transformations we obtain for

$$\widehat{p}(\lambda,t;\alpha,1) = \int_0^\infty e^{-\lambda x} p(x,t;\alpha,1)dx$$

the equation

$$\frac{\partial \widehat{p}(\lambda, t; \alpha, 1)}{\partial t} = -\lambda^{\alpha} \widehat{p}(\lambda, t; \alpha, 1).$$

Under the initial condition

$$\widehat{p}(\lambda, 0; \alpha, 1) = 1$$

it yields the Laplace characteristic function of the one-sided stable density

$$\widehat{p}(\lambda, t; \alpha, 1) = e^{-\lambda^{\alpha} t}.$$

Observe that if $\alpha = 1$ and $\theta = 1$ then equation (1.13) takes the form

$$\frac{\partial p(x, t; 1, 1)}{\partial t} = -\frac{p(x, t; 1, 1)}{\partial x}$$

describing the ballistic regime of motion with the constant unit velocity. Under the above initial condition we obtain:

$$p(x, t; 1, 1) = \delta(x - t).$$

With no doubts, the self-similarity property is more significant and more fundamental than finiteness of the variance of some random variable. Moreover, we generalize the process without losing any term of it. This allows us to suppose, that the stable laws different from Gaussian will play a more significant role in the near future than they do now. And in first turn, it concerns to the basement of thermodynamics.

1.2 Random flight models

1.2.1 *Continuous time random flights*

In order to make the following development of the stochastic approach, let us rewrite the Chapman-Kolmogorov equation (1.3) in the form

$$p\,(x, t) = \int_{-\infty}^{\infty} p(\xi, \tau) p(x - \xi, t - \tau) d\xi, \quad 0 < \tau < t. \tag{1.14}$$

Now, we suppose that the change of the walker position during short time-interval τ is not a continuous but a jump-like (or *flight*) process such that

$$p(\xi, \tau) = [1 - \mu\tau]\delta(\xi) + \mu\tau p(\xi) + o(\tau),$$

Inserting this transition probability into above equation and passing to the limit $\Delta t \to 0$, one can represent the equation as

$$\frac{\partial p(x, t)}{\partial t} = -\mu p(x, t) + \mu \int_{-\infty}^{\infty} p(\xi) p(x - \xi, t) d\xi.$$

Its solution obeying the initial condition

$$p(x, 0) = \delta(x)$$

is of the form

$$p(x,t) = e^{-\mu t}\delta(x) + \mu \sum_{j=1}^{\infty} \frac{(\mu t)^{j-1}}{(j-1)!} e^{-\mu t} p^{j*}(x). \tag{1.15}$$

In order to get a more general model of the flight process called the *Continuous Time Random Flights* (CTRF), we observe that

$$Q_0(t) \equiv e^{-\mu t}$$

can be interpreted as a probability $\mathrm{Prob}(T > t)$, and $q_0(t) \equiv \mu e^{-\mu t} = -dQ_0(t)/dt$ as a pdf for the random waiting time T. In addition,

$$\frac{\mu(\mu t)^{j-1}}{(j-1)!} e^{-\mu t} = q_0^{j\star}(t),$$

therefore, equation (1.15) takes the form

$$p(x,t) = Q_0(t)\delta(x) + \sum_{j=1}^{\infty} q_0^{j\star}(t) p^{j*}(x). \tag{1.16}$$

Taking instead of $q_0(t) = \mu e^{-\mu t}$ an arbitrary one-sided density $q(t)$, we obtain the desired generalization

$$p(x,t) = Q(t)\delta(x) + \sum_{j=1}^{\infty} q^{j\star}(t) p^{j*}(x). \tag{1.17}$$

The first term corresponds to the case when the first waiting time T_1 exceeds the observation time t. The rest sum in Eq. (1.17) relating to the rest path of the trajectory can be transformed in the following way:

$$\sum_{j=1}^{\infty} q^{j\star}(t) p^{j*}(x) = \int_0^t dt' q(t') \int_{-\infty}^{\infty} dx' p(x') p(x-x', t-t').$$

As a result we obtain instead of equation (1.17) the following one:

$$p(x,t) = Q(t)\delta(x) + \int_0^t dt' q(t') \int_{-\infty}^{\infty} dx' p(x') p(x-x', t-t'). \tag{1.18}$$

This equation can be commented by simple physical reasons. To do this, let us divide the coordinate space into small cubic elements ΔV_j centered at points $\mathbf{x}_j$ and prescribe the coordinate to the particle during all period of time T when the particle moves inside ΔV_j. This time depends on the point of entering into the element, on the direction and absolute value of the particle velocity. If we can not control these parameters we should consider T as a random variable. After T seconds, the particle leaves ΔV_j and enters one of the neighboring elements ΔV_k. The coarse-grained coordinate $\mathbf{x}_j$ changes instantaneously to $\mathbf{x}_k$ and the process repeats again. The probability $P_{ij}(t)$ to find the particle inside ΔV_j t seconds later as it entered ΔV_i obeys the following equation:

$$P_{ij}(t) = \int_t^\infty q(t')dt'\delta_{ij} + \int_0^t dt' q(t') \sum_k p_{ik} P_{kj}(t-t').$$

Passing to continuous limit,

$$P_{ij}(t) \sim p(\mathbf{x}_j - \mathbf{x}_i, t)dV_j, \; p_{ik} \sim p(\mathbf{x}_k - \mathbf{x}_i)dV_k,$$

we arrive at the three-dimensional analogue of the one-dimensional CTRF equation (1.18):

$$p(\mathbf{x}, t) = Q(t)\delta(\mathbf{x}) + \int_0^t dt' q(t') \int d\mathbf{x}' p(\mathbf{x}') p(\mathbf{x} - \mathbf{x}', t - t'). \tag{1.19}$$

It is easy to check that the solution of the equation obeys the normalizing condition

$$\int p(\mathbf{x}, t) d\mathbf{x} = 1.$$

In reality, this equation is nothing but just the conservation probability law, all physics is hidden in transition pdfs $q(t)$ and $p(\mathbf{x})$.

Let us come back to the one-dimensional equation (1.17). Recall that

$$Q(t) = \text{Prob}(T > t) = \int_t^\infty q(t')dt'$$

and it is easy to check that the solution of the equation obeys the normalizing condition

$$\int_{-\infty}^\infty p(x, t)dx = 1.$$

Eq. (1.18) is the master equation for CTRF process. It can be used for description of processes with any kinds of transition pdfs $p(x)$ and $q(t)$. For example, inserting

$$p(x) = \delta(x - 1)$$

and passing from pdfs to probabilities

$$\int_{n-\varepsilon}^{n+\varepsilon} p(x, t)dx = P_n(t), \; \varepsilon \in (0, 1),$$

we arrive at the integral equation for a renewal process

$$P_n(t) = Q(t)\delta_{n0} + (1 - \delta_{n0}) \int_0^t dt' q(t') P_{n-1}(t - t'),$$

counting the number of events (jumps) of the CTRF process.

1.2.2 *Counting process for number of jumps*

Consider a random sequence of times $\{T_j\} = T_1, T_2, T_3, ...$ with independent intervals $\tau_j = T_j - T_{j-1}$ identically distributed according to common cumulative distribution function

$$\Psi(t) = \text{Prob}(\tau < t) = \int_0^t \psi(t)dt.$$

Let $\Psi_n(t)$ be multi-fold convolution of $\Psi(t)$:

$$\text{Prob}\left(\sum_{j=1}^{n} T_j < t\right) = \Psi_n(t) = \int_0^t \Psi_{n-1}(t-t')d\Psi(t'), \qquad \Psi_1(t) = \Psi(t).$$

We will interpret this sequence as a sequence of instantaneous jumps of a particles being in rest between them. The probability distribution for a random number of jumps performed by this particle during time interval $(0,t)$ is expressed through $\Psi_n(t)$ as

$$W(n,t) \equiv \text{Prob}(N(t) = n) = \Psi_n(t) - \Psi_{n+1}(t).$$

We consider two cases.

• Case 1. Without using any special expression for $\Psi(t)$, we assume only that the random variable τ has a finite variance σ^2. The central limit theorem says that

$$\Psi_n(t) \sim \Phi(z_n), \quad , \quad n \to \infty,$$

where

$$z_n = \frac{t - nm_1}{\sigma\sqrt{n}}, \quad m_1 = \langle \tau \rangle$$

and

$$\Phi(z) = \frac{1}{\sqrt{2\pi}} \int_{-\infty}^{z} \exp\left(-\frac{x^2}{2}\right) dx$$

is the normal (Gaussian) cumulative distribution function. In the long-time asymptotics we arrive at the normal distribution

$$W(n,t) \sim \Phi(z_n) - \Phi(z_{n+1}) \sim \frac{1}{\sqrt{2\pi(t/m_1)(\sigma/m_1)^2}} \exp\left[-\frac{(n-t/m_1)^2}{2\,(t/m_1)(\sigma/m_1)^2}\right]$$

with the mean $\langle N(t) \rangle = t/m_1$ and relative fluctuations

$$\frac{\Delta(t)}{\langle N(t) \rangle} \equiv \frac{\sqrt{\langle N^2(t) \rangle - \langle N(t) \rangle^2}}{\langle N(t) \rangle} \sim \frac{\sigma/m_1}{\sqrt{t/m_1}} \to 0, \qquad t \to \infty.$$

• Case 2. Now, let

$$1 - \Psi(t) \sim \frac{1}{\Gamma(1-\nu)}(ct)^{-\nu}, \quad c > 0, \quad t \to \infty.$$

Evidently, the values $\nu > 2$ are related to the previous case. When $1 < \nu < 2$,

$$\Psi_n(t) \sim G^{(\nu,\theta_{\alpha_1})}(t_n), \quad t_n = c\,\frac{t - nm_1}{n^{1/\nu}}, \quad n \to \infty,$$

where $G^{(\nu,\theta_\nu)}(z_n)$ is the stable distribution function with the characteristic exponent $\nu \in (1,2)$ and extremal skewness $\theta_\nu = 2/\nu - 1$. One can show that in this case the mean value $\langle N(t)\rangle \sim t/m_1$ as above, but the relative fluctuations decay more slowly:

$$\frac{\Delta(t)}{\langle N(t)\rangle} \propto (t/m_1)^{-1/\nu} \to 0, \quad t \to \infty.$$

When $0 < \alpha_1 < 1$, the Generalized Limit Theorem yields

$$\Psi_n(t) \sim G_+(z_n;\nu), \quad z_n = ctn^{-1/\nu},$$

and therefore

$$W(n,t) \sim g_+\left(\frac{ct}{n^{1/\nu}};\nu\right)\frac{ct}{\nu n^{1+1/\nu}}, \quad t \to \infty.$$

Recall, the Laplace image of the stable density $g_+(t;\nu)$, $0 < \nu < 1$, has the form

$$\int_0^\infty e^{-st} g_+(t;\nu)dt = e^{-s^\nu}$$

and its moments converge for all real orders $\mu < \nu$:

$$\int_0^\infty t^\mu g_+(t;\nu)dt = \frac{\Gamma(1-\mu/\nu)}{\Gamma(1-\mu)}.$$

Observe that $g_+(t;\nu) \to \delta(t-1)$, if $\nu \to 1$.

The moments of random numbers of jumps during $(0,t)$

$$\langle N^\mu(t)\rangle = \sum_{n=0}^\infty n^\mu W(n,t)$$

are expressed in long time asymptotics as

$$\langle N^\mu(t)\rangle \sim \int_0^\infty (ct/\xi)^{\nu\mu} g_+(\xi;\nu)d\xi = \frac{\Gamma(1+\mu)}{\Gamma(1+\nu\mu)}(ct)^{\nu\mu}.$$

(see for detail [Uchaikin & Zolotarev (1999)]). As a consequence, we have

$$\frac{\Delta(t)}{\langle N(t)\rangle} = \sqrt{\frac{\nu\,[\Gamma(\nu)]^2}{\Gamma(2\nu)} - 1} = \text{const} \neq 0.$$

Fig. 1.4 represents the limit fluctuations dependence on the characteristic exponent ν. Physically, this distribution is related to such a long duration that contains a very large number of jumps.

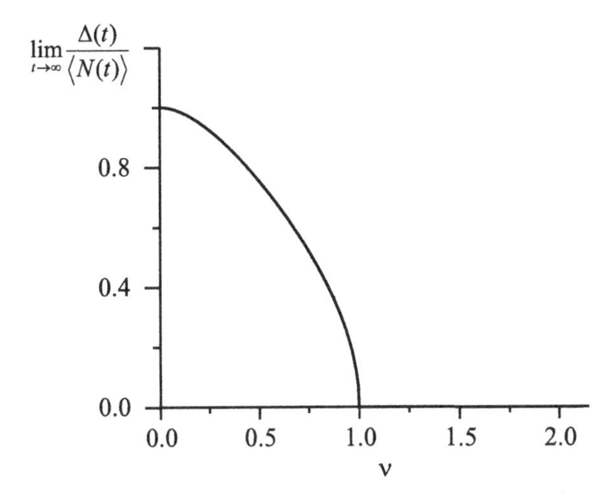

Figure 1.4 Limit relative fluctuations versus ν.

1.2.3 *The Poisson process*

In order to pass from the counting process to the well-known Poisson one, it is enough to take

$$q(t) = \mu e^{-\mu t}:$$

$$P_n(t) = e^{-\mu t}\delta_{n0} + (1-\delta_{n0})\int_0^t dt'\mu e^{-\mu t'}P_{n-1}(t-t').$$

After differentiation with respect to time, this integral equation takes the well-known differential form:

$$\frac{dP_n(t)}{dt} = -\mu P_n(t) + (1-\delta_{n0})\mu P_{n-1}(t), \qquad P_n(0+) = \delta_{n0}. \tag{1.20}$$

This equation can also be rewritten in the non-homogeneous form:

$$\frac{dP_n(t)}{dt} = -\mu P_n(t) + (1-\delta_{n0})\mu P_{n-1}(t) + \delta(t)\delta_{n0}, \qquad P_n(0-) = 0.$$

1.2.4 *The Fractional Poisson process*

The waiting time distribution density $q(t) = \mu e^{-\mu t}$ of the Poisson process obeys the ordinary differential equation

$$\frac{dq(t)}{dt} + \mu q(t) = \mu\delta(t).$$

In fractional Poisson process [Repin & Saichev (2000); Laskin (2003)], function $q(t) = \psi_\nu(t)$ obeys the fractional equation

$${}_0\mathsf{D}_t^\nu\psi_\nu(t) + \mu\psi_\nu(t) = \mu\delta(t).$$

The solution can be represented in two forms:

$$\psi_\nu(t) = -\frac{d}{dt}\mathrm{Prob}(T > t), \quad \mathrm{Prob}(T > t) = E_\nu(-\mu t^\nu), \tag{1.21}$$

where

$$E_\nu(z) = \sum_{n=0}^{\infty} \frac{z^n}{\Gamma(\nu n + 1)}$$

is the Mittag-Leffler function, and

$$\psi_\nu(t) = \frac{1}{t}\int_0^\infty e^{-x}\phi_\nu(\mu t/x)dx, \tag{1.22}$$

where

$$\phi_\nu(\xi) = \frac{\sin(\nu\pi)}{\pi[\xi^\nu + \xi^{-\nu} + 2\cos(\nu\pi)]}.$$

In [5], Eq. (1.21) is expressed through the two-parameter Mittag-Leffler function

$$\psi_\nu(t) = \mu t^{\nu-1} E_{\nu,\,\nu}(-\mu t^\nu), \tag{1.23}$$

$$E_{\alpha,\,\beta}(z) = \sum_{n=0}^{\infty} \frac{z^n}{\Gamma(\alpha n + \beta)}.$$

It is also convenient for the numerical calculation of the densities (see Fig. 1.5).

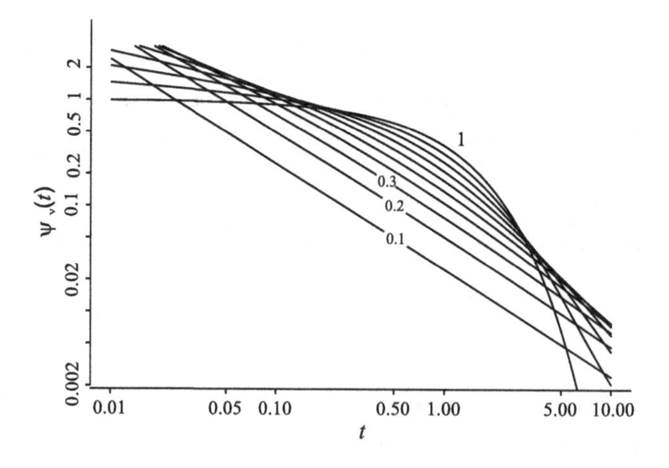

Figure 1.5 The fPp waiting time distribution densities (1.23) for $\mu = 1$ and $\nu = 0.1(0.1)1$.

1.2.5 *Simulation of waiting times*

The following result solves the problem of simulation of random waiting times.

The random variable T determined above has the same distribution as

$$T' = \frac{|\ln U|^{1/\nu}}{\mu^{1/\nu}} S_+(\nu), \tag{1.24}$$

where $S_+(\nu)$ is a random variable distributed according to $g_+(\tau;\nu)$ and U is independent of $S(\nu)$, is a uniformly distributed in $[0,1]$ random variable.

Making use of the formula of total probability, let us represent the cumulative function (1.21) in the following form

$$\mathrm{Prob}(T > t) = \int_0^\infty \mathrm{Prob}(T > t|\tau) g_+(\tau;\nu) d\tau,$$

where

$$\mathrm{Prob}(T > t|\tau) = e^{-\mu t^\nu/\tau^\nu}$$

is the conditional distribution. This means that

$$\mathrm{Prob}(T > t|\tau) = \mathrm{Prob}(U < e^{-\mu t^\nu/\tau^\nu}) = \mathrm{Prob}\left(\frac{|\ln U|^{1/\nu}}{\mu^{1/\nu}}\tau > t\right),$$

or

$$T|_\tau \stackrel{d}{=} \frac{|\ln U|^{1/\nu}}{\mu^{1/\nu}}\tau.$$

Because τ is a fixed possible value of $S(\nu)$, we obtain expression (1.24) for unconditional interarrival time.

Thus, the random variable T can be simulated in the following way

$$T \stackrel{d}{=} \frac{|\ln U_1|^{1/\nu}}{\mu^{1/\nu}} \frac{\sin(\nu\pi U_2)[\sin((1-\nu)\pi U_2)]^{1/\nu-1}}{[\sin(\pi U_2)]^{1/\nu}[\ln U_3]^{1/\nu-1}}, \tag{1.25}$$

where U_1, U_2 and U_3 are independent uniformly distributed on [0,1] random numbers. This conclusion follows from the Kanter algorithm for simulating $S(\nu)$ [Kanter, 1975].

Note that when $\nu \to 1$ this algorithm reduces to the standard rule of simulating random numbers with exponential distribution:

$$T \stackrel{d}{=} \frac{|\ln U|}{\mu}.$$

1.3 Some properties of the fractional Poisson process

1.3.1 *The nth arrival time distribution*

Let $T^{(n)}$, $n = 1, 2, 3, \ldots$, be the nth arrival time of a renewal process

$$T_n = T^{(1)} + T^{(2)} + \cdots + T^{(n)}$$

and $q^{*n}(t)$ be its probability density:

$$q^{*n}(t) = \underbrace{q * q * \cdots * q}_{n \text{ times}}(t).$$

Here, $T^{(j)}$'s are mutually independent copies of interarrival random times T and symbol $*$ denotes the convolution operation:

$$q * q(t) \equiv \int_0^t q(t-\tau)q(\tau)d\tau.$$

For the standard Poisson process,

$$q_0^{*n}(t) = \mu \frac{(\mu t)^{n-1}}{(n-1)!} e^{-\mu t},$$

and according to the Central Limit Theorem

$$\Psi^{(n)}(t) \equiv (\sqrt{n}/\mu) q^{*n}(n/\mu + t\sqrt{n}/\mu) \Rightarrow \frac{1}{\sqrt{2\pi}} e^{-t^2/2}, \ n \to \infty.$$

As numerical calculations show, $\Psi^{(n)}(t)$ practically reaches its limit form already by $n = 10$ (Fig. 1.6).

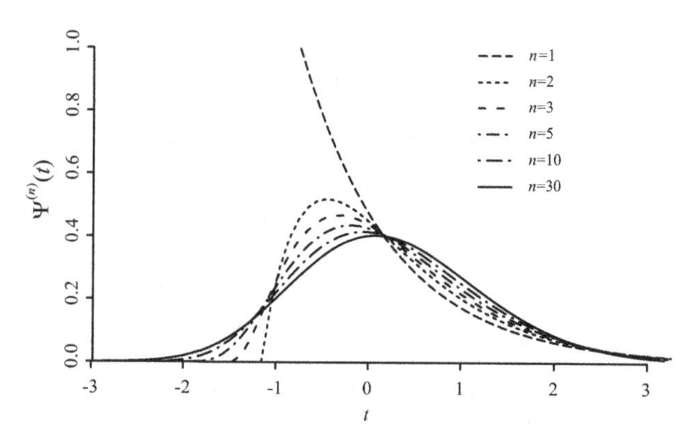

Figure 1.6 Rescaled arrival time distributions for the standard Poisson process ($\nu = 1, n = 1, 2, 3, 5, 10, 30$).

In case of the fPp,

$$\langle T \rangle = \int_0^\infty \psi_\nu(t) t dt = \infty$$

and the Central limit theorem is not applicable. Applying the Generalized Limit Theorem (see, for example, [Uchaikin & Zolotarev (1999)]), we obtain:

$$\Psi_\nu^{(n)}(t) \equiv \left(\frac{n}{\mu}\right)^{1/\nu} \psi_\nu^{*n}\left(t\left(\frac{n}{\mu}\right)^{1/\nu}\right) = n^{1/\nu} \overset{\circ}{\psi}{}_\nu^{*n} (tn^{1/\nu}) \Rightarrow g_+(t;\nu),\ n \to \infty,$$

where

$$\overset{\circ}{\psi}_\nu (t) = \psi_\nu(t)|_{\mu=1} = t^{\nu-1} E_{\nu,\,\nu}(-t^\nu).$$

Computing this multiple integrals can be performed by Monte Carlo technique. Taking $\mu = 1$ and observing that $\Psi_\nu^{(n)}(t)$ is the probability density of the renormalized sum $(T_1 + T_2 + \cdots + T_n)/n^{1/\nu}$ of independent random variables, distributed according to $\overset{\circ}{\psi}_\nu (t)$, one could directly simulate this sum by making use of the algorithm given by Eq. (1.25) and construct the corresponding histogram. However, the left tail of the densities is too steep for this method, and we applied some modification of the Monte Carlo method based on the partial analytical averaging of the last term.

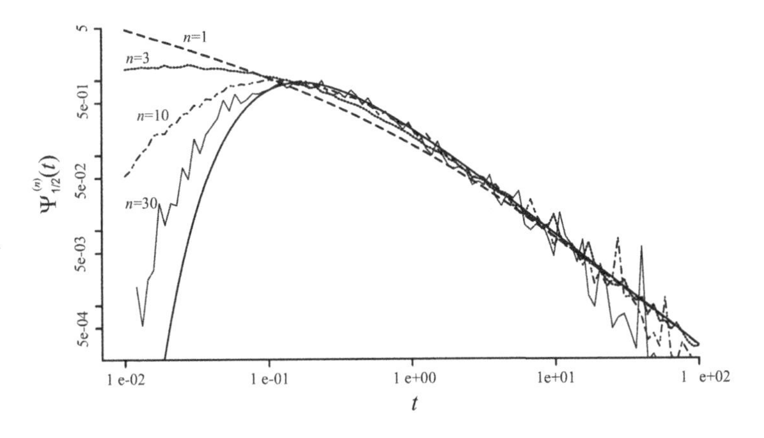

Figure 1.7 Rescaled arrival time distributions for fPp ($\nu = 1/2; n = 1, 3, 10$, and 30).

By making use of this modification, we computed the distributions $\Psi_\nu^{(n)}(t)$ for various n and ν. An example of these results is represented in Fig. 1.7.

1.3.2 *The fractional Poisson distribution*

Now we consider another random variable: the number of events (pulses) $N(t)$ arriving during the period t. According to the theory of renewal processes

$$p_n(t) \equiv \text{Prob}(N(t) = n) = \text{Prob}\left(\sum_{j=1}^{n} T_j > t\right) - \text{Prob}\left(\sum_{j=1}^{n+1} T_j > t\right), \quad n = 0, 1, 2, \ldots$$

and the following system of integral equations for $p_n(t)$ takes place:

$$p_n(t) = \delta_{n0} \int_t^\infty \psi_\nu(\tau)d\tau + [1-\delta_{n0}] \int_0^t \psi_\nu(t-\tau)p_{n-1}(\tau)d\tau, \quad n = 0, 1, 2, \ldots$$

After the Laplace transform with respect to time, we obtain

$$s^\nu \widehat{p}_n(s) = -\mu\widehat{p}_n(s) + \mu\widehat{p}_{n-1}(s) + s^{\nu-1}\delta_{n0}, \quad n = 0, 1, 2, \ldots, \quad \widehat{p}_{-1} = 0.$$

The inverse Laplace transform yields:

$$ {}_0\mathsf{D}_t^\nu p_n(t) = \mu[p_{n-1}(t) - p_n(t)] + \frac{t^{-\nu}}{\Gamma(1-\nu)}\delta_{n0}, \; 0 < \nu \le 1. \tag{1.26}$$

This is the master equation system for the fractional Poisson process. When $\nu \to 1$ it becomes the well known system for the standard Poisson process:

$$\frac{dp_n(t)}{dt} = \mu[p_{n-1}(t) - p_n(t)] + \delta(t)\delta_{n0}. \tag{1.27}$$

System (1.26) produces for the generating function

$$g(u,t) \equiv \sum_{n=0}^\infty u^n p_n(t) \tag{1.28}$$

the following equation:

$$ {}_0\mathsf{D}_t^\nu g(u,t) = \mu(u-1)g(u,t) + \frac{t^{-\nu}}{\Gamma(1-\nu)}. \tag{1.29}$$

When $\nu \to 1$ it becomes the well known equation for the standard Poisson process:

$$\frac{dg(u,t)}{dt} = \mu(u-1)g(u,t) + \delta(t). \tag{1.30}$$

Comparing (1.26) with (1.27) and (1.29) with (1.30), one can observe that the equations for standard processes are generalized to the equations for correspondent fractional processes by means of replacement of the operator d/dt with ${}_0\mathsf{D}_t^\nu$ and of right side the term $\delta(t)$ with $t^{-\nu}/\Gamma(1-\nu)$.

The solution to Eq. (1.29) is of the form

$$g(u,t) = E_\nu(\mu(u-1)t^\nu) \equiv \sum_{n=0}^\infty \frac{a^n}{\Gamma(\nu n+1)}(u-1)^n, \; a = \mu t^\nu.$$

Applying the binomial formula to each term of the sum and interchanging the summations, one can rewrite it as the series

$$g(u,t) = \sum_{n=0}^\infty u^n \left[\frac{a^n}{n!}\sum_{m=0}^\infty \frac{(m+n)!(-a)^m}{m!\Gamma(\nu(mk+n)+1)}\right]. \tag{1.31}$$

Comparing (1.31) with (1.28) yields

$$p_n(t) = \frac{a^n}{n!}\sum_{m=0}^\infty \frac{(m+n)!}{m!}\frac{(-a)^m}{\Gamma((m+n)\nu+1)}.$$

This distribution, which becomes the Poisson one, when $\nu = 1$, can be considered as its fractional generalization, called *fractional Poisson distribution.* The corresponding mean value and variance are given by

$$\langle N(t)\rangle = \frac{\mu t^\nu}{\Gamma(\nu+1)}$$

and

$$\sigma^2(t) = \langle N(t)\rangle\{1 + \langle N(t)\rangle[2^{1-2\nu}\nu\mathrm{B}(\nu, 1/2) - 1]\},$$

where

$$\mathrm{B}(\alpha_1, \alpha_2) = \int_0^1 x^{\alpha_1-1}(1-x)^{\alpha_2-1}dx$$

is the beta-function.

1.3.3 *Limit fractional Poisson distributions*

In case of the standard Poisson process, the probability distribution for random number $N(t)$ of events follows the Poisson law with $\langle N(t)\rangle = \mu t = \overline{n}$, which approaches to the normal one at large $\overline{n}$. Introducing normalized random variable $Z = N(t)/\overline{n}$ and quasicontinuous variable $z = n/\overline{n}$, one can express the last fact as follows:

$$f(z;\overline{n}) = \overline{n}\frac{\overline{n}^{\overline{n}z}}{\Gamma(\overline{n}z+1)}e^{-\overline{n}} \sim \sqrt{\frac{\overline{n}}{2\pi}}\exp\left\{-\frac{(z-1)^2}{2/\overline{n}}\right\}$$

as $\overline{n} \to \infty$. In the limit case $\overline{n} \to \infty$ the distribution of Z becomes degenerated:

$$\lim_{\overline{n}\to\infty} f(z;\overline{n}) = \delta(z-1).$$

Considering the case of fPp, we pass from the generating function to the Laplace characteristic function

$$g(u,t) = E_\nu(\mu t^\nu(u-1)) = E_\nu(\overline{n}\Gamma(\nu+1)(u-1)).$$

Introducing a new parameter $\lambda = -\overline{n}\ln u$ we get

$$\langle u^{N(t)}\rangle = \mathsf{E}e^{-\lambda Z} = E_\nu(\overline{n}\Gamma(\nu+1)(e^{-\lambda/\overline{n}}-1)).$$

At large $\overline{n}$ relating to large time t,

$$\langle e^{-\lambda Z}\rangle \equiv \int_0^\infty e^{-\lambda z}f_\nu(z)dz \sim E_\nu(-\lambda'), \quad \lambda' = \lambda\Gamma(\nu+1).$$

Comparison of this equation with formula (6.9.8) of the book [Uchaikin & Zolotarev (1999)]:

$$E_\nu(-\lambda') = \nu^{-1}\int_0^\infty \frac{e^{-\lambda' x}}{x^{1+1/\nu}}g_+(x^{-1/\nu};\nu)dx =$$

$$= \int_0^\infty e^{-\lambda z} \frac{[\Gamma(\nu+1)]^{1/\nu}}{\nu\, z^{1+1/\nu}} g_+\left(\frac{z^{-1/\nu}}{[\Gamma(\nu+1)]^{-1/\nu}};\nu\right) dz$$

shows that the random variable Z has the non-degenerated limit distribution at $t \to \infty$:

$$f_\nu(z;\overline{n}) \to f_\nu(z) = \frac{[\Gamma(\nu+1)]^{1/\nu}}{\nu\, z^{1+1/\nu}} g_+\left(\frac{z^{-1/\nu}}{[\Gamma(\nu+1)]^{-1/\nu}};\nu\right) \qquad (1.32)$$

with moments

$$\langle Z^k \rangle = \frac{[\Gamma(1+\nu)]^k \Gamma(1+k)}{\Gamma(1+k\nu)}.$$

By making use of series for $g_+(x;\nu)$, we obtain

$$f_\nu(z) = \sum_{k=0}^{\infty} \frac{(-z)^k}{k!\Gamma(1-(k+1)\nu)[\Gamma(\nu+1)]^{k+1}}.$$

When $z \to 0$,

$$f_\nu(z) \to f_\nu(0) = \frac{1}{\Gamma(1+\nu)\Gamma(1-\nu)} = \frac{\sin(\nu\pi)}{\nu\pi}.$$

It is also worth to note, that $\langle Z^0 \rangle = 1$, $\langle Z^1 \rangle = 1$ and $\langle Z^2 \rangle = 2\nu \mathrm{B}(\nu, 1+\nu)$, so that the limit relative fluctuations are given by

$$\delta_\nu \equiv \sigma_{N(t)}/\langle N \rangle = \sqrt{2\nu \mathrm{B}(\nu, 1+\nu) - 1}.$$

In particular cases

$$\delta_0 = 1, \quad \delta_1 = 0, \quad \delta_{1/2} = \sqrt{\pi/2} - 1.$$

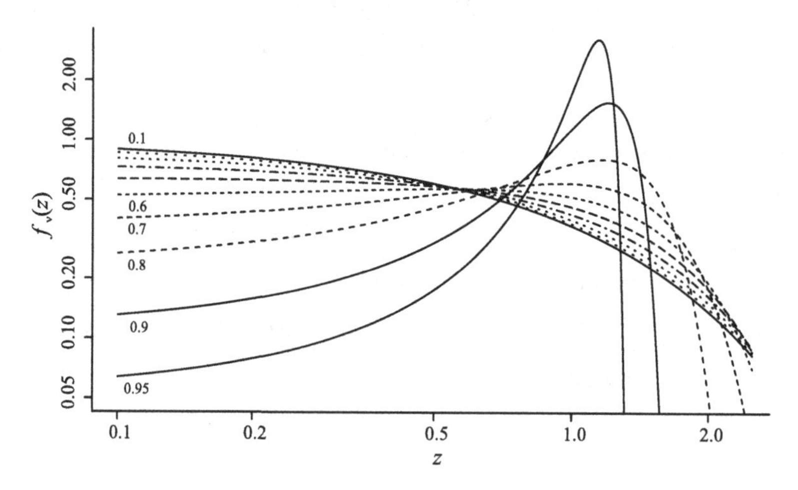

Figure 1.8 Limit distributions (1.32) for $\nu = 0.1(0.1)0.9$ and 0.95.

For $\nu = 1/2$, one can obtain an explicit expression for $f_\nu(z)$:

$$f_{1/2}(z) = \frac{2}{\pi} e^{-z^2/\pi}, \ z \geq 0.$$

The family of this limit distributions are plotted in Fig. 1.8.

1.3.4 *Fractional Furry process*

Let us move on to the branching processes and consider the simplest case, when each particle converts into two identical ones at the end of its waiting time, distributed with density $\psi_\nu(t)$. The process begins with one particle at $t = 0$ and the first arrival time has the same distribution density $\psi_\nu(t)$. When $\nu = 1$, the process is called the Furry process (Fp), therefore, in case of $\nu < 1$ we can call it the fractional Furry process (fFp). The following integral equations govern the fFp:

$$p_n(t) = \delta_{n1} \int_t^\infty \psi_\nu(\tau)d\tau + [1-\delta_{n0}-\delta_{n1}] \int_0^t \psi_\nu(t-\tau) \sum_{k=1}^{n-1} p_k(\tau)p_{n-k}(\tau)d\tau, \quad n = 1, 2, \ldots$$

Following the same way as before, we obtain

$${}_0\mathsf{D}_t^\nu p_n(t) = \mu \left[\sum_{k=1}^{n-1} p_k(t)p_{n-k}(t) - p_n(t) \right] + \frac{t^{-\nu}}{\Gamma(1-\nu)} \delta_{n1}, \; 0 < \nu \le 1.$$

The solution of this equation in case of $\nu = 1$ is well known: it is represented by the geometrical distribution

$$p_n(t) = e^{-\mu t} \left[1 - e^{-\mu t} \right]^{n-1}, \; n = 1, 2, 3, \ldots$$

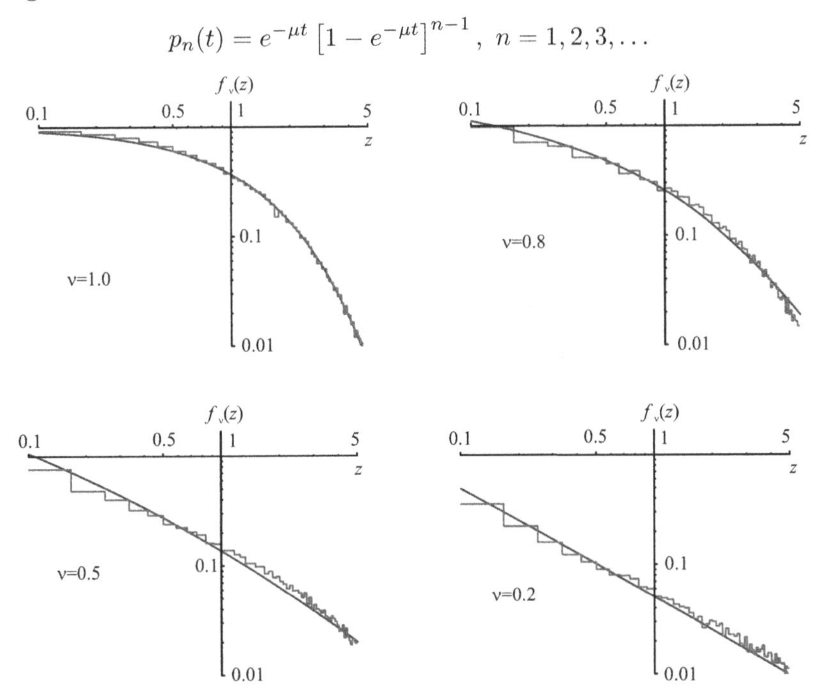

Figure 1.9 Monte Carlo calculation of $f_\nu(z)$ for $t = 5$ and $\nu = 1.0,\ 0.8,\ 0.5,\ 0.2$ (histograms) by comparison with hypothetical distribution (1.33) (smooth lines).

As to fFp for $\nu < 1$, we did not manage to derive the corresponding distribution from the fractional equation in a closed analytical form. The reason of the trouble lies in nonlinearity of the equation in case of branching. The only characteristics, the mean number of particles at time t has been found and expressed through the Mittag-Leffler function:

$$\langle N(t)\rangle = E_\nu(\mu t^\nu).$$

All other results have been obtained by means of Monte Carlo simulation using the algorithm described above.

Observe, that in contrast to the fPp, the limit distribution of the normalized random variable Z in case of fFp is not degenerated. In particular, for the standard Furry process

$$f(z) = \lim_{\overline{n}\to\infty} \overline{n} p_{\overline{n}Z}(\mu^{-1}\ln\overline{n}) = e^{-z}.$$

One could to suppose that in fractional case the "standard exponential function" is replaced with its fractional analogue

$$f_\nu(z) = z^{\nu-1}E_{\nu,\nu}(-z^\nu). \tag{1.33}$$

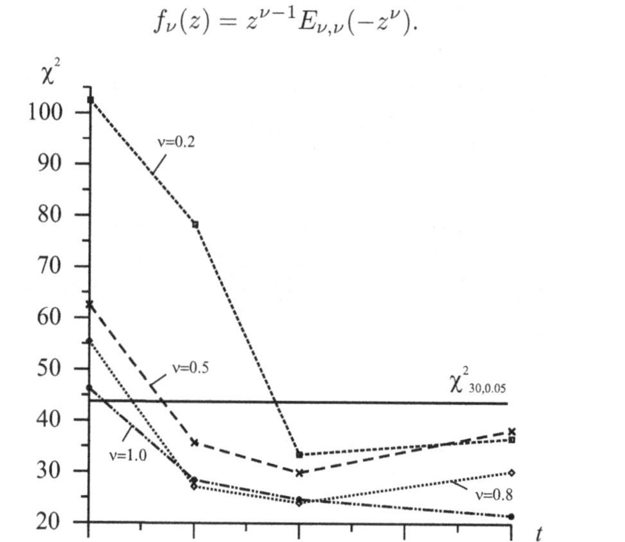

Figure 1.10 χ^2 Goodness-of fit Test.

Direct comparison of Monte Carlo data with formula (1.33) (Fig. 1.9) allows to propose that they coincide at large t, and the χ^2 goodness of fit analysis confirms this hypothesis (Fig. 1.10).

1.3.5 *Time-fractional equation*

The infinitesimal evolution $dP_n(t)$ of the probability distribution during elementary interval $(t, t+dt)$ for the Poisson process depends on distribution $P_n(t)$ and doesn't depend on the prehistory $P_n(t')$, $t' < t$. This property is interpreted as the *absence of memory* in the Poisson process. It seems to be natural for natural processes and one could expect that the "natural" CTRW equation should have a form

$$p(x,t) = e^{-\mu t}\delta(x) + \int_0^t dt' \mu e^{-\mu t'} \int_{-\infty}^{\infty} dx' p(x') p(x - x', t - t').$$

However, this form is valid when the rate μ for different traps are of the same value. In case of traps with different characteristics $\mu_1, \mu_2, \mu_3, \dots$ (disordered media) when

$$p(x,t;\mu_1,\mu_2,\mu_3,\dots) = e^{-\mu_1 t}\delta(x) + \int_0^t dt' \mu_1 e^{-\mu_1 t'} \int_{-\infty}^{\infty} dx' p(x') p(x - x', t - t'; \mu_2, \mu_3, \dots),$$

we have to average the equation over all possible values of $\mu_1, \mu_2, \mu_3, \dots$ for traps meeting the particle:

$$\langle p(x,t;\mu_1,\mu_2,\mu_3,\dots)\rangle = \langle e^{-\mu_1 t}\rangle \delta(x)$$

$$+ \int_0^t dt' \left\langle \mu_1 e^{-\mu_1 t'} \int_{-\infty}^{\infty} dx' p(x') p(x - x', t - t'; \mu_2, \mu_3, \dots) \right\rangle.$$

On assumption that μ_j are independent identically distributed random variables we obtain for the averaged pdf $\overline{p}(x,t) \equiv \langle p(x,t;\mu_1,\mu_2,\mu_3,\dots)\rangle$ the following equation:

$$\overline{p}(x,t) = \langle e^{-\mu_1 t}\rangle \delta(x) + \int_0^t dt' \left\langle \mu_1 e^{-\mu_1 t'} \right\rangle \int_{-\infty}^{\infty} dx' p(x') \overline{p}(x - x', t - t').$$

Replacing $e^{-\mu t}$ by $\langle e^{-\mu_j t}\rangle$ with distributed rates μ_j leads to a new regime called dispersive. Dispersive processes in disordered solids can be caused by multiple trapping of carriers in randomly distributed localized states (see [Zvyagin (1984); Madan & Shaw (1988)]). On the assumption that the localized states below the mobility edge fall off exponentially with energy, one can arrive at the waiting time distribution of power law type with the exponent α depending on temperature T.

Using designations

$$\widetilde{p}(k) = \int_{-\infty}^{\infty} e^{ikx} p(x) dx, \quad \widehat{q}(\lambda) = \int_0^{\infty} e^{-\lambda t} \overline{q}(t) dt, \quad \overline{p}(k,\lambda) = \int_{-\infty}^{\infty}\int_0^{\infty} dt e^{ikx - \lambda t} \overline{p}(x,t) dx dt,$$

we obtain the algebraic equation for the Fourier-Laplace transform of the process pdf:

$$\overline{p}(k,\lambda) = \lambda^{-1}[1 - \widehat{q}(\lambda)] + \widehat{q}(\lambda)\widetilde{p}(k)\overline{p}(k,\lambda).$$

Let us rearrange the equation by introducing transformations

$$\widehat{Q}(\lambda) = 1 - \widehat{q}(\lambda), \quad \widetilde{P}(k) = 1 - \widetilde{p}(k):$$

$$[\widehat{Q}(\lambda) + \widetilde{P}(k) - \widehat{Q}(\lambda)\widetilde{P}(k)]\overline{p}(k,\lambda) = \lambda^{-1}\widehat{Q}(\lambda).$$

This equation obtained from (1.18) without any approximations is a starting point for passing to differential equations.

We are going to consider a thermodynamic limit ($x \to \infty$, $t \to \infty$, $xt^{-\gamma} =$ const), consequently, we deal with infinitely small k, λ and $\widehat{Q}(\lambda)$, $\widetilde{P}(k)$ and the product $\widehat{Q}(\lambda)\widetilde{P}(k)$ can be neglected. Writing $\overline{f}(k,\lambda)$ for the leading asymptotic part of the solution $\overline{p}(x,t)$, we obtain:

$$[\widehat{Q}(\lambda) + \widetilde{P}(k)]\overline{f}(k,\lambda) = \lambda^{-1}\widehat{Q}(\lambda). \tag{1.34}$$

When $\langle T\rangle$ and $\langle X^2\rangle$ exist, we obtain

$$\widehat{Q}(\lambda) \sim \lambda\langle T\rangle, \quad \widetilde{P}(k) \sim (k^2/2)\langle X^2\rangle$$

(without drift) the normal diffusion equation with diffusivity $K = \langle X^2\rangle/2\langle T\rangle$. But if T has long tail distribution with infinite expectation,

$$\text{Prob}(T > t) \propto t^{-\omega}, \quad 0 < \omega < 1,$$

then the limit equation takes the form

$$_0\mathsf{D}_t^\omega f(x,t) = K\triangle f(x,t) + \delta(x)\frac{t^{-\omega}}{\Gamma(1-\omega)}. \tag{1.35}$$

Combining these reasoning with those in the preceding subsection, we obtain the time-space fractional equation:

$$_0\mathsf{D}_t^\omega f(x,t;\alpha,\omega,\theta) = K\triangle^{(\alpha/2,\theta)} f(x,t;\alpha,\omega,\theta) + \delta(x)\frac{t^{-\omega}}{\Gamma(1-\omega)}. \tag{1.36}$$

As shown in [?], the solution of the last time fractional equation is linked with the solution of the time-integer equation:

$$_0\mathsf{D}_t f(x,t;\alpha,1,\theta) = K\triangle^{(\alpha/2,\theta)} f(x,t;\alpha,1,\theta) + \delta(x)\delta(t)$$

via the following relation:

$$f(x,t;\alpha,\omega,\theta) = \frac{t}{\omega}\int_0^\infty f(\tau^{-1/\alpha}x;\alpha,1,\theta)g_+(t\tau^{-1/\omega};\omega)\tau^{-1/\alpha-1/\omega-1}d\tau$$

leading to the fractional stable density

$$f(x,t;\alpha,\omega,\theta) = (Kt^\omega)^{-1/\alpha}q((Kt^\omega)^{-1/\alpha}x;\alpha,\omega,\theta). \tag{1.37}$$

Eq. (1.36) can also be rewritten in the form

$$\frac{\partial f(x,t;\alpha,\omega,\theta)}{\partial t} = {}_0\mathsf{D}_t^{1-\omega}K\triangle^{(\alpha/2,\theta)} f(x,t;\alpha,\omega,\theta) + \delta(x)\delta(t). \tag{1.38}$$

Observe that when $\omega = 1$, Eqs. (1.36) and (1.38) relate to the Lévy motion. In case $\omega \neq 1$ we deal with the *subordinated Lévy motion*. One of interpretation of subordinated processes is based on involving concept of *operational time*, passing to which transforms Markovian process to its non-Markovian counterpart (see [Feller (1971)]).

Trajectories of the process for different values of α and ω are presented in Fig. 1.11.

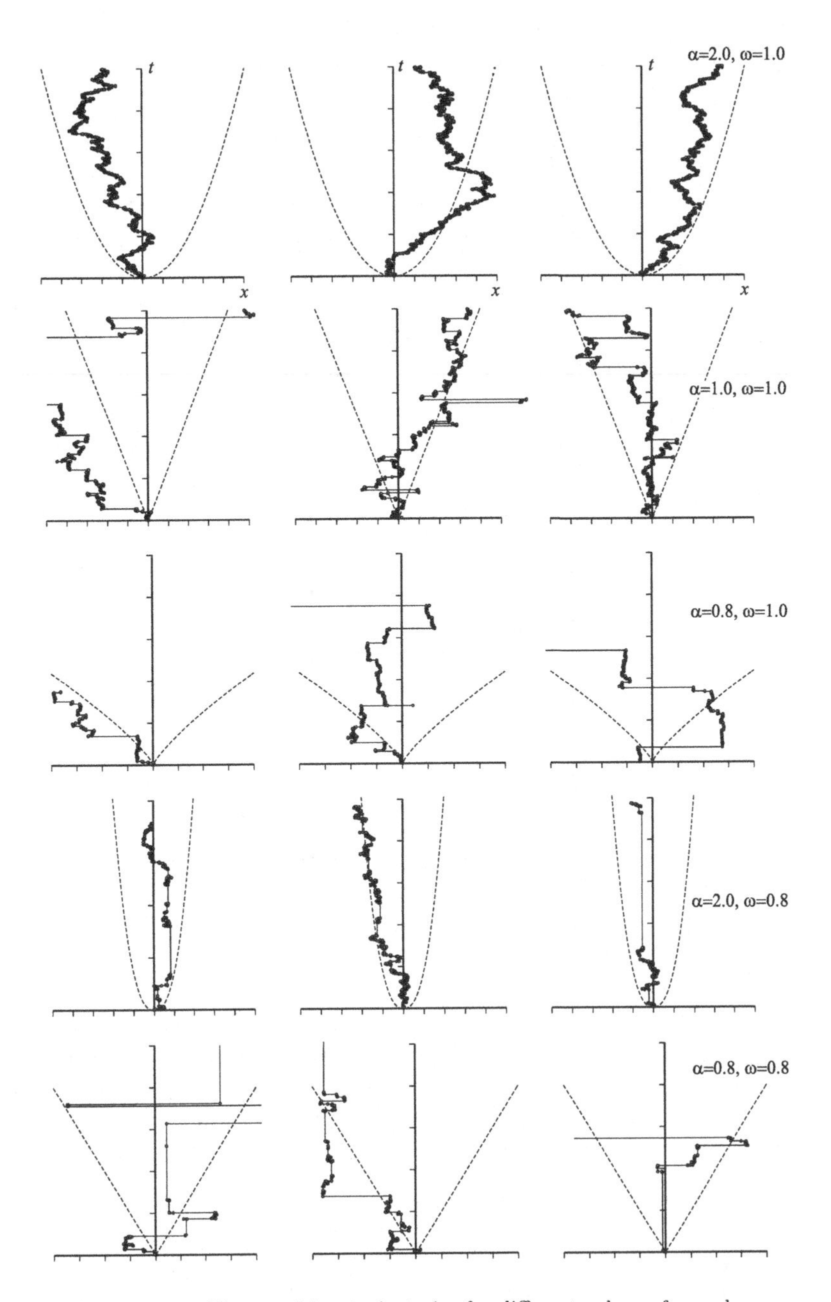

Figure 1.11 Time-position trajectories for different values of α and ω.

1.4 Random flights on a one-dimensional Lévy-Lorentz gas

1.4.1 *One-dimensional Lévy-Lorentz gas*

The term "anomalous diffusion" relates to the case when the size $\Delta(t)$ of a diffusion packet grows with time slower or faster then in the normal (Gaussian) case when $\Delta(t) \propto t^{1/2}$. Numerous anomalous phenomena have been investigated for a few last decades. It is well known that despite of variety of specific mechanisms generating normal diffusion process in various physical phenomena its main features can be obtained from the CTRF-scheme as time $t \to \infty$. This elevated hopes that the CTRF model can also describe a large number of different anomalous processes without consideration their specific mechanisms. Many works develop and improve this model (see for example [Montroll & Weiss (1965)][Schlesinger et al. (1993)],[Saichev & Zaslavsky (1997)] . We will call the process *fractional diffusion* (FD) in order to distinguish it from another process – *diffusion on fractals* (DF) which will be considered below. Whereas in the first case two consecutive jump lengths are independent in the second case they are correlated. Fig. 1 elucidates the sense of the difference: when the walker goes back it meet with the same atoms which were visited when it went forward. The back jump has the same length as the forward jump, this is a cause of the correlations between consecutive free paths.

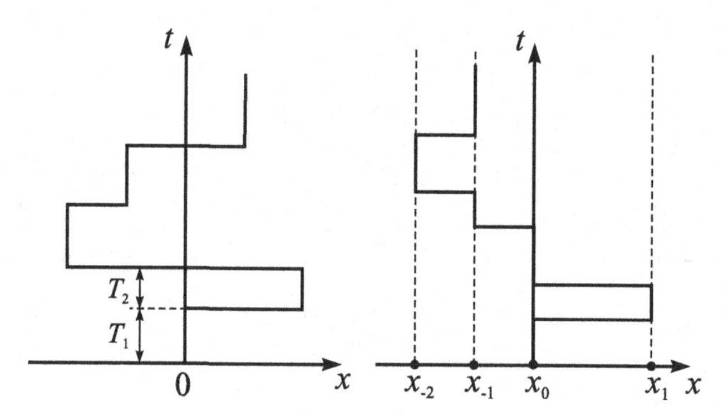

Figure 1.12 Fractional diffusion (left panel) and diffusion on a one-dimensional fractal (right panel). In the second case path lengths are correlated.

We consider the one-dimensional case when the substance is concentrated at points $\{X_j\} = ..., X_{-2}, X_{-1}, X_0 = 0, X_1, X_2, ...$ (atoms), randomly placed on x-axes. In order to construct the statistical ensemble we suppose, that $X_j - X_{j-1} = R_j$ are independent identically distributed random variables with a common distribution function

$$F(x) = \mathrm{Prob}(R < x) = \int_0^x f(y)dy.$$

On this assumption, the random values X_j form a correlated random sequence being composed of two independent sequences $X_1, X_2, ...$ and $X_{-1}, X_{-2}, ...$ with a common initial point $X_0 = 0$. This model is called the *one-dimensional Lorentz gas* [Barkai *et al.*, 2000].

Let $F_n(x)$ be the n-fold convolution of the distribution function $F(x)$:

$$F_{n+1}(x) = \int_0^x F_n(x-y)dF(y), \qquad F_1(x) = F(x)$$

and let $N_+(x)$ be the random number of atoms belonging to the interval $(0, x]$. Then

$$W(n,x) \equiv \text{Prob}(N_+(x) = n) = F_n(x) - F_{n+1}(x).$$

The similar relation takes place for the number $N_-(x)$ of atoms in the interval $[-x, 0)$. Note that the total number of atoms in the interval $[-x, x]$ is the sum $N(x) = N_+(x) + N_-(x) + 1$. Thus, choosing different functions $F(x)$ we obtain random sets of various kinds. Let $F(x)$ obey the asymptotic relation

$$1 - F(x) \sim \frac{A}{\Gamma(1-\alpha)} x^{-\alpha}, \quad A > 0, \quad x \to \infty.$$

For $\alpha > 2$ the variance of R is finite and we have asymptotically homogeneous medium. For $\alpha < 2$ the variance is infinite and we have a qualitatively another kind of medium called *Lévy-Lorentz gas* [Barkai *et al.*, 2000].

This a case of a *stochastic fractal Lorentz gas with fractal dimension* α embedded in a one-dimensional space. On the contrary to regular media stochastic fractals don't possess the property of self-averaging: while for a regular medium we have

$$\langle f(N(x), x)\rangle \to f(\langle N(x)\rangle, x), \quad x \to \infty,$$

for a fractal Lorentz gas we obtain

$$\langle f(N(x), x)\rangle \to \int_0^\infty f(N_1 x^\alpha z, x) w(z,\alpha) dz = \int_0^\infty f((c_\alpha x/\xi)^\alpha, x) g_+(\xi; \alpha) d\xi, \quad x \to \infty.$$

See for details [Uchaikin, 2004].

1.4.2 *The flight process on the fractal gas*

Now we consider a random sequence of moments $\{T_j\} = T_1, T_2, T_3, ...$ when the walker performs jumps. Let the sequence is built in the same manner as the sequence $X_1, X_2, X_3, ...$ namely $0 < T_1 < T_2 < T_3 < ...$, and $T_1, T_2 - T_1, T_3 - T_2, ...$ are mutually independent and identically distributed random variables with a common probability distribution function $Q(t) = \text{Prob}(T_{j+1} - T_j < t)$.

The random process $K(t)$ is defined in the same manner as the $N_+(x)$. Assuming in particular

$$1 - Q(t) \sim \frac{B}{\Gamma(1-\beta)} t^{-\beta}, \quad t \to \infty, \quad 0 < \beta < 1,$$

we obtain

$$W(k,t)dk \sim g_+(\tau_k; \beta) d\tau_k = w(z, \beta) dz$$

where

$$\tau_k = t c_\beta n^{-1/\beta}, \quad z = k/\langle K(t)\rangle, \quad \langle K(t)\rangle = \frac{1}{\Gamma(1+\beta)} (t/c_\beta)^\beta \equiv K_1 t^\beta.$$

Thus the times of jumps $T_1, T_2, T_3, ...$ form a fractal set on the time axis with fractal dimensionality β.

The test particle appears at time $t = 0$ at the origin $x_0 = 0$ and stays there up to time $T_1 > 0$. At time $t = T_1$, it performs an instantaneous jump to one of the neighbouring atoms: to x_{-1} with probability q_+ or to x_1 with probability q_-. It stays there up to time T_2 and then jumps again with the same probabilities to one of neighbours and so on.

This algorithm generates an ensemble of the particle trajectories $X_\theta(t|\{x_j\})$ on the set of fixed atoms $\{x_i\}$, where $\theta = q_+ - q_-$ is a skewness of the walking. Averaging over this ensemble will be denoted by the overbar:

$$F_\theta(x,t|\{x_j\}) \equiv \text{Prob}(X_\theta(t|\{x_j\}) < x) = \overline{H\left(x - X_\theta(t|\{x_j\}\right))}.$$

Let $J_\theta(t)$ be a random (positive or negative) number of the atom where the particle stays at the moment t

$$X_\theta(t|\{x_j\}) = x_{J_\theta(t)}.$$

Taking into account monotonicity of the function x_j $(j = ..., -1, 0, 1, ...)$ we obtain

$$F_\theta(x,t|\{x_j\}) = \overline{H(n(x) - J_\theta(t))}$$

where $n(x)$ obeys the equation

$$x_n \leq x \leq x_{n+1}, \quad n = ..., -1, 0, 1, ...$$

Further,

$$J_\theta(t) = \sum_{j=1}^{K(t)} U_j$$

where the independent random variables $U_j = \pm 1$ with probabilities $q_\pm$ and $K(t)$ is the random number of jumps up to time t (U_j and K are independent of each other).

1.4.3 *Propagators*

According to the central limit theorem

$$\text{Prob}(J_\theta < n|K = k) \sim \Phi\left((n - k\theta)/\sqrt{(1-\theta^2)k}\right), \qquad k \to \infty.$$

Then

$$\text{Prob}(J_\theta < n) \sim \overline{\Phi}\left((n - K\theta)/\sqrt{(1-\theta^2)K}\right)$$

and as a result

$$F_\theta(x,t|\{x_j\}) \sim \sum_{k=1}^{\infty} \Phi\left((n(x) - k\theta)/\sqrt{(1-\theta^2)k}\right) W(k,t), \qquad t \to \infty$$

where

$$W(k,t) = \text{Prob}(K(t) = k).$$

We meet here the subdiffusion behavior which is described by the distribution function

$$F_\theta(x,t|\{x_j\}) \sim \int_0^\infty \Phi\left((n(x)-\theta(c_\beta t/\tau)^\beta)/\sqrt{(1-\theta^2)(c_\beta t/\tau)^\beta}\right) g_+(\tau;\beta)d\tau.$$

In the symmetric case $\theta = 0$ and

$$F_0(x,t|\{x_j\}) \sim \Psi^{(2,\beta)}\left(n(x)(c_\beta t/\tau)^{-\beta/2}\right), \quad x \to \infty$$

where

$$\Psi^{(2,\beta)}(x) = \int_{-\infty}^{x} \psi^{(2,\beta)}(x)dx$$

is the symmetric fractional stable distribution function.

In the extreme case $\theta = 1$, the walker can not jump to the negative direction of the axis,

$$\lim_{\theta\to 1} \Phi\left((n(x)-\theta(c_\beta t/\tau)^\beta)/\sqrt{(1-\theta^2)(c_\beta t/\tau)^\beta}\right) = H\left(n(x)-(c_\beta t/\tau)^\beta\right)$$

and the probability distribution function takes the form

$$F_1(x,t|\{x_j\}) = \int_{c_\beta t[n(x)]^{-1/\beta}}^{\infty} g_+(\tau;\beta)d\tau = 1 - G_+\left(c_\beta t[n(x)]^{-1/\beta};\beta\right).$$

The third, final stage of the problem solution is averaging the conditional distribution function $F_\theta(x,t|\{X_j\})$ over all possible arrangements of atoms $\{X_j\}$. We will denote this operation by angle brackets as above:

$$F_\theta(x,t) = \langle F_\theta(x,t;\{X_j\})\rangle = \left\langle \overline{H(x-X_\theta(t|\{X_j\}))}\right\rangle.$$

The inversion property

$$p_\theta(x,t) \equiv \partial F_\theta(x,t)/\partial x = p_{-\theta}(-x,t)$$

allows us to focus on the positive semiaxis containing the subsequence $\{X_j\}_+$. As a result, the conditional distribution function $F_\theta(x,t|\{X_j\})$ can be decomposed into two terms

$$F_\theta(x,t|\{X_j\}) = F_\theta(0,t) + \Delta F_\theta(x,t|\{X_j\}_+),$$

the first of which is independent of $\{X_j\}$ and the second one depends only on $\{X_j\}_+$.

Taking it into account, one can write

$$F_\theta(x,t) \sim \int_0^\infty \left\langle \Phi\left(\frac{N_+(x)-\theta(c_\beta t/\tau)^\beta}{\sqrt{(1-\theta^2)(c_\beta t/\tau)^\beta}}\right)\right\rangle g_+(\tau;\beta)d\tau =$$

$$= \int_0^\infty\int_0^\infty \Phi\left(\frac{(c_\alpha x/\xi)^\alpha-\theta(c_\beta t/\tau)^\beta}{\sqrt{(1-\theta^2)(c_\beta t/\tau)^\beta}}\right) g_+(\xi;\alpha)g_+(\tau;\beta)d\xi d\tau.$$

Some results concerning asymmetric walking are presented in Fig. 1.13.

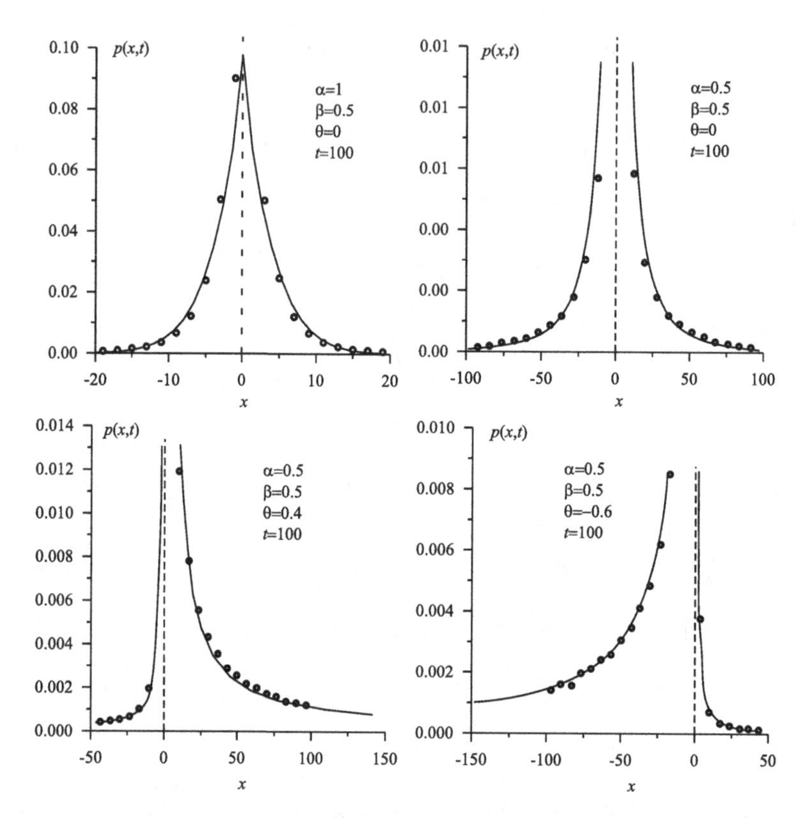

Figure 1.13 Density functions $p(x,t)$ for the diffusion on a one-dimensional fractal. Points are the results of Monte Carlo simulations, lines are analytical solutions.

1.4.4 *Fractional equation for flights on fractal*

Function $W(m,t,\beta)$ satisfies the equation

$$ {}_0\mathsf{D}_t^{\beta} W(m,t,\beta) = -c_\beta^\beta \frac{\partial W(m,t,\beta)}{\partial m} + \delta(m)\frac{t^{-\beta}}{\Gamma(1-\beta)}. $$

The conditional density $p(n|m)$ in accordance with central limit theorem

$$ p(n|m) = \frac{1}{\sqrt{2\pi m(1-\theta^2)}} \exp\left[-\frac{(n-m\theta)^2}{2m(1-\theta^2)}\right] $$

is the solution of the equation

$$ \frac{\partial p(n|m)}{\partial m} = \frac{1-\theta^2}{2}\frac{\partial^2 p(n|m)}{\partial n^2} - \theta\frac{\partial p(n|m)}{\partial n}, $$

where m and n are considered as continuous variables. Multiplying the equation for $W(m,t,\beta)$ by $p(n|m)$, integrating it on variable m ant taking into account

$$ p(n,t) = \int p(n|m)W(m,t,\beta)dm, $$

we obtain

$$ {}_0\mathsf{D}_t^\beta p(n,t) = -c_\beta^\beta \int p(n|m)\frac{\partial p(m,t)}{\partial m}dm + \frac{t^{-\beta}}{\Gamma(1-\beta)}p(n,0). $$

Integrating by parts and using the equation for $p(n|m)$ yields

$$ {}_0\mathsf{D}_t^\beta p(n,t) = c_\beta^\beta \left[\frac{1-\theta^2}{2}\frac{\partial^2 p(n,t)}{\partial n^2} - \theta\frac{\partial p(n,t)}{\partial n}\right] + \frac{t^{-\beta}}{\Gamma(1-\beta)}\delta(n). $$

Let's multiply this equation by $p(x|n)$ and integrate it on n. Taking into account

$$ p(x,t) = \int\limits_0^\infty p(x|n)p(n,t)dn, \qquad \frac{\partial p(x|n)}{\partial n} = c_\alpha^{-\alpha}\mathsf{D}_{x,0+}^\alpha p(x,n) + \delta(x)\delta(n), $$

as a result we obtain the DF-equation

$$ {}_0\mathsf{D}_t^\beta p(x,t) = \theta C\ {}_0\mathsf{D}_x^\alpha p(x,t) + c_\alpha^{-\alpha}\ \frac{C}{2}(1-\theta^2)\ {}_0\mathsf{D}_x^{2\alpha} p(x,t) + \delta(x)\frac{t^{-\beta}}{\Gamma(1-\beta)}. $$

The last equation describes random walks on a one-dimensional Lévy-Lorentz gas. The equation contains only Riemann-Liouville derivatives and does not contain the Feller fractional differential operator which is in the fractional diffusion equation.

Three important conclusions can be extracted from obtained results.

1. The fractal media does not possess the self-averaging property:

$$ p\Big(x,t|\{x_j\}\Big) \neq \Big\langle p(x,t|\{X_j\})\Big\rangle, \quad t\to\infty. $$

2. The DF-packet growth in width as $t^{\beta/2\alpha}$ i.e. much slower then the corresponding FD-packet the width of which $\sim t^{\beta/\alpha}$. This is the effect of neighbouring atoms playing the role of some kind of traps (see Fig. 1.12).

3. The DF- and FD- packet forms essentially differ from each other (see Fig. 1.13) but both of them are expressed through the stable distribution densities and obey fractional equations. The explicit expressions are brought above.

1.5 Subdiffusion

1.5.1 *Integral equations of diffusion in a medium with traps*

Usually by anomalous diffusion one means a random walk process involving a particle whose diffusion packet width $\Delta(t)$ [i.e. the width of the distribution density $\rho(x,t)$ with the initial condition $\rho(x,0)=\delta(x)$] grows in time according to the law

$$ \Delta(t) \propto t^\nu, \quad t\to\infty, \tag{1.39} $$

where the exponent ν differs from $1/2$, a value that corresponds to normal diffusion. When $\nu > 1/2$, we have *superdiffusion*, and when $\nu < 1/2$, we have *subdiffusion* (see [Bouchaud & Georges (1990); Isichenko (1992); Klages et al. (2008)]). The first type of anomalous diffusion is associated with anomalously long particle paths in the medium, ξ , while the second is associated with anomalously long times that the particle spends in a trap, τ.

If we assume that the random variables ξ_i and τ_i are mutually independent, subdiffusion can be described by integral equations. If we want the subdiffusion equation to look like an ordinary diffusion equation, we must consider equations in fractional derivatives [Wyss (1986); Nonnenmacher & Nonnenmacher (1989); Zaslavsky (1994); Chukbar (1995); Compte (1996)]. Schneider and Wyss (1989), Glöckle and Nonnenmacher (1993), and West et al. (1997) used Fox functions [Fox C. (1961)] to represent the solutions of such equation, but since Fox functions are peculiarly ill-suited to numerical problems, such representations are no better than Fourier-Laplace or Mellin transforms. The densities of the spatial distribution of a subdiffusive particle have yet to be found numerically.

The idea to seek the solution of subdiffusion equations in terms of stable laws emerged on the basis of two facts that are not well known to physicists. The first is that the Gaussian distribution is just one of the representatives of the infinite family of stable laws, whose common property is that all these laws describe limiting distributions of sums of independent random quantities, with each sum being normalized in a special way [Lévy (1965); Gnedenko & Kolmogorov (1954); Zolotarev (1983); Uchaikin & Zolotarev (1999)]. The Gaussian distribution emerges as the limiting distribution only if the terms have finite or logarithmically divergent variances. The second fact is that there exists a relation between stable laws and Fox functions making a bridge to other special functions [Schneider (1987)].

Since different approaches to the problem of anomalous diffusion invoke different variants of the equations, we begin with a complete description of the model under consideration, based on integral equations.

We use the model in which a particle can be in one of two states: of ordinary diffusion (state 1), or of rest (state 0) [Uchaikin (1999)]. Subdiffusion is a process in which the particle state changes successively at random moments in time. We assume that the random time intervals within which the particle is in one of the two states are mutually independent and distributed with densities $q_1(\tau)$ and $q_0(\tau)$, respectively.

The distribution of the time a particle stays in a trap, $q_0(\tau)$, which is determined by the trapping mechanism and the statistical spread of the trap properties, will not be nailed down for now. As for $q_1(\tau)$, we only assume that the mean time interval between leaving a trap and arriving at the next trap is finite:

$$\langle \tau_1 \rangle = \int_0^\infty \tau q_1(\tau) d\tau < \infty. \tag{1.40}$$

The medium is assumed spatially homogeneous and timeinvariant.

Let $p(x, t)$ be the spatial distribution of the probability in the case of continuous diffusion. In N-dimensional space,

$$p(x, t) = (4\pi Dt)^{-N/2} e^{-\frac{x^2}{4Dt}}, \quad x \in \mathrm{R}^N, \tag{1.41}$$

where D is the diffusion coefficient. Next, by $\rho_0(x,t)$ we denote the particle distribution at time t, where the particles history begins at $t=0$ when it lands in the trap at $x=0$, and by $\rho_1(x,t)$ we denote the particle distribution at time t, where the particles history begins at $t=0$ when it leaves the trap at $x=0$. These two distributions are related by a pair of integral equations:

$$\rho_0(x,t) = Q_0(t)\delta(x) + \int_0^t d\tau q_0(\tau)\rho_1(x,t-\tau), \tag{1.42}$$

$$\rho_1(x,t) = Q_1(t)p(x,t) + \int_0^t d\tau q_1(\tau)p(x,\tau) * \rho_0(x,t-\tau), \tag{1.43}$$

where

$$Q_i(t) = \int_t^\infty q_i(\tau)d\tau,$$

and $*$ denotes spatial convolution:

$$p(x,\tau) * \rho_0(x,t-\tau) = \int p(x',\tau)\rho_0(x-x',t-\tau)d^N x'. \tag{1.44}$$

The system of equations (1.42) and (1.43) normally describes a more general class of processes, since it holds for an arbitrary distribution $p(x,t)$. In particular, instead of using the diffusion regime (1.41) in the interval between two traps, we can use the ballistic regime or, say, the superdiffusion regime. In this paper we limit ourselves to the study of solutions of Eqs. (1.42) and (1.43) with distribution (1.41). We select the distribution of the time a particle spends in a trap, $q_0(\tau)$, in a form that ensures that the subdiffusion regime prevails.

Necessary and sufficient condition for subdiffusion

Let us find the condition that the distribution $q_0(\tau)$ must meet so that the model leads to subdiffusion (1.39) with $\nu < 1/2$. Introducing, for the sake of brevity, the notation

$$s_i(t) = \int |x|^2 \rho_i(x,t)d^N x,$$

and using (1.42) and (1.43), we obtain

$$s_0(t) = \int_0^t d\tau q_0(\tau)s_1(t-\tau), \tag{1.45}$$

$$s_1(t) = Q_1(t)s(t) + \int_0^t d\tau q_1(\tau)[s(\tau) + s_0(t-\tau)], \tag{1.46}$$

where

$$s(t) = \int |x|^2 p(x,t) d^N x = at, \quad a = 2ND. \tag{1.47}$$

Taking the Laplace transform of Eqs. (1.45) and (1.46),

$$s_i(\lambda) = \int_0^\infty e^{-\lambda t} s_i(t) dt,$$

we obtain a system of algebraic equations for the components

$$s_0(\lambda) = q_0(\lambda) s_1(\lambda), \quad s_1(\lambda) = K(\lambda) + q_1(\lambda) s_0(\lambda),$$

where

$$K(\lambda) = \int_0^\infty dt\, e^{-\lambda t} \Big[Q_1(t) s(t) + \int_0^t d\tau\, q_1(\tau) s(\tau) \Big].$$

Its solution has the form

$$s_0(\lambda) = \frac{q_0(\lambda) K(\lambda)}{1 - q_0(\lambda) q_1(\lambda)}, \tag{1.48}$$

$$s_1(\lambda) = \frac{K(\lambda)}{1 - q_0(\lambda) q_1(\lambda)}, \tag{1.49}$$

Using (1.47), we can transform $K(\lambda)$ into

$$K(\lambda) = -a \frac{d}{d\lambda} \int_0^\infty Q_1(t) e^{-\lambda t} dt \; - \frac{a}{\lambda} \frac{dq_1 \lambda}{d\lambda} =$$

$$= -a \frac{d}{d\lambda} \frac{1 - q_1(\lambda)}{\lambda} + \frac{a}{\lambda} \frac{d}{d\lambda} \Big[1 - q_1(\lambda) \Big] = \frac{a[1 - q_1(\lambda)]}{\lambda^2}. \tag{1.50}$$

According to Tauberian theorem, $s_i(t) \sim A_i t^\alpha, \quad t \to \infty$, implies

$$s_i(\lambda) \sim \Gamma(\alpha+1) A_i \lambda^{-\alpha-1}, \quad \lambda \to 0 \tag{1.51}$$

and conversely [Feller (1967)]. By virtue of (1.40) we have

$$1 - q_1(\lambda) \sim \bar{\tau}_1 \lambda, \quad Q_1(\lambda) = \frac{1 - q_1(\lambda)}{\lambda} \sim \bar{\tau}_1 \tag{1.52}$$

so that $K(\lambda) \sim a\bar{\tau}_1/\lambda$ as $\lambda \to 0$. Substituting (1.51) into (1.48) and (1.49) and solving the resulting equations for $1 - q_0(\lambda)$, we find the necessary condition for subdiffusion:

$$1 - q_0(\lambda) \sim b\lambda^\alpha, \quad \lambda \to 0, \quad b = \frac{a\bar{\tau}_1}{\Gamma(\alpha+1) A}, \quad \alpha < 1, \tag{1.53}$$

with $A_1 = A_2 = A$ (the asymptotic behavior of the width of the subdiffusion packet is independent of the initial particle state). By virtue of the reciprocity of Tauberian theorem, the condition (1.52) is also sufficient.

To reformulate the condition (1.52) for the distribution density $q_0(\tau)$, we again turn to Tauberian theorem and apply it to the function $Q_0(t)$, with the result that

$$Q_0(\lambda) = \frac{1 - q_0(\lambda)}{\lambda}.$$

We obtain

$$Q_0(t) = \int_t^\infty q_0(\tau)d\tau \sim Bt^{-\alpha}, \quad t \to \infty,$$

where

$$B = \frac{a\bar{\tau}_1}{[\Gamma(1-\alpha)]^2 A} \tag{1.54}$$

or, for the density,

$$q_0(t) \sim \alpha B t^{-\alpha-1}, \quad t \to \infty. \tag{1.55}$$

Thus, in the model considered, subdiffusion emerges if and only if the distribution of the times particles stay in traps exhibits asymptotic behavior of the power-law type (1.54) with an exponent $\alpha < 1$. This means, in particular, that the average time a particle stays in a trap is infinite:

$$\int_0^\infty \tau q_0(\tau)d\tau = \infty, \quad \alpha < 1.$$

If it is finite,

$$\int_0^\infty \tau q_0(\tau)d\tau = \bar{\tau}_0,$$

the asymptotic behavior of $q_0(\lambda)$ is

$$q_0(\lambda) \sim 1 - \bar{\tau}_0\lambda, \quad \lambda \to 0. \tag{1.56}$$

Substituting (1.52) and (1.56) into (1.49), we see that in this case

$$s_1(\lambda) \sim \frac{a}{\lambda^2[1 + \bar{\tau}_0/\bar{\tau}_1]}, \quad \lambda \to 0$$

with the result that the effect of traps reduces to a variation in the diffusion coefficient,

$$D \to \frac{D}{1 + \bar{\tau}_0/\bar{\tau}_1},$$

and that the temporal variation of the mean square $s_1(t)$ remains linear. It can be shown that the diffusion packet in this case remains Gaussian.

1.5.2 *Differential equations of subdiffusion*

Now, let's go back to Eqs. (1.42) and (1.43) and take Fourier and Laplace transforms with respect to position and time, respectively:

$$\rho_i(k,\lambda) = \int_0^\infty dt \int d^N x \exp\{-\lambda t + ikx\} p_i(x,t), \; k \in \mathbb{R}^N.$$

This yields

$$\rho_0(k,\lambda) = Q_0(\lambda) + q_0(\lambda)\rho_1(k,\lambda),$$

$$\rho_1(k,\lambda) = Q_1(\lambda + Dk^2) + q_1(\lambda + Dk^2)\rho_0(k,\lambda).$$

The solution of this system has the form

$$\rho_0(k,\lambda) = \frac{Q_0(\lambda) + q_0(\lambda)Q_1(\lambda + Dk^2)}{1 - q_0(\lambda)q_1(\lambda + Dk^2)} \tag{1.57}$$

and

$$\rho_1(k,\lambda) = \frac{Q_1(\lambda + Dk^2) + Q_0(\lambda)q_1(\lambda + Dk^2)}{1 - q_0(\lambda)q_1(\lambda + Dk^2)}. \tag{1.58}$$

Combining conditions (1.53-1.55) with Eqs. (1.57) and (1.58), we obtain an expression for the leading asymptotic terms,

$$\rho^{as}(k,\lambda) = \frac{\lambda^\alpha}{\lambda[D'k^2 + \lambda^\alpha]}, \quad D' = \bar{\tau}_1 D/b, \tag{1.59}$$

which is independent of the initial state. We postpone taking the inverse transform to next subsection. Here we write the above relationship in three equivalent forms:

$$\lambda^\alpha \rho^{as}(k,\lambda) = -D'k^2 \rho^{as}(k,\lambda) + \lambda^{\alpha-1}, \tag{1.60}$$

$$\lambda \rho^{as}(k,\lambda) = -D'k^2 \lambda^{1-\alpha} \rho^{as}(k,\lambda) + 1. \tag{1.61}$$

$$\rho^{as}(k,\lambda) = -D'k^2 \lambda^{-\alpha} \rho^{as}(k,\lambda) + \lambda^{-1}. \tag{1.62}$$

As is known, on a suitable class of functions the Laplace transform $F(\lambda)$ of the Riemann-Liouville fractional derivative

$$F(t) = \frac{d^\mu f(t)}{dt^\mu} \equiv \frac{1}{\Gamma(-\mu)} \int_0^t (t-\tau)^{-\mu-1} f(\tau) d\tau, \; \mu < 1, \tag{1.63}$$

is related to $f(\lambda)$ of the differentiable function $f(t)$ [Samko et al. (1987); Miller & Ross (1993)]:

$$F(\lambda) = \lambda^\mu f(\lambda). \tag{1.64}$$

When $\mu < 0$ the expression (1.63) is a fractional integral of order $|\mu|$. Using this notation in the inverse FourierLaplace transform of Eqs. (1.601.62), we obtain an equation in fractional derivatives that describes the asymptotic behavior of the subdiffusion process::

$$\frac{\partial^\alpha \rho^{as}}{\partial t^\alpha} = D' \bigtriangledown^2 \rho^{as} + \frac{t^{-\alpha}}{\Gamma(1-\alpha)}\delta(x), \tag{1.65}$$

$$\frac{\partial \rho^{as}}{\partial t} = D' \bigtriangledown^2 \frac{\partial^{1-\alpha} \rho^{as}}{\partial t^{1-\alpha}} + \delta(x)\delta(t), \tag{1.66}$$

$$\rho^{as} = D' \bigtriangledown^2 \frac{\partial^{-\alpha} \rho^{as}}{\partial t^{-\alpha}} + \delta(x). \tag{1.67}$$

These equations have a general solution with the Fourier Laplace transform found earlier, which is represented by Eq. (1.59).

Note that the special case of Eq. (1.65) corresponding to $\alpha = 1/2$ was obtained by Nigmatullin (1984) in connection with diffusion in fractal structures of the Koch-tree type, which models porous and disordered media. The one-dimensional analog of Eq. (1.66) was written out by Compte (1996), and the integral equation (1.67) was solved by Schneider and Wyss (1989). We now discuss their solution.

1.5.3 *Subdiffusion distribution density*

Let $D' = 1$ in Eq. (1.67) and write the latter

$$\rho(x,t) = \delta(x) + \frac{1}{\Gamma(\alpha)} \int_0^t (t-\tau)^{\alpha-1} \bigtriangledown^2 \rho(x,t), \quad 0 < \alpha < 1. \tag{1.68}$$

Equation (1.68) was studied by Schneider and Wyss (1989), who expressed its solution as a function of the distance $r = |x|$ in terms of the Fox functions:

$$\rho(r,t) = \alpha^{-1}\pi^{-N/2}r^{-N} \times$$

$$\times H_{12}^{20}\left((r/2)^{2/\alpha} t^{-1} \middle| \begin{matrix} (\quad 1 \quad , \quad 1 \quad) \\ (N/2, 1/\alpha), (1, 1/\alpha) \end{matrix} \right). \tag{1.69}$$

They also found the explicit form of the Mellin component in r:

$$\rho(s,t) = \int_0^\infty r^{s-1}\rho(r,t)dr = 2^{s-N-1}\pi^{-N/2}t^{\alpha(s-N)/2}$$

$$\times \frac{\Gamma(s/2)\Gamma\left((s-N)/2\right)}{\alpha\Gamma\left(\alpha(s-N)/2\right)}. \tag{1.70}$$

We establish another form of the solution that relates the form to stable distributions. This will make it possible not only to carry out a qualitative analysis but also to understand the physics of the solution.

We write (1.59) as

$$\rho^{as}(k,\lambda) = \lambda^{\alpha-1}\int_0^\infty e^{-[D'k^2+\lambda^\alpha]y}dy, \quad \alpha<1 \tag{1.71}$$

and also write the inverse Laplace transform:

$$\rho^{as}(k,t) = \int_0^\infty dye^{-D'k^2y}(2\pi i)^{-1}\int_\gamma d\lambda\lambda^{\alpha-1}e^{\lambda t-\lambda^\alpha y}.$$

Evaluating the innermost integral by parts, we obtain

$$\rho^{as}(k,t) = \frac{t}{\alpha}\int_0^\infty dye^{-D'k^2y}y^{-1}(2\pi i)^{-1}\int_\gamma e^{\lambda t-\lambda^\alpha y}d\lambda.$$

In the innermost integral we transform to the new variable $s = y^{1/\alpha}\lambda$,

$$\rho^{as}(k,t) = \alpha^{-1}t\int_0^\infty dye^{-D'k^2y}y^{-1-1/\alpha}\left[(2\pi i)^{-1}\int_\gamma \exp\left(sy^{-1/\alpha}t - s^\alpha\right)ds\right].$$

The expression in square brackets is a one-sided stable density with characteristic exponent $\alpha<1$:

$$g_+(t;\alpha) = (2\pi i)^{-1}\int_\gamma e^{st-s^\alpha}ds. \tag{1.72}$$

Thus,

$$\rho^{as}(k,t) = \alpha^{-1}t\int_0^\infty dye^{-D'k^2y}y^{-1-1/\alpha}g_+\left(y^{-1/\alpha}t;\alpha\right).$$

Introducing the integration variable $\tau = y^{-1/\alpha}t$, we find that

$$\rho^{as}(k,t) = \int_0^\infty d\tau e^{-D'k^2t^\alpha/\tau^\alpha}g_+(\tau;\alpha).$$

Finally, taking the inverse Fourier transform, we obtain

$$\rho^{as}(x,t) = (D't^\alpha)^{-N/2}\,\Psi_N^{(\alpha)}\left(|x|/\sqrt{D't^\alpha}\right), \tag{1.73}$$

where

$$\Psi_N^{(\alpha)}(r) = (4\pi)^{-N/2}\int_0^\infty d\tau e^{-r^2\tau^\alpha/4}\tau^{N\alpha/2}g(\tau;\alpha), \quad \alpha<1 \tag{1.74}$$

is a function of the distance and depends on two parameters: the subdiffusion exponent α and the dimensionality N of the space.

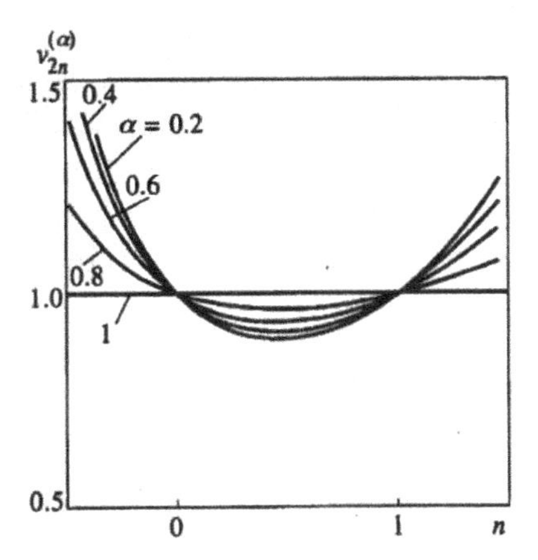

Figure 1.14 Ratio of dimensionless moments, $\nu_{2s}^{(\alpha)}$, for $\alpha = 0.2$; 0.4; 0.6; 0.8 and $\alpha = 1$.

Note that both (1.59) and (1.71) have the same meaning at the limit $\alpha = 1$ and lead to normal diffusion with the same coefficient D'. This means that the function $\Psi_N^{(\alpha)}(r)$ can be redefined so that it holds at $\alpha = 1$:

$$\Psi_N^{(\alpha)}(r) = (4\pi)^{-N/2} e^{-r^2/4}.$$

The subdiffusion distribution in the form (1.73) can be defined as an ordinary diffusion distribution from simple probabilistic considerations based on the GLT. Ignoring the particle dwell time in the diffusion state in our calculations of the distribution of the number of trap events over the observation time $t \to \infty$, we find that

$$p_n \approx Q_0^{(n)}(t) - Q_0^{(n+1)}(t) = G_+\left((nB^*)^{-1/\alpha} t; \alpha\right) - G_+\left([(n+1)B^*]^{-1/\alpha} t; \alpha\right),$$

$$Q_0^{(n)}(t) = \int_t^\infty q_0^{(n)}(t)dt,$$

$$q_0^{(n)}(t) = \int_0^t q_0^{(n-1)}(t - t')q_0(t')dt'.$$

Representing the argument of the subtracted function in the form

$$[(n+1)B^*]^{-1/\alpha} t = [nB^*]^{-1/\alpha} t - [nB^*]^{-1/\alpha} t(n\alpha)^{-1}$$

and expanding in series, we obtain the asymptotic expression

$$p_n \sim [nB^*]^{-1/\alpha} t(n\alpha)^{-1} g_+\left([nB^*]^{-1/\alpha} t; \alpha\right), \quad t \to \infty.$$

When n is fixed, the conditional distribution of the coordinates of a particle can be expressed in terms of the ordinary diffusion density as $\rho(x, t|\nu) \sim p(x, n/\mu)$.

Here the random diffusion time is replaced by the mean value ν/μ for understandable reasons. Averaging over the number of continuous- diffusion events, $\rho(x,t) = \sum_n \rho(x,t|n)p_n$, and replacing summation over n by integration with respect to the variable $\tau = [nB^*]^{-1/\alpha} t$, we obtain the distribution (1.73).

A convenient way to compare our solution with the one obtained by Schneider and Wyss (1989) is to compare the Mellin transforms of the two solutions. According to (1.73),

$$\rho^{as}(s,t) = (1/2)\pi^{-N/2}(4D't^\alpha)^{(s-N)/2}\Gamma(s/2)\int_0^\infty \tau^{(N-s)\alpha/2}g_+(\tau;\alpha)d\tau.$$

Expressing the surviving integral in terms of gamma functions and comparing the result with (1.70) at $D' = 1$, we see that the solutions are identical.

1.5.4 *Analysis of subdiffusion distributions*

In this section we discuss some properties of the solutions, examine their asymptotic behavior at small and large distances, and discuss the results of numerical calculations.

We begin with the spatial moments, which can be explicitly expressed in terms of the Mellin transforms discussed above:

$$\langle |x|^{2n}\rangle = \frac{2\pi^{N/2}}{\Gamma(N/2)}\int_0^\infty r^{2n+N-1}\rho^{as}(r,t)dr = \frac{\Gamma(n+1)\Gamma(N/2+n)}{\Gamma(\alpha n+1)\Gamma(N/2)}(4D't^\alpha)^n.$$

The second moment

$$\langle |x|^2\rangle = \frac{2ND'}{\Gamma(\alpha+1)}t^\alpha, \tag{1.75}$$

which is the same as the one calculated by Schneider and Wyss (formula (1.14)), increases with time in proportion to t^α, $0 < \alpha < 1$, a hallmark of subdiffusion. The ratio

$$\mu_{2n}^{(\alpha)} = \langle |x|^{2n}\rangle / \langle |x|^2\rangle^n = \frac{2^n\Gamma(n+1)\Gamma(N/2+n)\left[\Gamma(\alpha+1)\right]^n}{N^n\Gamma(\alpha n+1)\Gamma(N/2)},$$

which represents dimensionless moments of orders higher than the second, does not depend on time. This suggests that the shape of the distribution remains unchanged @incidentally, this follows immediately from Eq. (1.73). At $\alpha = 1$ the ratio yields the dimensionless moments of the normal distribution:

$$\mu_{2n}^{(1)} = \frac{2^n\Gamma(N/2+n)}{N^n\Gamma(N/2)}.$$

Figure 1 depicts the ratio of the dimensionless absolute moments of order $2s$ $(-1/2 < s < 3/2)$

$$\nu_{2s}^{(\alpha)} = \mu_{2s}^{(\alpha)}/\mu_{2s}^{(1)} = \Gamma(s+1)\left[\Gamma(\alpha+1)\right]^s/\Gamma(\alpha s+1),$$

which characterizes the difference between the shape of the subdiffusion distribution $\Psi_N^{(\alpha)}(r)$ for $\alpha < 1$ and the normal distribution $\Psi_N^{(1)}(r)$. Close to zero ($s < 0$), and at large distances ($s > 1$), $\Psi_N^{(\alpha)}(r)$ exceeds the normal distribution, while in the transitional region ($0 < s < 1$) the opposite is true.

Moving on to analyze the shape of the distribution $\Psi_N^{(\alpha)}(r)$, we note first and foremost that by differentiating (1.74) with respect to r we can easily obtain a relationship between the distributions in N- and $N+1$-dimensional space:

$$\Psi_{N+2}^{(\alpha)}(r) = -\frac{1}{2\pi r}\frac{df_N^{(\alpha)}(r)}{dr}. \tag{1.76}$$

Let $\mathbf{r} = (x_1, \ldots, x_N)$ be an N-dimensional vector, so that $r = \sqrt{x_1^2 + \cdots + x_N^2}$. Integrating (1.74) with respect to the variables $x_{n+1}, \ldots, x_N$, where $1 < n < N$, we obtain

$$\int dx_{n+1} \ldots \int dx_N \Psi_N^{(\alpha)}\left(\sqrt{x_1^2 + \cdots + x_N^2}\right) = \Psi_n^{(\alpha)}\left(\sqrt{x_1^2 + \cdots + x_n^2}\right),$$

which means that the behavior of the projection of a random point in subdiffusive motion in N-dimensional space onto an n-dimensional subspace is described by an n-dimensional subdiffusion equation with the same characteristic exponent α. In this respect the situation is similar to that in normal diffusion, but there is an important difference. In normal diffusion, the coordinates X_1 and X_2 a diffusing particle are mutually independent, while in subdiffusion the joint distribution of the particles,

$$\mathrm{Prob}(X_1 \in dx_1, X_2 \in dx_2) = (4\pi)^{-1}\int_0^\infty d\tau e^{-(x_1^2+x_2^2)\tau^\alpha/4}\tau^\alpha g_+(\tau;\alpha)dx_1 dx_2$$

does not reduce to the product $\mathrm{Prob}(X_1 \in dx_1)$ $\mathrm{Prob}(X_2 \in dx_2)$, with the result that the random coordinates X_1 and X_2 cease to be independent. The nature of their statistical dependence at small and large distances can be clarified by examining the asymptotics.

Equation (1.74) implies that in the one-dimensional ($N = 1$) case the distribution density at the origin exists:

$$\Psi_1^{(\alpha)}(0) = (4\pi)^{-1/2}\int_0^\infty \tau^{\alpha/2} g_+(\tau;\alpha)d\tau,$$

since $g_+(\tau;\alpha)$ has finite moments of order less than α. Using property of stable pdf [Uchaikin & Zolotarev (1999)], we obtain

$$\Psi_1^{(\alpha)}(0) = [2\Gamma(1-\alpha/2)]^{-1}.$$

In spaces with $N \geqslant 2$ the subdiffusion density in contrast to the normal diffusion density! has an integrable singularity. When $N = 2$ this singularity is logarithmic, as can be easily verified by splitting the integral in (1.74) into two parts, the transient and the asymptotic, and replacing the density $g_+(\tau;\alpha)$ in the latter by the leading term in the expansion:

$$\Psi_2^{(\alpha)}(r) \approx (4\pi)^{-1}\left\{\int_0^T e^{-r^2\tau^\alpha/4}\tau^\alpha g_+(\tau;\alpha)d\tau+\frac{1}{\pi}\Gamma(1+\alpha)\sin(\pi\alpha)\int_T^\infty e^{-r^2\tau^\alpha/4}\tau^{-1}d\tau\right\}.$$

When $r \to \infty$, the second term in this sum dominates, which leads to a logarithmic singularity:

$$\Psi_2^{(\alpha)}(r) \sim [4\pi\Gamma(1-\alpha)]^{-1}E_1(r^2T^{\alpha/4}) \sim [2\pi\Gamma(1-\alpha)]^{-1}|\ln r|, \quad r \to 0.$$

In a space with $N \geqslant 3$, the singularity at the origin is hyperbolic:

$$\Psi_N^{(\alpha)}(r) \sim \pi^{-1}(4\pi)^{-N/2}\Gamma(\alpha+1)\sin(\pi\alpha)\int_0^\infty e^{-r^2\tau^\alpha/4}\tau^{(N/2-1)\alpha-1}d\tau =$$

$$= (1/4)\pi^{-N/2}\left[\Gamma(N/2-1)/\Gamma(1-\alpha)\right]r^{-(N-2)}, \quad r \to 0. \tag{1.77}$$

Setting (1.75) $N = 1$ and substituting the asymptotic expression (1.77) into the left-hand side, we find that at:

$$\frac{d\Psi_1^{(\alpha)}(r)}{dr} \to \frac{-2}{\Gamma(1-\alpha)}, \quad r \to \infty.$$

the derivative of the function $x = 0$ has a finite discontinuity, i.e., rather than $\Psi_1^{(\alpha)}(x)$ being smooth as in normal diffusion!, the peak of the distribution in the one-dimensional case is a cusp. As $\alpha \to 1$, the derivative vanishes and the vertex becomes smooth.

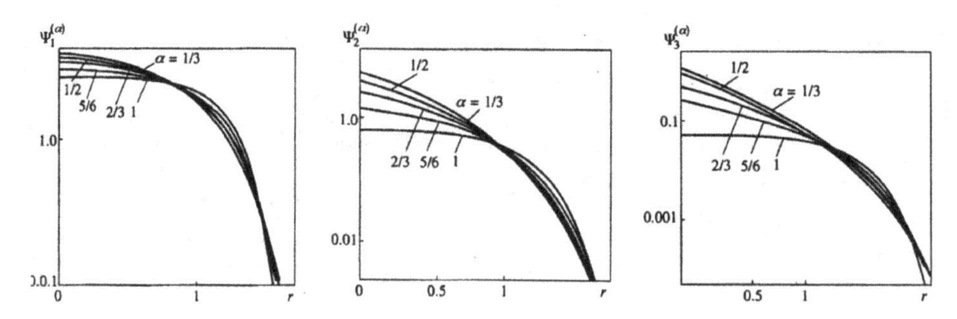

Figure 1.15 Left panel: one-dimensional distributions $\Psi_1^{(\alpha)}(r)$ for $\alpha = 2/6;\ 3/6;\ 4/6;\ 5/6;\ 1$. Middle panel: two-dimensional distributions $\Psi_2^{(\alpha)}(r)$ for the same values of α. Right panel: Three-dimensional distributions $\Psi_3^{(\alpha)}(r)$ for the same values of α.

When r is large, the exponential in the integrand of (1.74) rapidly decreases, so that to make an asymptotic estimate of the integral we use the expression, which

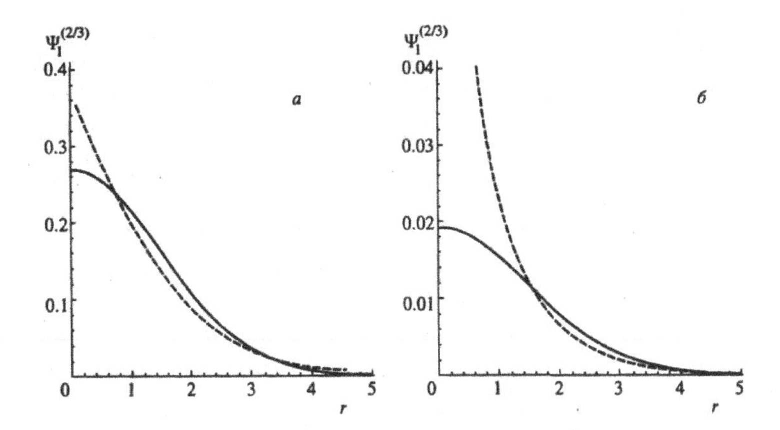

Figure 1.16 Comparison of subdiffusion distributions ($\alpha = 2/3$, solid curves) and normal distributions (dashed curves) with the same variance. Left panel: $N = 1$; right panel: $N = 3$.

approximates the stable density at small values of the argument. We calculate the resulting integral

$$\Psi_N^{(\alpha)}(r) \approx (4\pi)^{-N/2} A \int\limits_0^\infty e^{-r^2\tau^\alpha/4 - b\tau^{-\delta}} \tau^{(N\alpha/2-\gamma)} d\tau \qquad (1.78)$$

$$\varphi(\tau) = -r^2\tau^\alpha/4 - b\tau^{-\delta}. \qquad (1.79)$$

Using the condition $\varphi'(\bar{\tau}) = 0$, we find the position of the maximum of this function:

$$\bar{\tau} = \left(\frac{4b\delta}{\alpha r^2}\right)^{\frac{1}{\delta+\alpha}}. \qquad (1.80)$$

Proceeding in the usual way, we find that

$$\int\limits_0^\infty e^{-r^2\tau^\alpha - b\tau^{-\delta}} \tau^{(N\alpha/2-\gamma)} d\tau \sim \pi^{-N/2} A \bar{\tau}^{-\frac{N\alpha/2-\gamma}{\delta+\alpha}} \sqrt{2\pi/|\varphi''(\bar{\tau})|} e^{\varphi(\bar{\tau})}. \qquad (1.81)$$

Substituting (1.79) and (1.80) into (1.81), and the result into (1.78), we obtain the asymptotic expression (1.78) in the form

$$\Psi_N^{(\alpha)}(r) \sim (4\pi)^{-N/2} \frac{\alpha^{[(N+1)\alpha/2-1]/(2-\alpha)}}{\sqrt{2-\alpha}} \times$$

$$\times (r/2)^{-N(1-\alpha)/(2-\alpha)} \exp\left\{-(2-\alpha)\alpha^{\alpha/(2-\alpha)} (r/2)^{2/(2-\alpha)}\right\}. \qquad (1.82)$$

A comparison with exact calculations carried out below shows that for $\alpha > 1/2$, Eq. (1.82) provides a satisfactory approximation of the distribution over the entire region, except at small distances, and as $\alpha \to 1$ it turns into the normal distribution

$$\Psi_N^{(1)}(r) = (4\pi)^{-N/2} e^{-r^2/4},$$

i.e., it becomes exact.

Figures 1.15 and 1.16 provide an accurate idea of the shape of the distributions. They depict the subdiffusion distributions $f_N^{(\alpha)}(r)$ for several values of α, including the limit $\alpha = 1$, corresponding to normal diffusion [the variances of these distributions are different and depend on α according to (1.75)]. An important difference in the shape of subdiffusion distributions that sets them apart from normal distributions is the higher concentration of probability at both small and large distances. But if these features do not play a significant role in a specific problem, in the one-dimensional case with $\alpha > 1/2$ subdiffusion distributions can indeed be approximated by a Gaussian with subdiffusion variance, as demonstrated by Klimontovich (1995) (see Fig. 1.16). For spaces of higher dimensionality, the normal approximation fails to yield satisfactory results.

1.5.5 *Discussion*

The fact that there are different formulations of the anomalous-diffusion problem and different ways in which the results can be presented sometimes leads to a situation in which researchers fail to see the logical relations among them, even when the same problem is being studied. The approach developed in the present paperfrom the random walk model based on integral equations to the asymptotic part of solutions of these equations that satisfy equations in fractional derivativesmakes it possible not only to express the coefficients of anomalous diffusion in terms the characteristics of "elementary distributions", but also to establish a relationship among the solutions obtained by various means. That such a problem exists can easily be seen by comparing some of the papers devoted to anomalous diffusion [Bouchaud & Georges (1990); Schneider & Wyss (1989)] with Ref. [West et al. (1997)].

As noted above, Schneider and Wyss (1989) used the integral equation (1.67) with a multidimensional Laplacian, found the Mellin and Laplace transforms, expressed the solution in terms of Fox functions, found an approximate expression [that is exactly equivalent to Eq. (1.82)] for the density at large distances, and obtained in the one-dimensional case an exact expression for the density of the form (in our notation)

$$\Psi_1^{\alpha}(r) = \alpha^{-1} r^{-1-2/\alpha} g_+(r^{-2/\alpha}; \alpha/2). \tag{1.83}$$

Taking $N = 1$ and using Eq.(A3), we obtain the same formula, which advantageously differs from Eq.(1.74). Note that, according to the property (1.76), we can express the distributions in spaces with a large odd number of dimensions in terms of the density $g_+(t; \alpha/2)$ and its derivatives. As is well known, however, integration is preferable to differentiation in numerical calculations.

At $\alpha = 2/3$ distribution (1.83) can be expressed, according to (A4), in terms of modified Bessel functions of the second kind:

$$\Psi_1^{(2/3)}(r) = \frac{1}{3\pi}\sqrt{r}K_{1/3}\frac{2r^{3/2}}{\sqrt{27}}.$$

In Sec. 1.2.3.1 of their review, Bouchaud and Georges (1990) discuss the problem in which a particle hops suddenly fromone point to another with a density characterized by a finite value of the rms hop length, and by a distribution of time intervals between successive hops satisfying the condition (1.54). Using the Central Limit Theorem, they found the Laplace transform $\rho^{as}(x, \lambda)$. which is the same as the one obtained by Schneider and Wyss11 [Eqs. (2.8) and (2.10)], established the self-similar behavior of the distribution, i.e., introduced the function $\Psi_N(r)$, and once more derived Eq. (1.83) without any mention of Ref. [Schneider & Wyss (1989)].

The work [Schneider & Wyss (1989)] is mentioned in the Introduction to the paper by West et al. [West et al. (1997)] who nevertheless believed that they solved a different problem [not the one solved by Schneider and Wyss with the use of Eq. (1.66)]. They expressed their solution (only the one-dimensional case was considered) in terms of Fox functions and gave the approximate expression (1.82), derived earlier by Schneider and Wyss (1989). West et al. (1997) gave neither a general formula for the multidimensional case nor the exact solution (1.83) for the one-dimensional case, and the well known review of Bouchaud and Georges (1990) was not cited at all. It must also be noted that by writing the equation in the form

$$\frac{\partial \sigma_0(x,t)}{\partial t} = C\frac{\partial^2}{\partial x^2}\,\frac{\partial^\beta \sigma_0(x,t)}{\partial t^\beta} \tag{1.84}$$

West et al. (1997) used a nonstandard notation for the fractional derivative

$$\frac{\partial^\beta f(t)}{\partial t^\beta} = \frac{1}{\gamma(1-\beta)}\int\limits_0^t \frac{f(t')dt'}{(t-t')^\beta}$$

instead of the standard notation (1.63). As a result, Eq. (1.84) corresponds to Eq. (1.66) at $\beta - 1 = 1 - \alpha$, i.e. $\beta = 2 - \alpha$. With this modification, the results of West et al. (1997) are identical to their analogs in the cited papers and in the present paper, but the improperly defined order of the fractional derivative led them to believe that for $\alpha > 1$ their solution describes subdiffusion (see the remark to Eq. (1.82) and Figs. 2-5 of the cited work, which depict the distributions for $\alpha > 1$). In actuality, however, the paper [?] suggests, the parameter α cannot exceed unity in this problem: even if we put $\alpha > 1$ in the distribution (1.54) where it first appears, the transforms (1.57) and (1.58) will lead, due to (1.56), to an ordinary diffusion equation, i.e., to Eq. (1.66) with $\alpha = 1$. As Ref. [Zolotarev et al. (1999)] suggests, the superdiffusion regime is described by equations containing fractional derivatives with respect to the spatial variables (incidentally, the same is stated in the last section of Ref. [West et al. (1997)]).

The representation of subdiffusion distributions in terms of steady distributions (in contrast to Fox functions) appears to be more convenient, physically clear, and logically justified (in the sense of the limit theorem). The properties of stable distributions have been thoroughly studied and the densities have been tabulated, so that they can be added to the class of special functions [Schneider (1987)].

Chapter 2

Fractional kinetics of dispersive transport

Starting with remarks on significance of the phenomenological approach in natural sciences, we refer to such general properties of the phenomena under consideration as self-similarity of experimental data concerning macroscopic properties of relaxation and transport processes in disordered semiconductors[1]. Among elementary functions, only power functions manifest this property and the use of them directly leads to fractional equations. Further, we discuss identifying specific mechanisms and finding connection between macroscopic parameters and microscopic characteristics of the medium, some contradictions in the existing treatments of the time-of-flight experiments and their resolution in frameworks of the fractional differential approach.

2.1 Macroscopic phenomenology

2.1.1 *A role of phenomenology in studying complex systems*

The standard diffusion theory describing charge kinetics in ordered (for example, crystalline) materials is based on the normal (Gaussian) model. At the same time, transport processes often reveal non-Gaussian property observed in many disordered materials with various microscopic structures: amorphous hydrogenated Si [Hvam & Brodsky (1981); Madan & Shaw (1988)], amorphous selenium [Pfister (1976); Noolandi (1977)], amorphous chalcogenides [Kolomietz et al. (1978); Shutov et al. (1979)], in organic semiconductors, polymers, [Tyutnev et al. (2005); Bässler (1993); Tyutnev et al. (2006)], in porous solids [Averkiev et al. (2000, 2003); Kazakova et al. (2004)], in nanostructured materials [Choudhury et al. (2003)], in polycrystalline films [Ramirez-Bon et al. (1993)], liquid crystals [Boden et al. (1993)], etc. Processes of such kind are joined under the name *anomalous transport* (AT) or the *dispersive transport* (DT). Recall that there is a small difference between terms AT and DT. Saying about AT, we can mean an unusual value of diffusivity or its time- or space-dependence in the framework of the standard diffusion approach, whereas DT is

[1]Of course, in practice, this property is observed only in bounded regions.

interpreted as a complex process consisting of many normal processes with wide distributions of their characteristics. However, the long-tail power type distributions used in DT can hardly be approximated by short-tail normal distributions. The results appear rather cumbersome and do not look natural. Moreover, comparison of the various experimental results manifests the presence of *universal* DT properties which weakly depend on the detailed atomic and molecular structure of matter [Scher & Montroll (1975); Jonscher (1986)].

Scher and Montroll (1975) suggested to consider not normal processes but power-type (Lévy type) distributions themselves as elementary components of a new kinetics taking place in disordered semiconductors. This idea has proved to be very fruitful and was followed by a series of articles that used the CTRW model (CTRF, to be more precise) with waiting time distributions of Lévy type [Scher & Montroll (1975)]. This model gives results which are easy for mathematical representations and physical interpretations. The only disadvantage of this approach is the absence of a continuous transition between normal and anomalous models of transport. Embedding fractional derivatives in the theory [Nigmatullin (1984); Westerlund (1991); Metzler et al. (1998, 1999); Margolin & Berkowitz (2000); Barkai (2001a); Bisquert (2003, 2005); Uchaikin & Sibatov (2008)] removed this unwanted feature and opened opportunities for the development of normal and anomalous kinetics in the framework of unified mathematical formalism. In this book, we will consider a subclass of DT processes called *fractional dispersive transport* (FDT).

The fractional derivative formalism is sometimes criticized for lacking a clear physical interpretation of the proposed equations[2]. Indeed, the fractional equations are often introduced by replacement of integer-order derivatives in classic kinetic equations by fractional ones, and only after this, one tries to give physical explanation of such replacement. In other cases, the equations are derived from assumptions on a microscopic mechanism of a process.

(1) Equations with fractional time-derivative of order 1/2 were offered by Nigmatullin on the base of the frequency response function approach to non-Debye relaxation in systems with residual memory [Nigmatullin (1984)].
(2) Another way of deriving the equations is based on the asymptotic transition from the integral equations of the Scher-Montroll random walk model [Scher & Montroll (1975)] to the fractional equations (see details in [Saichev & Zaslavsky (1997); Uchaikin (1999)]).
(3) Metzler and Klafter (2000a) developed a three-dimensional description of subdiffusion with the transition to macroscopic fractional differential dynamics using the Langevin equation approach. They analyzed also non-Markovian processes by generalization of the Chapman-Kolmogorov equations and showed that the power-law distribution of waiting times asymptotically leads to a special case of

[2]Remarkably, that Dyhne [Dykhne et al. (2004)] initially expressing this reproach has finally arrived at the equation with fractional derivative.

the Klein-Kramers equation leading to fractional generalizations of the Rayleigh and Fokker-Planck equations.

(4) Fractional equations can also be obtained from the balance equations of the multiple trapping model with the exponential density of localized states (for more details, see [Bisquert (2005); Sibatov & Uchaikin (2007)]).

All the above methods are based on some special assumptions on the transport mechanism. These assumptions have to be justified for each particular material under particular experimental conditions. One can ask the question: Can a certain conclusion on the fractional differential character of the transport equation be made immediately from the experimental data?

From physical point of view, the dispersive transport may be explained by involving various mechanisms: multiple trapping of charge carriers into localized states distributed in the mobility gap, hopping conduction assisted by phonons, percolation through conducting states, etc. [Scher & Montroll (1975); Arkhipov et al. (1983); Zvyagin (1984); Tiedje (1984); Nikitenko (2011)]. The variety of approaches reflects a complexity of the systems and processes under consideration. For this reason, the construction of a consistent dispersive transport theory based on first principles is still an unsolved problem. From the other side, experimental data reveal a universal behavior of some important characteristics of the dispersive transport such as time-behaviour of transient photocurrent. This indicates predominance of statistical laws over dynamical ones. It is reasonable to believe that kinetic equations describing such transport processes must have the same form for different materials. The question arises: *is it possible to derive the kinetic equation directly from experimental data without any special assumptions about transport mechanism?* If so, the presupposed derivation may be considered as a phenomenological way to the *generic* equation of dispersive transport in disordered semiconductors.

In connection with this problem, it is very instructive to refer to the Heisenberg article [Heisenberg (1966)]. One of outstanding theoretical physicists of the XX century has defined a phenomenological theory as such a formulation of laws observed in physical phenomena, which does not attempt to completely reduce them to general fundamental "first principles", through which they could be understood. The phenomenological theories always played value-significant role in the development of physics. Referring to semi-empirical laws in meteorology, valency rules, interrelations between radii of atoms and ions, binding energies and excitation energies in chemistry, main interrelations in turbulent hydrodynamics, the Drude dispersive theory, phenomenological thermodynamics and Ptolemaic system in antique astronomy, Heisenberg wrote: "For technical and other applications, they were often more important than the apprehension of relations, and from a purely pragmatic point of view phenomenological theories can make the knowledge of the nature laws to a large extent even redundant."[3]

[3]Not giving an absolute sense to this statement of Heisenberg, we would like to emphasize his belief in the virtue of the phenomenological theories.

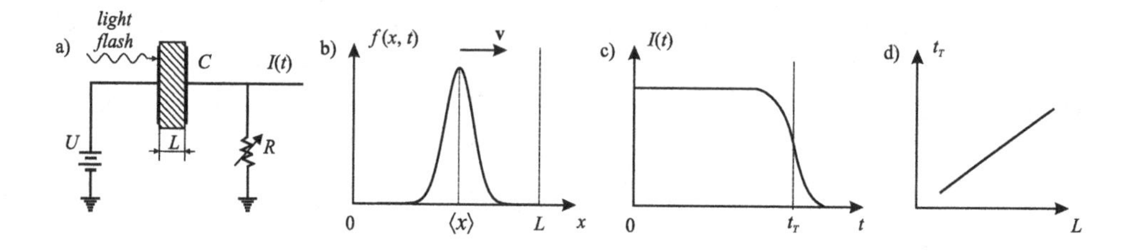

Figure 2.1 Experimental method for determining drift mobility: (a) electrical scheme of the experiment; (b) distribution of charge carriers in normal transport; (c) transient current curve in normal transport, and (d) time-of-flight linear dependence plotted vs. sample thickness.

2.1.2 *Universality of transient current curves*

Let us consider a classical "time-of-flight" experiment for determining the drift mobility of charge carriers. Electrons and holes are usually generated in a sample by a pulse of laser radiation from the side of the semitransparent electrode. The voltage applied to the electrodes is such that the corresponding electric field inside the sample is significantly stronger than the field of nonequilibrium charge carriers. The electrons (or holes, depending on the voltage sign) enter the semitransparent electrode, while holes (or electrons) drift to the opposite electrode. In the case of normal transport, drifting free carriers in the field E give rise to a rectangular photocurrent pulse:

$$I(t) \propto \begin{cases} \text{const}, \ t < t_T, \\ \quad 0, \quad t > t_T, \end{cases} \quad \alpha < 1. \tag{2.1}$$

where the time of flight t_T is given by drift velocity v and sample length L: $t_T = L/v$. Taken together, the scattering of delocalized carriers during the drift, trapping into localized states, and thermal emission of the carriers lead to packet spreading. Such a packet has a Gaussian shape with a mean value of $\langle x(t) \rangle \propto t$ and width $\Delta x(t) \propto \sqrt{t}$. In this case, the transient current $I(t)$ remains constant until the leading edge of the Gaussian packet reaches the opposite edge of the sample. The current decrease takes a time of $\Delta x / \langle v \rangle$. As a result, the right edge of the photocurrent pulse becomes smooth (Fig. 2.1). Such a picture is typical for most ordered materials.

However, when determining drift mobility in certain disordered (amorphous, porous, disordered organic, strongly doped, etc.) semiconductors, a specific signal of transient current $I(t)$, is observed, having two regions with the power-law behavior of $I(t)$ and an intermediate region:

$$I(t) \propto \begin{cases} t^{-1+\alpha}, \ t < t_T, \\ t^{-1-\alpha}, \ t > t_T, \end{cases} \quad \alpha < 1. \tag{2.2}$$

Exponent α, termed the *dispersion parameter*, depends on the medium characteristics and can vary with temperature. Parameter t_T is called *transient time* (or *time of flight*) in analogy with normal transient processes, but has a different physical sense. It has been shown experimentally [Scher & Montroll (1975); Pfister (1976)] that in the dispersive transport regime the following relationship takes place:

$$t_T \propto (L/U)^{1/\alpha}, \tag{2.3}$$

where U is the voltage.

As noted in Refs. [Scher & Montroll (1975); Pfister & Scher (1977)] the shape of the transient current signal in the reduced coordinates $\lg[I(t)/I(t_T)] - \lg[t/t_T]$ is virtually independent of the applied voltage and sample size. This property, inherent in many (but not all: see [Van Roosebroeck (1960)]), materials, is referred to as the property of shape universality of transient current curves (Fig. 2.2). Occurrence of these features in many disordered materials confirms the universality of transport properties. A large number of experimental observations of this universality were reported both in early and recent publications (see for details Refs. [Madan & Shaw (1988); Zvyagin (1984); Scher & Montroll (1975); Jonscher (1983, 1986)]).

Given dependencies (2.2-2.3), transient current curves automatically possess the asymptotic property of universality. In general, current (2.2) depends on size L, strength E etc. through coefficients A and B:

$$I(t) \sim \begin{cases} A(L,E,\alpha,\dots)\, t^{-1+\alpha},\ t < t_T, \\ B(L,E,\alpha,\dots)\, t^{-1-\alpha},\ t > t_T, \end{cases} \quad \alpha < 1. \tag{2.4}$$

Time of flight (transient time) t_T is found from the intersection of asymptotes:

$$I_T = A(L,E,\alpha,\dots)\, t_T^{-1+\alpha} = B(L,E,\alpha,\dots)\, t_T^{-1-\alpha}.$$

Hence, $t_T = (B/A)^{1/2\alpha}$. The property of asymptotic universality means that the function $I(\tau t_T)/I_T$ is independent of t_T. It is easy to see that this property holds for functions with asymptotes (2.2):

$$I(\tau t_T)/I_T \sim \begin{cases} \tau^{-1+\alpha},\ \tau < 1, \\ \tau^{-1-\alpha},\ \tau > 1, \end{cases} \quad \alpha < 1.$$

Observe, that the value of $I(t_T)$ does not equal I_T – the point (t_T, I_T) is determined by intersection of asymptotes of the transient current at small and large times.

Transient photocurrent in a sample of thickness L is defined as conduction current density averaged over thickness:

$$I(t) = \frac{1}{L}\int_0^L j(x,t)dx. \tag{2.5}$$

Evidently, the integral in the last formula is independent of L at the initial stage for $t \ll t_T$. According to relationship (2.3), one finds $t_T \propto L^{1/\alpha}$. Therefore, it is possible to write out for A, B and I_T the following relations:

$$A \propto L^{-1}, \qquad B \propto A\, t_T^{2\alpha} \propto L, \qquad I_T \propto L^{-1/\alpha}.$$

The current density $j(x,t)$ is proportional to the probability flux $q(x,t)$

$$j(x,t) = eNq(x,t),$$

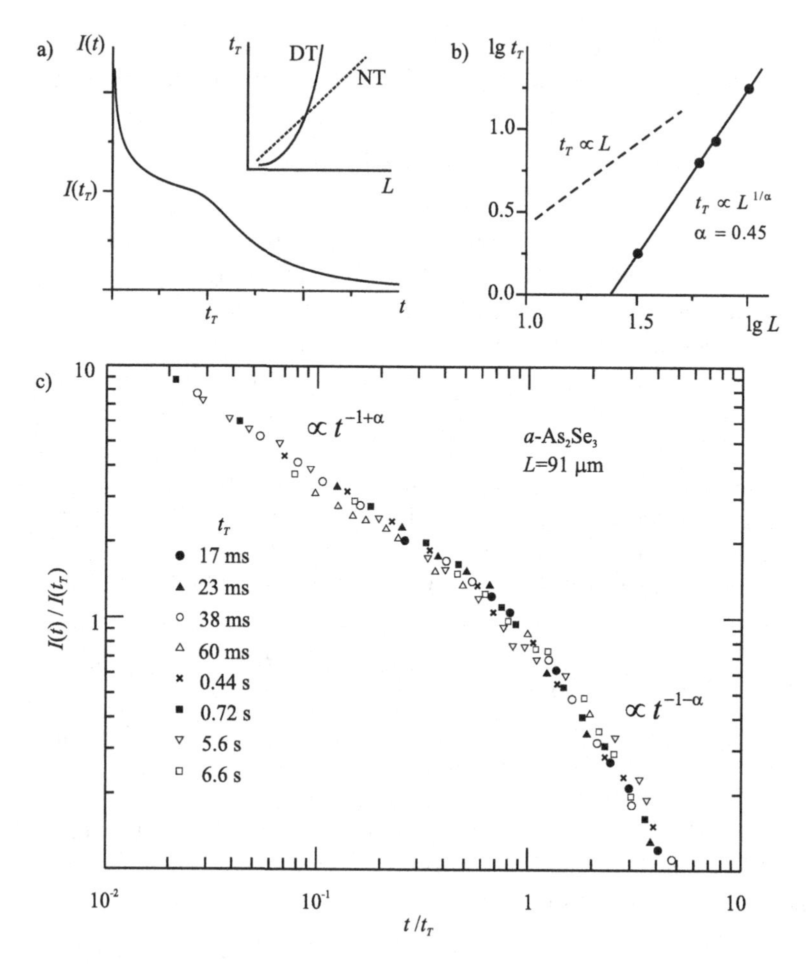

Figure 2.2 (a) Transient photocurrent relaxation curve; inset shows time of flight plotted vs. sample thickness for the cases of dispersive transport (DT) and normal transport (NT); (b) $\lg t_T - \lg L$ dependence; dots are the experimental data for a-As_2Se_3 from Ref. [Scher & Montroll (1975)], solid line is a power-law approximation; c) reduced time dependencies of photocurrent in a-As_2Se_3 in the log-log scale digitized from Ref. [Scher & Montroll (1975)].

where e is the charge of one carrier, and N is the number of photoinjected carriers per unit area of the illuminated electrode. The probability flux is equivalent to the first passage time distribution density $p(t|x)$:

$$q(x,t) \equiv p(t|x).$$

Product $p(t|x)dt$ is the probability that a walking particle (in our case, a charge carrier) will reach coordinate x within a time interval $(t, t+dt)$. The coordinate distribution density $p(x|t)$ of the walking particle and probability flux $p(t|x)$ are related by the probability conservation equation

$$\frac{\partial p(x|t)}{\partial t} + \frac{\partial p(t|x)}{\partial x} = \delta(x)\delta(t). \tag{2.6}$$

In a mathematical sense, the shape universality of transient current curves constitutes a self-similarity property at the time scale:

$$I(t;L_2) \approx (L_2/L_1)^{-1/\alpha} I\left(t(L_2/L_1)^{-1/\alpha};L_1\right), \tag{2.7}$$

where $I(t;L_1)$ and $I(t;L_2)$ are the time dependences of transient current in samples of thickness L_1 and L_2, respectively. According to relationships (2.2), the long-time asymptotes of transient current and the first passage time distribution density follow the power law with exponent α. By substituting first passage time density for conduction current density in formula (2.5), taking into account the shape universality of transient current curves (2.7), it is possible to show that functions $p(t|x)$ are self-similar at the time scales:

$$p(t|x_2) = (x_2/x_1)^{-1/\alpha} p\left(t(x_2/x_1)^{-1/\alpha} \mid x_1\right).$$

Thus, the property of shape universality of transient current curves with respect to variations in sample thickness and the external field intensity is self-similar over time (scaling) upon a change in parameter t_T.

2.1.3 *From self-similarity to fractional derivatives*

The time for a carrier to get an opposite semiconductor boundary is composed of random waiting times elapsed in layers transverse to the electric field. These times are mainly determined by the processes of localization in a disordered semiconductor, but one should note that percolation paths significantly affects the type of waiting time distribution.

Time $\tau(kx)$; $k \in \{1,2,3,\dots\}$ during which coordinate kx is reached is a random variable being the sum of k independent random time intervals during which a carrier passes through the layers $(0,x)$, $(x,2x)$, ..., $((k-1)x,kx)$: $\tau(kx) = \sum_{j=1}^{k} \tau_j(x)$. The corresponding distribution is expressed as a k-fold convolution of the passage time distributions for each layer:

$$p(t|kx) = k^{-1/\alpha} p(tk^{-1/\alpha}|x) = p^{*k}(t|x), \qquad p^{*k}(t|x) = \int_0^t dt\, p(t-t'|x)\, p^{*(k-1)}(t'|x).$$

Applying Laplace transformation

$$\widehat{p}(s|x) = \int_0^\infty dt\; p(t|x)\exp(-st),$$

we arrive at

$$\widehat{p}(s|kx) = \widehat{p}(k^{1/\alpha}s|x) = [\widehat{p}(s|x)]^k.$$

Taking the logarithm of the latter relation and raising it to the $1/\alpha$ power yields

$$\left[\ln \widehat{p}(s|kx)\,\right]^{1/\alpha} = \left[\ln \widehat{p}\left(k^{1/\alpha}s|x\right)\right]^{1/\alpha} = k^{1/\alpha}\left[\ln \widehat{p}(s|x)\,\right]^{1/\alpha}.$$

The latter equalities lead to

$$\left[\ln \widetilde{p}(s|x)\right]^{1/\alpha} = s\left(-\frac{x}{K}\right)^{1/\alpha},$$

where K is a certain constant. For $\widetilde{p}(s|x)$, we have the following expression

$$\widetilde{p}(s|x) = \exp\left(-\frac{x}{K}\,s^{\alpha}\right). \tag{2.8}$$

This function is the Laplace transform of the one-sided stable density with the characteristic exponent α (see, for instance, Refs. [Zolotarev (1983); Uchaikin & Zolotarev (1999); Sibatov & Uchaikin (2010)]):

$$p(t|x) = \frac{1}{2\pi i}\int\limits_C ds\; p(s|x)\exp(st) = \left(\frac{x}{K}\right)^{-1/\alpha} g_+\left(t\left(\frac{x}{K}\right)^{-1/\alpha};\alpha\right). \tag{2.9}$$

Thus, the first passage time distribution density represents the one-sided stable pdf with a characteristic exponent equaling the dispersion parameter.

Substitution of the expression for conduction current density

$$j(x,t) = eNp(t|x) = eN\left(\frac{x}{K}\right)^{-1/\alpha} g_+\left(t\left(\frac{x}{K}\right)^{-1/\alpha};\alpha\right). \tag{2.10}$$

into formula (2.5), leads to

$$I(t) = \frac{eKN\alpha}{L}t^{\alpha-1}\int_{\zeta_0}^{\infty}\zeta^{-\alpha}g_+(\zeta;\alpha)\,d\zeta, \quad \zeta_0 = t\,(L/K)^{-1/\alpha}. \tag{2.11}$$

Fig. 2.3a displays transient current curves calculated by formula (2.11) for different values of the dispersion parameter. Comparison with experimental data [Shutov et al. (1979); Enck & Pfister (1976)] for glassy As_2S_3 and a-As_2Se_3 is presented in Fig. 2.3b.

Asymptotic analysis of expression (2.11) shows that ζ_0 is small when $t \ll t_T$ (the expression for t_T will be derived below). Using the formula for moments of one-sided stable distributions [Uchaikin & Zolotarev (1999)], namely

$$\int\limits_0^{\infty}\zeta^{-\alpha}g_+(\zeta;\alpha)d\zeta = \frac{1}{\Gamma(1+\alpha)},$$

yields

$$I(t) \cong \frac{e\,KN}{L\,\Gamma(\alpha)}\,t^{\alpha-1}, \quad t < t_T. \tag{2.12}$$

At long time scales, $t \gg t_T$ and $\zeta \gg 1$, the density $g_+(\zeta;\alpha)$ behaves as

$$g_+(\zeta;\alpha) \cong \frac{\Gamma(1+\alpha)\sin\pi\alpha}{\pi\zeta^{\alpha+1}},$$

so

$$I(t) \cong \frac{eNL}{2K\,\Gamma(1-\alpha)}\,t^{-\alpha-1}, \quad t \gg t_T. \tag{2.13}$$

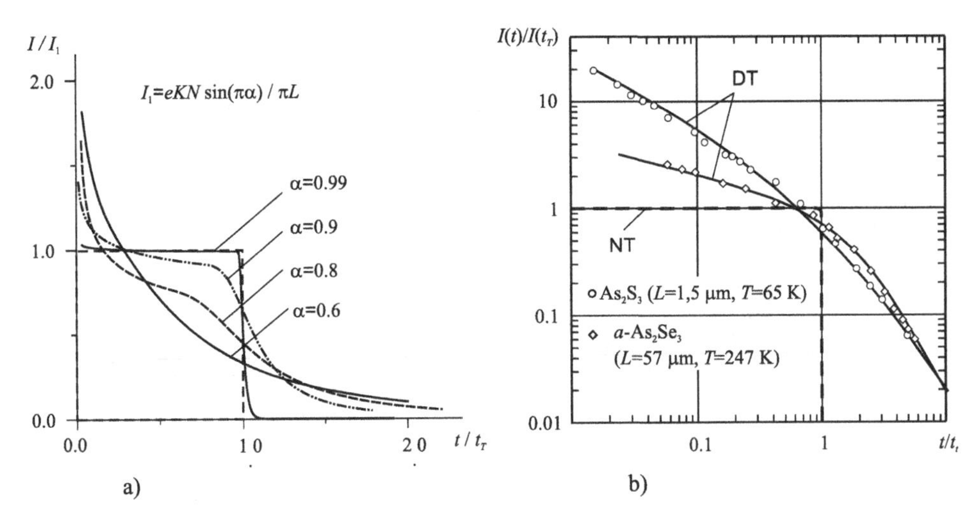

Figure 2.3 (a) Theoretical transient photocurrent curves for different α values. (b) Reduced transient current density: circles are the experimental data for glassy As_2S_3 [Shutov et al. (1979)], rhombi are the data for $a-As_2Se_3$ [Enck & Pfister (1976)], solid lines present calculations by formula (2.11) with adjustable parameters α and t_T.

Equating expressions (2.12) and (2.13), we find the following relation for parameter t_T:

$$t_T = \left(\frac{L}{K} \sqrt{\frac{\Gamma(\alpha)}{2\Gamma(1-\alpha)}} \right)^{1/\alpha}. \tag{2.14}$$

This formula implies that the relation $t_T \propto (L/E)^{1/\alpha}$ [Scher & Montroll (1975)] is experimentally confirmed at a fixed temperature. Drift mobility is defined as

$$\mu_D = \frac{L}{Et_T} = \frac{L}{E} \left(\frac{L}{K} \sqrt{\frac{\Gamma(\alpha)}{2\Gamma(1-\alpha)}} \right)^{-1/\alpha}. \tag{2.15}$$

It follows from Eqs. (2.6) and (2.9), that the carrier concentration $n(x,t)$ is given by

$$n(x,t) = Np(x|t) = -N\frac{\partial}{\partial x}\int_0^t p(t|x)dt = \frac{Nt}{\alpha K}\left(\frac{x}{K}\right)^{-1/\alpha-1} g_+\left(t\left(\frac{x}{K}\right)^{-1/\alpha}; \alpha\right). \tag{2.16}$$

Transient current is expressed through the total concentration of carriers according to the formula (see [Zvyagin (1984)] for details):

$$I(t) = \frac{e}{L}\frac{d}{dt}\int_0^L (x-L)\, n(x,t)dx.$$

Substituting carrier concentration (2.16), into this formula as a check, we come back to the formula (2.11).

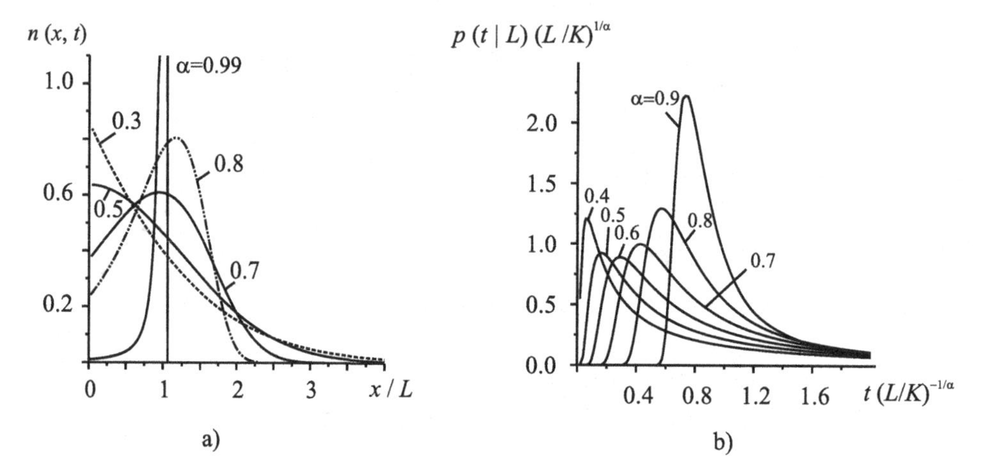

Figure 2.4 (a) Coordinate distribution densities in case of dispersive drift. (b) First passage time distribution densities.

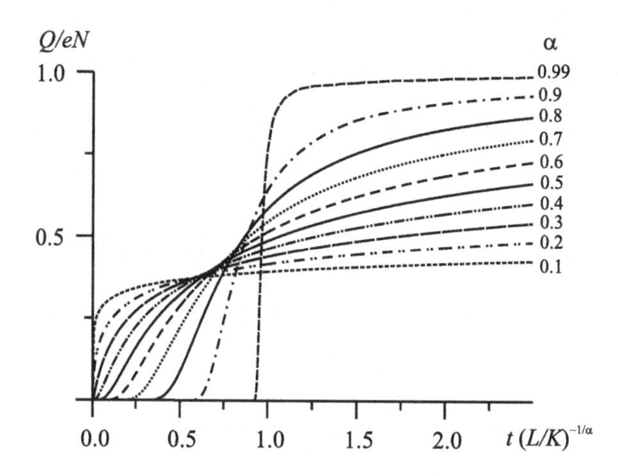

Figure 2.5 Time dependence of charge passed through the sample for different values of the dispersion parameter α.

The pdfs of the reduced coordinate and the first passage time in the case of dispersive transport in stationary homogeneous electric field are shown in Fig. 2.4. Plots in Fig. 2.4a and 2.4b were calculated by formulas (2.16) and (2.9), respectively. Time dependence of charge passed through the sample

$$Q(t) = \int_0^t j(L,t)dt = eN\ G_+\left(t\,(L/K)^{-1/\alpha};\alpha\right),$$

is presented in Fig. 2.5.

The transform of the first passage time pdf (2.8) satisfies the equation

$$\frac{s^\alpha}{K}\ \widehat{p}(s|x) + \frac{\partial}{\partial x}\ \widehat{p}(s|x) = \delta(x).$$

Taking into account that the inverse Laplace transformation gives

$$s^\alpha \widehat{p}(s|x) \mapsto \frac{1}{\Gamma(1-\alpha)} \frac{\partial}{\partial t} \int_0^t \frac{p(t'|x)}{(t-t')^\alpha} dt' = {}_0\mathsf{D}_t^\alpha p(t|x),$$

we arrive at the fractional equation for density $p(t|x)$:

$$K^{-1}\,{}_0\mathsf{D}_t^\alpha p(t|x) + \frac{\partial}{\partial x} p(t|x) = \delta(x)\delta(t).$$

Formula (2.10) implies that an analogous equation holds true for conduction current density and free carrier concentration.

The Laplace image of normalized concentration (2.16), taking the form:

$$\widehat{n}(x,s) = -N \frac{\partial}{\partial x} \frac{\widehat{p}(s|x)}{s} = -N \frac{\partial}{\partial x} \frac{1}{s} \exp\left(-\frac{x}{K} s^\alpha\right)$$

$$= N \frac{s^{\alpha-1}}{K} \exp\left(-\frac{x}{K} s^\alpha\right) = N \frac{s^{\alpha-1}}{K} \widehat{p}(s|x) \tag{2.17}$$

is the solution of the equation

$$s^\alpha \widehat{n}(x,s) + K \frac{\partial}{\partial x} \widehat{n}(x,s) = N s^{\alpha-1} \delta(x),$$

the inverse transformation of which yields the fractional equation for dispersive drift:

$${}_0\mathsf{D}_t^\alpha n(x,t) + K \frac{\partial}{\partial x} n(x,t) = \frac{N t^{-\alpha}}{\Gamma(1-\alpha)} \delta(x). \tag{2.18}$$

The universality of transient current curves and the power-law dependence of transient time on sample thickness lead to the conclusion that the first passage time (without additional assumptions as regards the transport mechanism) is distributed according to a one-dimensional stable law with the characteristic exponent being equal to the dispersion parameter. Such time distribution accounts for the presence of power-law asymptotics of transient current at long and short time scales. Moreover, conduction current density, free carrier concentration, and total carrier concentration are the solutions of fractional equations. It is also worth to note that the form of stable transient current density agrees with the Scher-Montroll hypothesis [Scher & Montroll (1975)] for power-law distribution of the times of carrier residence in localized states during dispersive transport.

2.1.4 *From transient current to waiting time distribution*

The waiting time pdf is related to important physical characteristics of both the transport process and the medium wherein this transport occurs. Such characteristics comprise the energy density of localized states in the case of multiple trapping and trap spatial distribution in the case of tunneling. Transport under certain physical conditions can be described using a phenomenological integral equation based on waiting time distribution density alone, even if the transport mechanism remains unknown. The question is how to deduce waiting time distribution from transient current curves.

The proposed algorithm is applicable on the following assumptions inherent in the standard time-of-flight experiment:

1. Distribution of traps in semiconductor sample is homogeneous.
2. The external electric field is high compared with the field of nonequilibrium carriers.
3. The carriers move in the positive direction of the x-axis.

The transient current is determined from formulas

$$I(t) = \frac{1}{L}\int_0^L j(x,t)dx, \qquad j(x,t) = eNp(t|x).$$

If the sample is mentally divided into identical layers transverse to the electric field, a carrier that crosses n layers has the coordinate roughly equal to the product of the number n by elementary layer width l: $x \approx nl$. Propagation time along coordinate x is the sum of n random residence times in individual layers; in distribution terms, it is written with the help of convolution

$$p(t|x) = \psi^{*n}(t),$$

where $\psi(t)$ is the residence time distribution density in each elementary layer. Substituting the latter expression into the formula for transient current gives

$$I(t) = \frac{eN}{L}\int_0^L \psi^{*n}(t)dx.$$

The Fourier transform converts the convolution of functions into the product of their Fourier images and makes the integral easy to calculate:

$$\widetilde{I}(\omega) = \frac{eN}{L}\int_0^L \left[\widetilde{\psi}(\omega)\right]^{x/l} dx = \frac{eN}{L}\int_0^L \exp\left(\frac{x}{l}\ln\widetilde{\psi}(\omega)\right) dx$$

$$= \frac{eNl}{L\ln\widetilde{\psi}(\omega)}\left[\exp\left(\frac{L}{l}\ln\widetilde{\psi}(\omega)\right) - 1\right]$$

The solution of this equation is written as

$$\ln\widetilde{\psi}(\omega) = -\frac{l}{L}W\left(-\frac{eN}{\widetilde{I}}\exp\left(-\frac{eN}{\widetilde{I}}\right)\right) - \frac{eNl}{L\widetilde{I}}, \tag{2.19}$$

where $W(x)$ is the Lambert function (the solution of the transcendent equation $W\exp(W) = x$).

Thus, by finding the Fourier transform of the transient current curve, calculating transformant (2.19) by formula $\widetilde{\psi}(\omega)$ and inverting it, we obtain the waiting time distribution density.

Knowing the waiting time distribution, it is possible to write out the integral transport equation for a regular homogeneous medium controlled by trapping in localized states. As before, it is assumed that the trap concentration in any region of a semiconductor is much higher than the concentration of the carriers involved in the transport. Free and localized carrier densities are related by the equation

$$n_t(\mathbf{r},t) = \tau_0^{-1} \int_0^t n_d(\mathbf{r},t')\, Q(t-t')\, dt', \tag{2.20}$$

where $Q(t) = 1 - \Psi(t) = \mathrm{Prob}(\tau > t)$ is the additional distribution function of carrier residence time in traps. Concentration $n_t(\mathbf{r},t)$ is independent of the localized state energy, and $Q(t)$ is the energy-averaged waiting time distribution function in localized states:

$$Q(t) = \left\langle \exp\left(-\omega_\varepsilon t\, \mathrm{e}^{-\varepsilon/kT}\right)\right\rangle = N_t^{-1} \int_0^{\varepsilon_g} \rho(\varepsilon) \exp\left(-\omega_\varepsilon t\, \mathrm{e}^{-\varepsilon/kT}\right) d\varepsilon.$$

The density of delocalized carriers can be expressed through the trapped carrier density with the help of the integral equation

$$n_d(\mathbf{r},t) = \tau_0 \int_0^t n_t(\mathbf{r},t')\, Q^*(t-t')\, dt', \tag{2.21}$$

where $Q^*(t)$ is an integral kernel that can be defined in terms of the Laplace transform:

$$\tilde{R}(s) = [\widehat{Q}(s)]^{-1}.$$

After substituting expression (2.20) or (2.21) into continuity equation (2.53) with the use of formula (2.52), we obtain transport equations for the mobile carrier concentration:

$$\frac{\partial n_d(\mathbf{r},t)}{\partial t} + \tau_0^{-1} \frac{\partial}{\partial t} \int_0^t n_d(\mathbf{r},t')\, Q(t-t')\, dt' + \mathrm{div}[\, \mu \mathbf{E}\, n_d(\mathbf{r},t) - D\nabla n_d(\mathbf{r},t)] = 0$$

and for the total concentration of nonequilibrium carriers:

$$\frac{\partial n(\mathbf{r},t)}{\partial t} + \tau_0 \int_0^t dt'\, R(t-t')\, \mathrm{div}[\, \mu \mathbf{E}\, n(\mathbf{r},t') - D\nabla n(\mathbf{r},t')] = 0, \quad n = n_t + n_d \approx n_t.$$

2.2 Microscopic backgrounds of dispersive transport

2.2.1 *From the Scher-Montroll model to fractional derivatives*

CTRF model, introduced in the famous paper [Scher & Montroll (1975)], provided the first detailed explanation of all the main patterns of current behavior in time-of-flight experiments with amorphous semiconductors.

The main assumptions of this model are as follows:

(1) The transport of charge carriers is a jump random process in which the walkers change their positions by instantaneous random jumps ("flights") at random instants of time.
(2) Carrier jumps are independent of one another, and time intervals between them (waiting times) are independent, identically distributed random variables τ.
(3) Waiting times are characterized by a broad power-law distribution:

$$\mathsf{Prob}(\tau > t) \propto t^{-\alpha}, \quad t \to \infty, \qquad 0 < \alpha < 1. \tag{2.22}$$

Power-law asymptotics of transient current (2.2) were for the first time explained by Shlesinger [Schlesinger (1974)], based on the CTRW model. The power-law waiting time distribution in the hopping transport model was deduced from *ab initio* calculations done by Tunaley [Tunaley (1972b)], and Scher and Lax [Scher & Lax (1972)] in 1972. The authors of Ref. [Scher & Montroll (1975)] simulated charge transfer in disordered semiconductors as carrier hopping within the model grid of localized states. The grid constitutes a regular cubic lattice, each cell of which contains randomly distributed sites (localized states). The waiting time till the next hopping depends on the distance to the nearest neighbor sites. As shown in Ref. [Scher & Montroll (1975)], the cell residence time distribution can be of the power law type due to spatial disorder of sites in a cell.

Assumption (2.22) is of fundamental importance because such a time distribution having an infinite mean value leads to anomalous transient current behavior, whereas any distribution with a finite mean waiting time, e.g., exponential distribution, produces a normal signal [?]. The reason, why the infinite mean value does not contradict the physical sense, was plausibly explained in Ref. [Tunaley (1972b)]. The divergence of integral

$$\lim_{t\to\infty} \int_0^t \psi(t')\, t'\, dt' = \infty$$

actually means only the impossibility of using $\langle T \rangle$ for calculations. Because

$$\lim_{t\to\infty} \int_t^\infty \psi(t') dt' = 0,$$

all observed realizations of T are finite, and there is no paradox.

In the Scher-Montroll model, waiting times τ (positive random quantities) are distributed according to the asymptotically power law, which means that at macroscopic scales much greater than the average distance between localized states the first passage time distribution, conduction current density, and concentration of delocalized carriers considered as functions of time should have the form of stable density distribution in agreement with formulas (2.9) and (2.10) obtained in the previous section.

The main characteristics of the CTRF model are pdfs of waiting times and path lengths. Let us denote densities of these distributions as $\psi(t)$ and $p(\mathbf{r})$. Coordinate distribution density $p(\mathbf{r}, t)$ of a particle executing the random walks and initially located at the origin of coordinates is defined in terms of the Fourier-Laplace transform [Montroll & Weiss (1965)]

$$\widehat{\widetilde{p}}(\mathbf{k},s)=\frac{1-\widetilde{\psi}(s)}{s}\frac{1}{1-\widehat{p}(\mathbf{k})\widetilde{\psi}(s)}, \tag{2.23}$$

where

$$\widehat{\widetilde{p}}(\mathbf{k},s)=\int\limits_{\mathbb{R}^3} d\mathbf{r}\; e^{i\mathbf{kr}}\int\limits_0^\infty dt\; e^{-st}p(\mathbf{r},t)$$

is the Fourier-Laplace transform of normalized particle concentration, $\widetilde{p}(\mathbf{k})$ is the Fourier transform of path distribution density, and $\widehat{\psi}(s)$ is the Laplace transform of waiting time distribution density. Substituting into Eq. (2.23) the asymptotic series expansion of the Laplace image of waiting time distribution density with the power-law tail

$$\widehat{\psi}(s)\sim 1-s^\alpha/c^\alpha, \quad s\ll c,$$

along with asymptotic expansion of the Fourier image of path distribution density:

$$\widetilde{p}(\mathbf{k})\sim 1+i\mathbf{c}_1\cdot\mathbf{k}-c_2k^2, \quad |\mathbf{k}|\ll 1/|\mathbf{c}_1|,$$

and applying the Tauberian theorem [Feller (1967)], we shall obtain

$$\widehat{\widetilde{p}}(\mathbf{k},s)=\frac{c^{-\alpha}s^{\alpha-1}}{-i\mathbf{c}_1\cdot\mathbf{k}+c_2k^2+s^\alpha/c^\alpha}.$$

Rewriting the last expression in the form

$$[s^\alpha-i\mathbf{c}_1\cdot\mathbf{k}c^\alpha+c_2c^\alpha k^2]\,\widehat{\widetilde{p}}(\mathbf{k},s)=s^{\alpha-1}$$

and applying inverse Fourier and Laplace transformations yield

$${}_0\mathsf{D}_t^\alpha p(\mathbf{r},t)+\mathbf{K}.\nabla p(\mathbf{r},t)-C\nabla^2 p(\mathbf{r},t)=\frac{t^{-\alpha}}{\Gamma(1-\alpha)}\delta(\mathbf{r}). \tag{2.24}$$

This equation in a one-dimensional case with $C=0$ coincides with equation (2.18) obtained in the previous section.

Thus, fractional equations and their solutions are consistent with the Scher-Montroll random walk model.

As shown in Sec. 2.1.3, fractional equations of dispersive transport for a time-of-flight experiment can be obtained based on the shape universality of transient current curves and the power-law dependence of the flight time on sample thickness. In the present section, we propose more general equations for a three-dimensional case taking into account the diffusion term.

2.2.2 Physical basis of the power-law waiting time distribution

The physical foundations of a broad power-law waiting time distribution (2.22) have been discussed in many studies. It was shown in Ref. [Silver & Cohen (1977)] that such distribution underlain by a multiple trapping mechanism may be a consequence of exponential energy distribution of localized states. The exponential form of the density of band tail states was extensively used in model calculations and discussions of the results of experiments on transport and luminescence in disordered semiconductors [Silver & Cohen (1977); Spear (1991)]. However, as mentioned in Refs. [Arkhipov et al. (1988); Arkhipov & Perova (1993)], the main features of carrier packet propagation controlled by carrier capture in localized states may be just as well retained in other, sufficiently broad energy distributions of traps. Transport in a purely hopping model occurs via tunneling between adjacent traps, and power-like waiting time distribution results from dispersion of distances between localized states [Tiedje (1984)]. Dispersive transport is observed in disordered semiconductors of a different nature. It is worthwhile to study the common mathematical principles accounting for anomalous transport in different physical models.

Random waiting time T in a given localized state is characterized by the probability:

$$\text{Prob}(T > t) = \exp(-t/\theta), \tag{2.25}$$

where the parameter θ is the mean residence time in a given trap.

In a model of transport controlled by multiple trapping (see, for instance, Refs. [Noolandi (1977); Arkhipov et al. (1983); Tiedje & Rose (1981)]) it is used in the form of Arrhenius law:

$$\theta = w_\varepsilon^{-1} \exp(\varepsilon/kT), \tag{2.26}$$

where w_ε is the carrier capture rate in a trap with energy ε.

In the tunnelling transport model proposed in Ref. [Tunaley (1972c)], this parameter is represented by the Sommerfeld-Bethe formula

$$\theta = \beta[\exp(\gamma d) - 1], \tag{2.27}$$

where d is the distance to the neighbor site, γ is the positive constant, and the parameter β is inversely proportional to the applied field strength.

In the Miller-Abrahams random grid model for hopping conductivity (see, e.g., Ref. [Shklovskiy & Efros (1979)]), the mean residence time is defined as

$$\theta = \left[\nu_0 \sum_j \exp\left(-\frac{2d_{ij}}{a} - \frac{\varepsilon_{ij}}{kT}\right)\right]^{-1}, \tag{2.28}$$

where ν_0 is the characteristic carrier hopping rate, a is the localization radius, d_{ij} is the distance between the i-th and j-th sites, and, ε_{ij} is the corresponding energy difference.

The lack of long-range order, dispersion of intertrap distances, and fluctuations of localized state energy in disordered semiconductors make use to consider parameter θ as a random variable. As follows from formulas (2.26-2.28) a weak dispersion of intertrap distances and/or localized state energies result in a broad distribution of this variable, that may have a "heavy" power-law tail.

Let us consider the case specified by formula (2.27). It is assumed in this model that carriers undergo one-dimensional motion along the positive direction of the x-axis, given by the direction of the applied field. Potential wells are randomly distributed along this axis. In a quasiclassical approximation, the distribution of random time τ during which a charge carrier resides at a selected site is given by probability (2.25).

On the assumption of uniform potential well distribution along the straight line, the probability of interest equals

$$\mathrm{Prob}(d > x) = \exp(-x/d_0),$$

hence, one arrives at

$$\mathrm{Prob}(\theta > t) = \mathrm{Prob}\left(\beta[\exp(\gamma d) - 1] > t\right) = \frac{1}{(1 + t/\beta)^{\alpha_1}}, \tag{2.29}$$

where

$$\alpha_1 = 1/\gamma d_0.$$

According to formula (2.25) conditional probability at a given θ equals:

$$\mathrm{Prob}(\tau > t|\theta) = \exp(-t/\theta),$$

and it follows from relationship (2.29) that θ has the distribution density

$$p_\theta(t) = \frac{\alpha_1}{\beta(1 + t/\beta)^{\alpha_1+1}}.$$

Then, the absolute (i.e., averaged over θ) probability for the residence time of the carrier in a site takes the form

$$\mathrm{Prob}(\tau > t) = \int_0^\infty \mathrm{Prob}(\tau > t\ |\theta = t')p_\theta(t')dt' \sim \alpha_1\Gamma(\alpha_1)\left(\frac{t}{\beta}\right)^{-\alpha_1}, \quad t \gg \beta. \tag{2.30}$$

As shown in Ref. [Scher & Montroll (1975)] the dispersive behavior may be a consequence of multiple trapping in localized states distributed by energies. The assumption that the density of states below the mobility edge exponentially falls off with energy may lead to formula (2.2) with the parameter α_1, depending on temperature T. In the case of exponential tail of the density of states, namely

$$\rho(\varepsilon) \propto \exp(-\varepsilon/\varepsilon_0),$$

assuming that the capture rate weakly depends on trap energy $w_\varepsilon \approx w_0$, we obtain

$$p_\theta(t) = -\frac{d}{dt}\mathrm{Prob}(\theta > t) = -\frac{d}{dt}\mathrm{Prob}(w_0^{-1}\exp(\varepsilon/kT) > t)$$

$$= -\frac{d}{dt}(w_0 t)^{-\alpha_1} = \alpha_1 w_0 (w_0 t)^{-\alpha_1 - 1},$$

where

$$\alpha_1 = kT/\varepsilon_0.$$

The absolute probability for the residence time of a carrier in the localized state is equal to

$$\text{Prob}(\tau > t) = \int_0^\infty \text{Prob}(\tau > t \mid \theta = t') p_\theta(t') dt' \sim \alpha_1 \Gamma(\alpha_1)(w_0 t)^{-\alpha_1}, \quad t \gg w_0^{-1}. \tag{2.31}$$

Naturally, energy and space distributions of localized states in most cases differ from exponential ones. Notice that exponential distribution has a methodological deficiency of being characterized by a single parameter: the mean value of this distribution equals the root-mean-square deviation. Varying this parameter, it is impossible to theoretically determine changes in transport properties during transition from an ordered to a disordered position of trapping centers.

Let us consider the multiple trapping mechanism at greater length. Take a gamma-distribution, instead of an exponential one, and write its normalized density in the form

$$\rho(\varepsilon) = \frac{\varepsilon_0}{\sigma_\varepsilon^2 \, \Gamma(\varepsilon_0^2/\sigma_\varepsilon^2)} \left(\frac{\varepsilon_0 \, \varepsilon}{\sigma_\varepsilon^2}\right)^{-1+\varepsilon_0^2/\sigma_\varepsilon^2} \exp\left(-\frac{\varepsilon_0 \, \varepsilon}{\sigma_\varepsilon^2}\right), \tag{2.32}$$

where ε_0 and σ_ε^2 are the mean level depth and mean square of energy fluctuations of localized states, respectively. Plots of gamma-distribution densities are depicted in Fig. 2.6. Parameter σ_ε represents the localized state distribution width and thereby characterizes the degree of semiconductor disorder. If σ_ε tends to zero, we have a crystalline semiconductor model with a single trapping level: $\rho(\varepsilon) \to \delta(\varepsilon - \varepsilon_0)$. If $\sigma_\varepsilon = \varepsilon_0$, we arrive at an exponential tail of the density of localized states, and $\rho(\varepsilon) = \varepsilon_0^{-1} \exp(-\varepsilon/\varepsilon_0)$.

$$p_\theta(t) = -\frac{d}{dt}\text{Prob}(\theta > t) = -\frac{d}{dt}\text{Prob}\left(w_0^{-1}\exp(\varepsilon/kT) > t\right)$$

$$= -\frac{d}{dt}\text{Prob}(\varepsilon > kT \ln w_0 t) = \rho(kT \ln w_0 t)\,\frac{kT}{t}.$$

Thus, for the distribution density of parameter θ, one obtains

$$p_\theta(t) = \frac{\varepsilon_0 kT}{w_0 \sigma_\varepsilon^2 \, \Gamma(\varepsilon_0^2/\sigma_\varepsilon^2)} \left(\frac{\varepsilon_0 kT}{\sigma_\varepsilon^2} \ln w_0 t\right)^{-1+\varepsilon_0^2/\sigma_\varepsilon^2} (w_0 t)^{-1-kT\varepsilon_0/\sigma_\varepsilon^2}.$$

Then, the probability for waiting time distribution is defined as

$$\text{Prob}(\tau > t) = \int_0^\infty \text{Prob}(\tau > t \mid \theta = t') p_\theta(t') dt' \sim u(t)(w_0 t)^{-\alpha_1}, \tag{2.33}$$

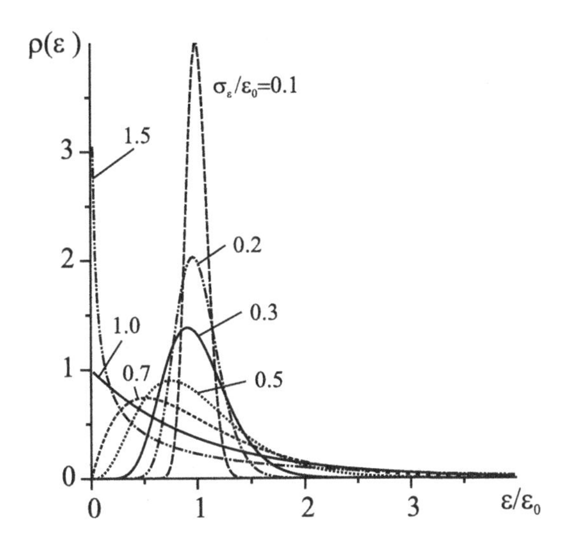

Figure 2.6 Gamma density for different values of $\sigma_\varepsilon/\varepsilon_0$.

$$t \gg w_0^{-1}, \quad \alpha_1 = kT\frac{\varepsilon_0}{\sigma_\varepsilon^2},$$

where $u(t)$ is a slowly varying function. Dispersive transport corresponds to $\alpha_1 < 1$. In accordance with expression (2.33), this condition is satisfied if the mean square σ_ε^2, of energy fluctuations of localized states, characterizing semiconductor disorder, is greater than product $kT\varepsilon_0$.

In the tunneling model proposed in Ref. [Tunaley (1972c)], for the dispersion parameter, one has

$$\alpha_1 = d_0/(\sigma_d^2\gamma),$$

where d_0 and σ_d^2 are the mean value and root-mean-square deviation of random distance between adjacent sites; $\sigma_d^2 > d_0/\gamma$, the semiconductor satisfies the dispersive transport condition. In other words, the successive tunneling is dispersive when fluctuations squared of the intertrap distance exceed the mean hopping length multiplied by the half-radius of wave function localization.

2.2.3 *Multiple trapping regime*

In the multiple trapping model, charge carriers are separated into quasi-free (delocalized) carriers and captured in traps (localized). Relationship of their concentrations $n_d(\mathbf{r},t)$ and $n_t(\mathbf{r},t;\varepsilon)$ is expressed by

$$n(\mathbf{r},t) = n_f(\mathbf{r},t) + \int_0^\infty n_t(\mathbf{r},t;\varepsilon)\, d\varepsilon. \tag{2.34}$$

Undergoing scattering by phonons and irregularities, delocalized carriers drift in an external electric field $\mathbf{E}$ with an average velocity $\mathbf{v} = \mu\mathbf{E}$, where μ is the mobility of quasi-free carriers. In framework of the dispersive transport, the vast majority of carriers captured traps, and only free carriers are responsible for the charge transfer.

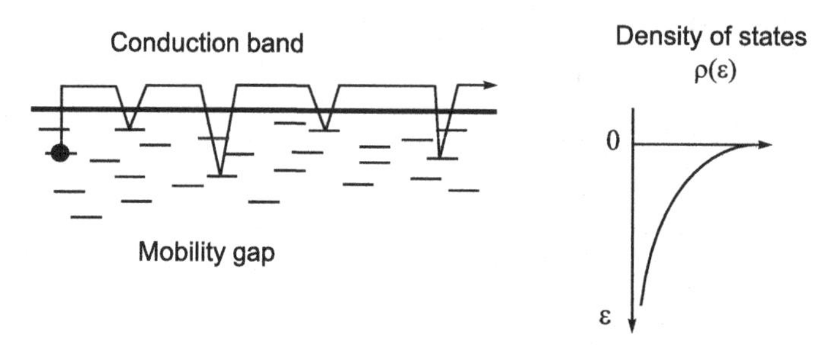

Figure 2.7 Scheme of multiple trapping.

The continuity equation looks as usual:

$$\frac{\partial n(\mathbf{r},t)}{\partial t} + \mathrm{div}\left\{ \mu\mathbf{E}\; n_d(\mathbf{r},t) - D\nabla n_d(\mathbf{r},t)\right\} = 0. \tag{2.35}$$

On assumption on poor trap population, the capture-emission kinetic equation is written in the form

$$\frac{\partial n_t(\mathbf{r},t;\varepsilon)}{\partial t} = w_\varepsilon \rho(\varepsilon) n_d(\mathbf{r},t) - \frac{N_c}{N_t}\; w_\varepsilon e^{-\varepsilon/kT} n_t(\mathbf{r},t;\varepsilon).$$

Being solved with respect to $n_t(\mathbf{r},t;\varepsilon)$, this equation leads to the following interrelation between the concentrations of free and trapped carriers:

$$n_t(\mathbf{r},t;\varepsilon) = \int_{-\infty}^{t} n_d(\mathbf{r},\tau)\; w_\varepsilon \rho(\varepsilon) \exp\left\{ -w_\varepsilon \frac{N_c}{N_t} e^{-\varepsilon/kT}(t-\tau)\right\} d\tau. \tag{2.36}$$

Inserting it in Eq. (2.34) and then in Eq. (2.35), we obtain the integro-differential equation

$$\frac{\partial n_d(\mathbf{r},t)}{\partial t} + \frac{\partial}{\partial t}\int_{-\infty}^{t} n_d(\mathbf{r},\tau)\; Q(t-\tau)d\tau$$
$$+\mathrm{div}\left\{ \mu\mathbf{E}\; n_f(\mathbf{r},t) - D\nabla n_f(\mathbf{r},t)\right\} = 0, \tag{2.37}$$

with the kernel Q determined by integrating over trap energies:

$$Q(t) = \int_0^\infty w_\varepsilon \exp\left\{ -w_\varepsilon t\frac{N_c}{N_t} e^{-\varepsilon/kT}\right\} \rho(\varepsilon)d\varepsilon.$$

In the case of a weak energy dependence of trapping coefficient $w_\varepsilon \approx w_0$ and exponential density of states

$$\rho(\varepsilon) = \varepsilon_0^{-1}\exp(-\varepsilon/\varepsilon_0),$$

we obtain:

$$Q(t) = \frac{w_0}{\varepsilon_0}\int_0^\infty \exp\left\{ -w_0 t\frac{N_c}{N_t}\; e^{-\varepsilon/kT}\right\} \exp(-\varepsilon/\varepsilon_0)d\varepsilon.$$

Making the change of integration variable $\xi = w_0 t\;(N_c/N_t)\; e^{-\varepsilon/kT}$, yields

$$Q(t) = \frac{w_0\alpha}{(w_0 t N_c/N_t)^\alpha}\int_0^{w_0 t\cdot N_c/N_t} e^{-\xi}\xi^{\alpha-1}d\xi \sim \frac{w_0\alpha\Gamma(\alpha)}{(w_0 N_c/N_t)^\alpha}\, t^{-\alpha}, \quad t\to\infty, \quad \alpha = \frac{kT}{\varepsilon_0}.$$

2.2.4 *Hopping conductivity*

If the phonon energy is not enough to activate the carrier mobility, or if the mobility band is absent (as it takes place in majority of organic semiconductors), the transport occurs via hopping of carriers between localized states (Fig. 2.8).

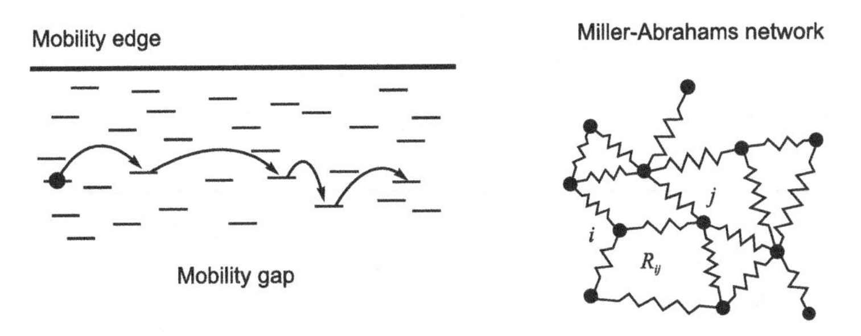

Figure 2.8 Scheme of hopping and the Miller-Abrahams resistor network.

To simulate the hopping conduction in disordered Semiconductors, the Miller-Abrahams model of random network [Shklovskiy & Efros (1979)] is often used. The use of this model reduces the problem to calculations of the resistance of such a network, each node of which is coincides with one of the traps distributed over the sample. Assuming that the resistance R_{ij} between pairs of the nodes is inversely proportional to the rate of electron transitions from one trap to another ν_{ij}^0 and denoting the distance between i-th and j-th nodes by r_{ij}, we write the transition rate as

$$\nu_{ij} = \nu_0 \exp\left(-2r_{ij}/a - \varepsilon_{ij}/k_0 T\right),$$

where

$$\varepsilon_{ij} = \frac{1}{2}(|\varepsilon_i - \varepsilon_j| + |\varepsilon_i - \mu| + |\varepsilon_j - \mu|),$$

a is the radius of the wave function, μ the Fermi energy, ν_0 a positive constant, k_0 the Boltzmann constant, T temperature. The resistance between two nodes is defined by

$$R_{ij} = R_{ij}^0 \exp(\xi_{ij}), \qquad \xi_{ij} = \frac{2r_{ij}}{a} + \frac{\varepsilon_{ij}}{k_0 T}.$$

The mean value of electron trapping time is inversely proportional to the sum of rates of transitions to other localized states.

The most important features of hopping conductivity have been understood on the basis of the numerical simulations for different densities of states and spatial distributions of traps. Considering the equilibrium conductivity, Grünewald & Thomas (1979); O'Shaughnessy & Procaccia (1985) came to the conclusion that the main contribution to the conductivity is made by electrons, performing jumps near the characteristic energy level. An analytical formula was obtained for the energy showing that it does not depend on the position of the Fermi level. Independently of them, Monroe (1985) considered the energy relaxation of electrons by hopping over the band tail states. The author showed that the electrons starting from the mobility edge make a few jumps and quickly relax in energy, but the type of relaxation changes when the characteristic energy (*transport level*) ε_{tr} is reached. The hopping process near and below the transport level is similar to the multiple trapping relaxation. The transport level can be considered as an analog to the mobility edge. This level separates conductive states from traps. Only states with energies less than ε_{tr} should be interpreted as traps [Baranovskii et al. (2002); Nikitenko (2011)].

Note that in the frame of the hopping conductivity model with an exponential density of states in energy, we again arrive at the fractional diffusion equation and the dispersion parameter is proportional to α on the temperature (see for detail [Uchaikin & Sibatov (2008)]).

2.2.5 *Bässler's model of Gaussian disorder*

The Scher-Montroll (1975) approach and the Arkhipov-Rudenko (1982) theory predict a transition to the Gaussian regime, when the dispersion parameter α tends to 1. In the framework of multiple trapping and thermoactivated hops, transition to the normal statistics is observed with temperature increasing. However, it should be noted that the model for hopping transport in organic semiconductors predicts the transition to the normal transport with increasing a sample thickness or decreasing a voltage applied to the sample during the transit time experiments [Bässler (1993)]. In other words, a change in transport statistics can be due to changes in macroscopic large-scale parameters. For small flight times, i.e. small values of sample thickness and/or high voltages, the normalized transient current curves are almost universal, and correspond to the dispersion mode of transport. In samples with greater thickness, or at lower voltages, a plateau on the curves of $I(t)$ is observed [Bässler (1993); Tyutnev et al. (2005)], which indicates the Gaussian mode of transfer. This phenomenon demonstrating the spatio-temporal scale effect, relates to the case with many low molecular weight, molecular-doped and conjugated polymers and can be described in terms of the theory of quasi-equilibrium transport [Nikitenko (2006)].

Bäsler's model assumes that the energy distribution of hopping centers involved in tunnel-activation transfer is described by the Gaussian function. It is important to know the form of waiting time distribution for such energy distribution of hopping centers. Results of numerical calculations are presented in Fig. 2.9 for two energy distributions of localized states: exponential with $\varepsilon_0 = 2.5kT$ (diamonds), and Gaussian with the same ε_0 (circles). The solid line represents the power law of fractional order $\alpha = 0.4$. The dotted line corresponds to the power law with $\alpha = 0.45$ truncated by the exponential function $e^{-t/(625\ t_0)}$. Because a waiting time distribution for Bässler's model does not have pure power tail, we can not apply the standard fractional differential approach at all scales. Moreover, all

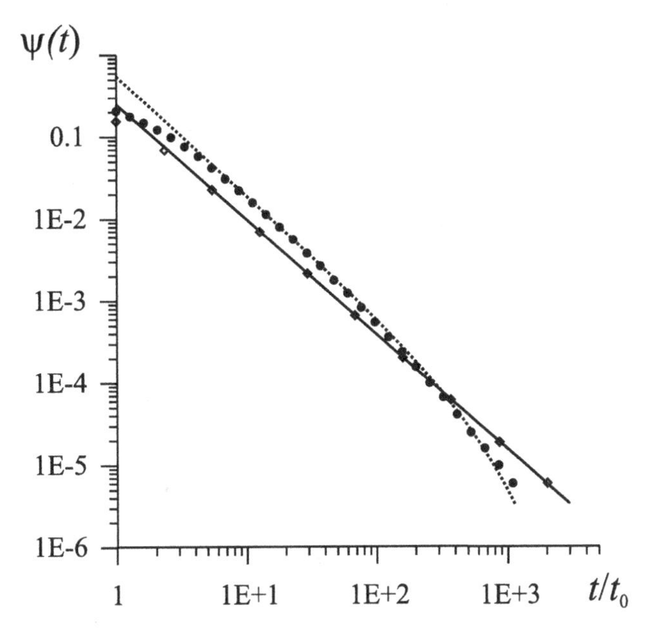

Figure 2.9 Numerically calculated pdf of waiting times for exponential (diamonds) and Gaussian (circles) distribution of hopping center energies. Solid line represents a power law function, dotted line is an approximation

moments of sojourn times are finite, and normal transport regime has to be observed at large times. As Fig. 2.9 demonstrates, the exponentially truncated power law is a good approximation to waiting time pdf obtained in frames of Bässler's transport model. No doubt, truncation of long-tailed distributions of waiting times can arise due to not only Gaussian shape of density of states. From this point of view, the generalized Scher-Montroll theory with truncated power pdf of waiting times is more general than the model of Gaussian disorder.

A truncated power-law distribution provides the transition from anomalous (non-Gaussian) to the normal (Gaussian) transport with increasing the thickness of a sample under investigation. It is worth to note that waiting time distributions and transient current curves obtained from the multiple trapping model are in agreement with the results of direct simulation for the hopping mechanism ([Hartenstein et al. (1996)]). This means that in not too strong electric fields, the observed macroscopic manifestations of both mechanisms are indistinguishable, despite their significant physical difference. By opinion of Hartenstein et al. (1996), the cause of this lies in the existence of the transport energy level in both cases.

Detailed analysis of the localization time distributions shows that the complementary cumulative function $Q(t) = \mathrm{Prob}(T > t)$ in the framework of Gaussian disorder model is better described by an inverse power function multiplied by a stretched exponential one. It is also important that the index of the stretched exponential function is not arbitrary: it is twice smaller than the power law index

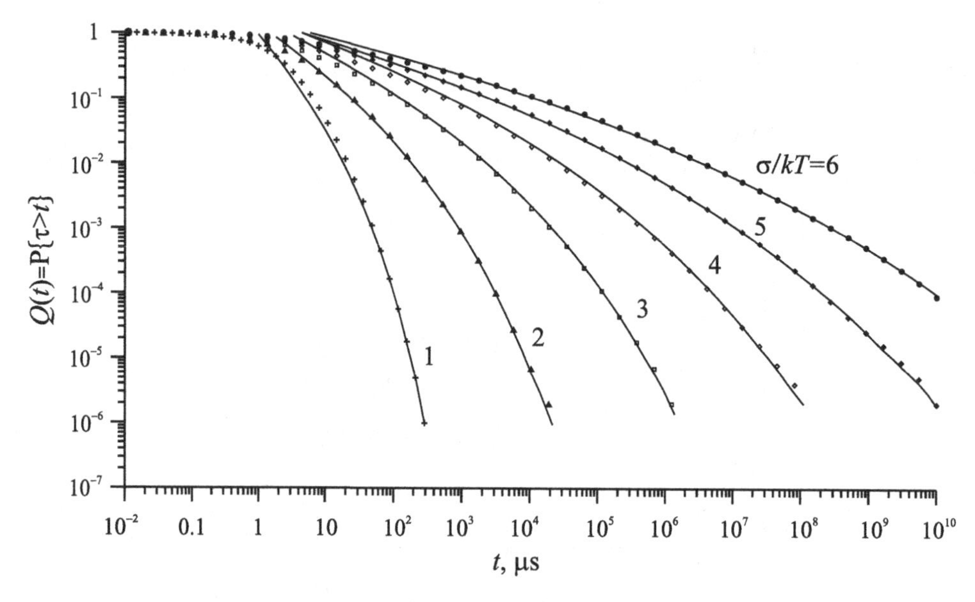

Figure 2.10 Complementary distribution function of waiting times in localized states in the Gaussian disorder model of hopping conductivity. Points are the results of Monte-Carlo calculations, lines are functions $At^{-kT/\sigma}\exp\left(-(\sigma t/4kT)^{kT/2\sigma}\right)$.

$\alpha_1 = kT/\sigma$:

$$Q(t) = \left\langle \exp\left(-\nu_0 t\ \mathrm{e}^{-\varepsilon/kT}\right)\right\rangle = \int\limits_{-\infty}^{+\infty} g(\varepsilon)\exp\left(-\nu_0 t\ \mathrm{e}^{-\varepsilon/kT}\right) d\varepsilon$$

$$\approx Q^{\mathrm{as}}(t) \equiv At^{-kT/\sigma}\exp\left(-(\sigma t/4kT)^{kT/2\sigma}\right).$$

Fig. 2.10 manifests a good agreement of the analytical formula (with parameters given in Table) and Monte Carlo simulation results.

Parameters of asymptotical distribution of waiting times.

$\hat{\sigma} = \sigma / kT$	$\alpha = kT/\sigma$	$\gamma = kT/2\sigma$	A
1	1,000	0,5000	1,50
2	0,500	0,2500	3,00
3	0,333	0,1667	4,19
4	0,250	0,1250	4,81
5	0,200	0,1000	4,90
6	0,167	0,0833	5,01

The table above contains parameters for asymptotic representation of localization time distribution Q^{as} in the framework of the Gaussian disorder model.

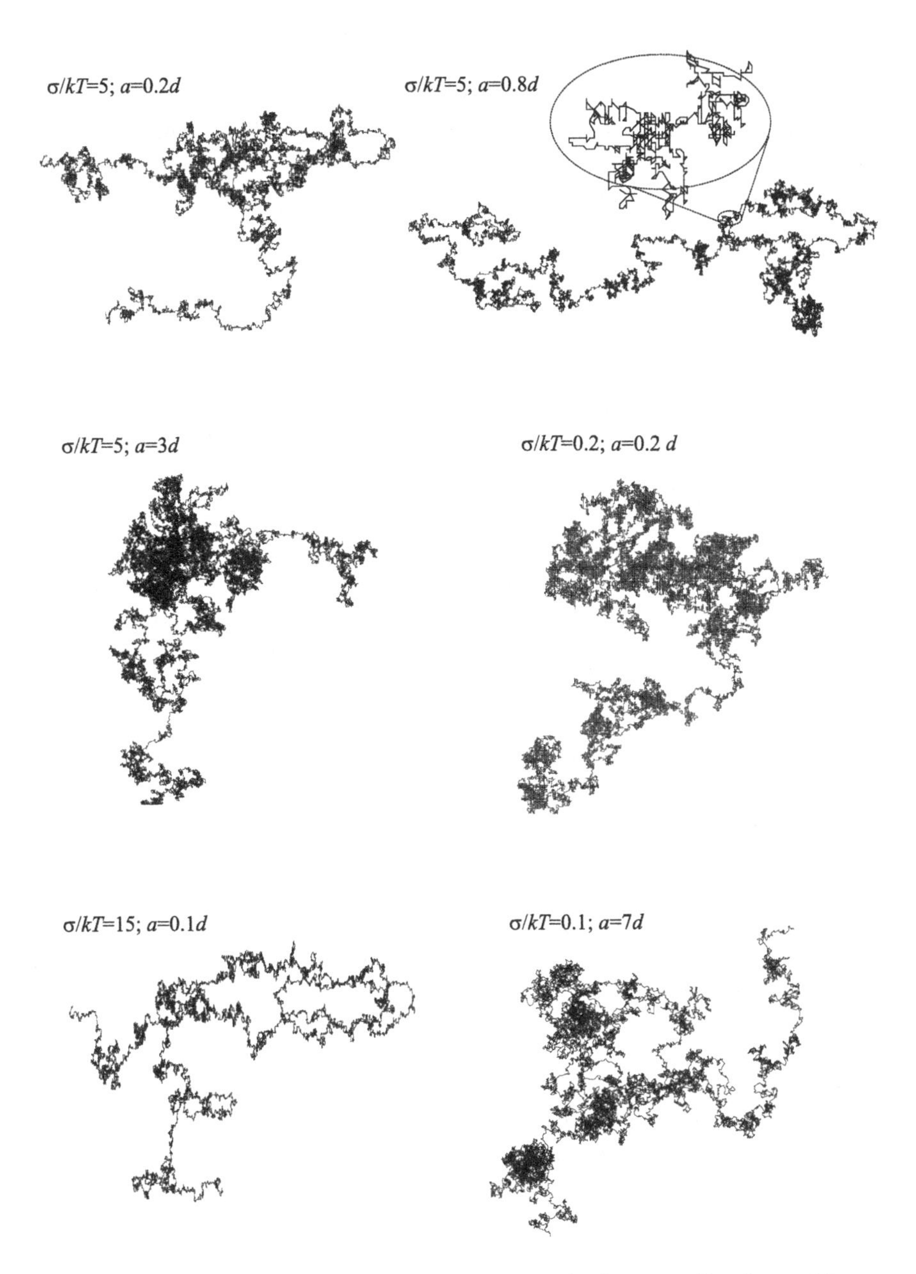

Figure 2.11 Trajectories of hopping diffusion in the Gaussian disorder model.

Analyzing the hopping conduction through exponentially distributed localized states, [Grünewald & Thomas (1979); O'Shaughnessy & Procaccia (1985)], came to the conclusion about dominance of hops near a certain energy level. Shapiro and Adler got an analytical formula for this level and showed that it is independent of the Fermi energy. Monroe (1985) considering the hopping relaxation demonstrated that the carriers starting with the mobility edge energy relax deeper into the tail of the state density, and the relaxation nature changes significantly when the carriers reach the transport level ε_t. The nature of the relaxation and transport in this case is similar to the properties of the multiple trapping transport. The transport level plays a role of the mobility edge [Nikitenko (2006)].

2.3 Fractional formalism of multiple trapping

2.3.1 *Prime statements*

In the multiple trapping model, carriers are divided into delocalized and trapped ones. The delocalized carriers are scattered on phonons and inhomogeneities, and drift in the external electric field $\mathbf{E}$ with an average velocity $\mathbf{v} = \mu\mathbf{E}$, where μ is mobility. As noted in Ref. [Tiedje (1984)], the overwhelming majority of carriers involved in dispersive transport are captured in traps, but free carriers are actually responsible for the transport. In the classical theory of crystalline semiconductors, most kinetic equations are written for the concentration of free carriers. The hopping model implies zero concentration of free carriers; therefore, the equation for them makes sense only in the case of a multiple trapping mechanism. Let us use a mathematical trick to avoid derivation for each mechanism of a separate generalized drift-diffusion equation for total concentration. In the hopping transport model we turn the number of free charges to zero and velocity v to infinity.

We assume that particles are transported in a regular medium, i.e., particles in identical macroscopic volumes of a material are involved in a roughly similar mean number of capture events. Trap density is much higher than the number density of carriers involved in the transport process. The transition rate $w_{M\to R}(\mathbf{r},t)$, from motion to rest per unit volume, i.e., the rate of capture events, is proportional to the concentration of delocalized carriers:

$$w_{M\to R}(\mathbf{r},t) = \tau_0^{-1}\, n_d(\mathbf{r},t).$$

Here, τ_0 is the mean time of particle's motion between two entrapments to the localized state; in the hopping model, $\tau_0 \to 0$. The total concentration of carriers is the sum of concentrations of mobile, $n_d(\mathbf{r},t)$ and trapped, $n_t(\mathbf{r},t)$ carriers:

$$n(\mathbf{r},t) = n_d(\mathbf{r},t) + n_t(\mathbf{r},t),$$

in this case, the relation

$$n_t(\mathbf{r},t) = \int_0^t \tau_0^{-1} n_d(\mathbf{r},t')\,[1 - \Psi(t-t')]\,dt', \qquad (2.38)$$

is fulfilled, where $\Psi(t)$ is the distribution function of random residence time in the localized state (see Section 2).

From physical point of view, the long time asymptotics ($t \to \infty$) means that during time interval t the carrier (an electron or a hole) undergoes a large number of localization-delocalization events, that is $t \gg w_0^{-1}N_t/N_c$. Reducing $Q(t)$ to the form

$$Q(t) \sim \frac{\alpha\pi w_0}{(w_0 N_c/N_t)^\alpha \sin\pi\alpha}\,\frac{t^{-\alpha}}{\Gamma(1-\alpha)} = w_0\frac{(c_\alpha t)^{-\alpha}}{\Gamma(1-\alpha)},$$

$$c_\alpha = \frac{w_0 N_c}{N_t}\left(\frac{\sin\pi\alpha}{\pi\alpha}\right)^{1/\alpha},$$

and inserting it in Eq.(2.37), we arrive at the equation with fractional time-derivative which order is defined as $\alpha = kT/\varepsilon_0$:

$$\frac{\partial n_d(\mathbf{r},t)}{\partial t} + w_0 c_\alpha^{-\alpha}\ {}_0\mathsf{D}_t^\alpha\ n_d(\mathbf{r},\tau) + \operatorname{div}\left\{\ \mu\mathbf{E}\ n_f(\mathbf{r},t) - D\nabla n_f(\mathbf{r},t)\right\} = 0. \quad (2.39)$$

In case $\alpha > 1$, the term with fractional derivative becomes negligible at long time region and Eq. (2.39) takes the normal form of the standard Fokker-Planck equation:

$$\frac{\partial n_d(\mathbf{r},t)}{\partial t} + \operatorname{div}\left\{\ \mu\mathbf{E}\ n_f(\mathbf{r},t) - D\nabla n_f(\mathbf{r},t)\right\} = 0. \quad (2.40)$$

Otherwise, when $\alpha < 1$, we observe an opposite interrelation leading to the subdiffusion form of this equation:

$$w_0 c_\alpha^{-\alpha}\ {}_{-\infty}\mathsf{D}_t^\alpha\ n_d(\mathbf{r},\tau) + \operatorname{div}\left\{\ \mu\mathbf{E}\ n_f(\mathbf{r},t) - D\nabla n_f(\mathbf{r},t)\right\} = 0. \quad (2.41)$$

As follows from Eq. (2.36), the trapped carrier concentration can be found from free carrier counterpart by solving the differential equation

$$w_0^{-1}\frac{\partial n_t(\mathbf{r},t)}{\partial t} = c_\alpha^{-\alpha}\ {}_{-\infty}\mathsf{D}_t^\alpha\ n_d(\mathbf{r},\tau).$$

Let us come back to the time-flight experiment. The mobile carriers are generated in the near-electrode layer of the semiconductor at $t = 0$,

$$n_d(x,t=0) = n(x,t=0) = N\delta(x).$$

Thus, the concentration of nonequilibrium (quasi-free and localized) carriers obeys the equation

$$\frac{\partial n(x,t)}{\partial t} + {}_0\mathsf{D}_t^{1-\alpha}\left\{\frac{\mu E c_\alpha^\alpha}{w_0}\frac{\partial n(x,t)}{\partial x} - \frac{D c_\alpha^\alpha}{w_0}\frac{\partial^2 n(x,t)}{\partial x^2}\right\} = N\delta(x)\delta(t).$$

Action of fractional integration operator ${}_0\mathsf{I}_t^{1-\alpha}$ reduces it again to

$${}_0\mathsf{D}_t^\alpha n(x,t) + \frac{\mu E c_\alpha^\alpha}{w_0}\frac{\partial n(x,t)}{\partial x} - \frac{D c_\alpha^\alpha}{w_0}\frac{\partial^2 n(x,t)}{\partial x^2} = N\delta(x)\frac{t^{-\alpha}}{\Gamma(1-\alpha)},$$

but now we know connection of kinetic coefficients with physical parameters of multiple trapping model.

In the previous section, we established a relationship between universality of the transition current curves and the fractional generalization of the Fokker-Planck equation. In the model of multiple trapping considered in this section, we arrive at a fractional equation and consequently, at the universal current curves on assumption of exponential energy distribution of localized states.

2.3.2 *Multiple trapping regime and Arkhipov-Rudenko approach*

In 1982, Arkhipov and Rudenko proposed an approach in the framework of the concept of multiple trapping that yields analytical results for the description of dispersive transport [Arkhipov & Rudenko (1982); Rudenko & Arkhipov (1982)]. The Arkhipov-Rudenko theory is rather popular and widely used in ongoing research (see, e.g., Refs. [Kaczer et al. (2005); Nikitenko & Tyutnev (2007)]). Here is a quotation from their work, giving an idea of this approach [Arkhipov & Rudenko (1982)]: "The carriers make transitions between the conduction state and localized levels. This process finally results in equilibrium between the mobile and immobile carrier fractions. Before thermal equilibrium is established, the energy interval, in which the localized states are distributed, may be divided into two regions: (1) The regions of shallow traps is situated between the mobility edge and some boundary energy $\varepsilon_*(t)$. In this region the equilibrium between the conduction state and traps is established at some time t. The region of deep traps lies below the boundary energy $\varepsilon_*(t)$. No carriers are realized from these traps until the time t. The concentration of carriers captured by the deep traps changes only due to trapping. Consequently we may neglect the 'release term'." Although this approach is approximate, it fairly well describes the main features of transient current behavior in time-of-flight experiments. Specifically, it explains the power-law asymptotics of transient current for exponential density of states, power-law dependence of drift mobility on sample thickness, electric field strength, etc. [Rudenko & Arkhipov (1982)]. In later studies, the field-stimulated diffusion was taken into account (see [Arkhipiv & Nikitenko (1989); Nikitenko (2011)]). It has been shown that in the non-equilibrium case, one can neglect usual diffusion coefficient in the transport equation, because the field-stimulated diffusion coefficient is much greater. With the application of the concept of the transport level, this approach is adapted for hopping conduction in the works Nikitenko (see [Nikitenko (2006)] and references therein).

To understand the main ideas of the approach, let us consider equation obtained in [Arkhipov & Rudenko (1982)] without the field-stimulated diffusion. By dividing localized states into shallow and deep ones, an equation for nonequilibrium transport with time-dependent diffusion coefficient D and mobility μ was derived [Arkhipov & Rudenko (1982)]:

$$\frac{\partial n(x,t)}{\partial t} + \mu_1(t)E\frac{\partial n(x,t)}{\partial x} - D_1(t)\frac{\partial^2 n(x,t)}{\partial x^2} = -\lambda(t)[n(x,t) - n(x,0)], \tag{2.42}$$

where time functions are defined by the relationships

$$\mu_1(t) = \frac{1}{1+1/\theta(t)}\,\mu, \qquad D_1(t) = \frac{1}{1+1/\theta(t)}\,D, \qquad \lambda(t) = \frac{1}{1+1/\theta(t)}\cdot\frac{1}{\tau(t)},$$

$$\frac{1}{\theta(t)} = \int\limits_0^{\varepsilon_*(t)} d\varepsilon\,\frac{\rho(\varepsilon)}{N_c}\exp\left(\frac{\varepsilon}{kT}\right), \qquad \frac{1}{\tau(t)} = \int\limits_{\varepsilon_*(t)}^{\infty} d\varepsilon\; c(\varepsilon)\rho(\varepsilon), \tag{2.43}$$

and the demarcation level position is found by solving the equation

$$c(\varepsilon_*)N_c \exp(-\varepsilon_*/kT)\, t = 1. \tag{2.44}$$

In the case of a weak energy dependence of the capture coefficient, $c(\varepsilon) \approx c_0$, the demarcation level equals $\varepsilon_* = kT\ln(\nu_0 t)$, where $\nu_0 = N_c\, c_0$ is the mean frequency of escape attempts.

Arkhipov and Rudenko (1982), Arkhipov et al. (1988) distinguish a strongly nonequilibrium transport regime in which most carriers are captured in deep traps (below the demarcation level) and are unlikely to be released by the instant of time t. This regime corresponds to the values of dispersion parameter $\alpha < 0.5$. The the master equation of *τ-approximation* has the form

$$n_d(\mathbf{r},t) = \frac{\partial}{\partial t}[\tau(t)\; n(\mathbf{r},t)], \tag{2.45}$$

where the function $\tau(t)$ is given by the expression

$$\tau(t) = \left[\nu_0 \int\limits_{kT\ln(\nu_0 t)}^{\infty} d\varepsilon\; \rho(\varepsilon)\right]^{-1},$$

Here, $\rho(\varepsilon)$ is the density of localized states below the band edge.

In this regime, the contribution of carriers occupying shallow traps can be neglected. This is equivalent to neglecting the time derivative in equation (2.42) (see Ref. [Arkhipov & Rudenko (1982)] for details):

$$n(x,t) + \mu E\tau(t)\, \frac{\partial}{\partial x}\, n(x,t) - D\tau(t)\, \frac{\partial^2}{\partial x^2}\, n(x,t) = n(x,0). \tag{2.46}$$

For the initial condition $n(x,0) = N\delta(x)$, the solution is the exponential function:

$$n(x,t) = \frac{N}{\mu E\tau(t)} \exp\left(-\frac{x}{\mu E\tau(t)}\right). \tag{2.47}$$

Solutions calculated by relationship (2.45), and their comparison with the solutions of the fractional equation will be presented below.

As follows from Eq. (2.45), the delocalized carrier density has the form

$$n_d(x,t) = Nx\, \frac{\tau'(t)}{[\mu E\tau(t)]^2} \exp\left(-\frac{x}{\mu E\tau(t)}\right), \tag{2.48}$$

and consequently, the transient current is defined as follows:

$$I(t) = \frac{e\mu E}{L}\int\limits_0^L dx\, n_d(x,t) = eN\mu EL^{-1}\tau'(t)\left[1 - \left(1 + \frac{L}{\mu E\tau(t)}\right)\exp\left(-\frac{L}{\mu E\tau(t)}\right)\right].$$

For the exponential energy distribution of band tail states and a weak energy dependence of the capture coefficient, function $\tau(t)$ takes the form [Arkhipov et al. (1988); Emelyanova & Arkhipov (1998)]:

$$\tau(t) = \frac{N_c}{N_t}\, \nu_0^{-1}\, (\nu_0 t)^{\alpha}. \tag{2.49}$$

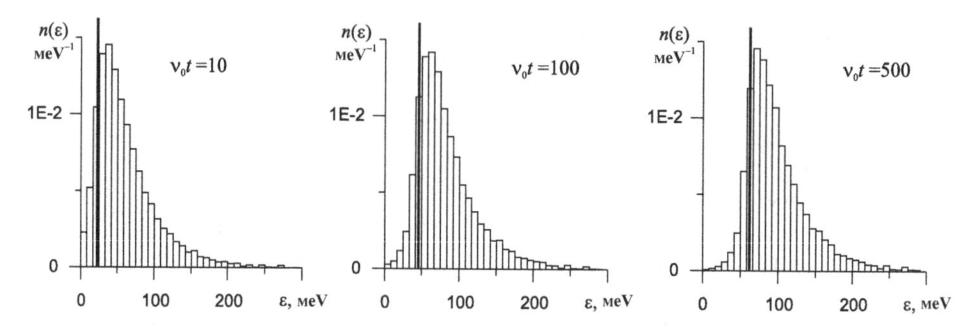

Figure 2.12 Evolution of the energetic distribution of carriers in localized states, $\varepsilon_0 = 40$ meV, $kT = 10$ meV.

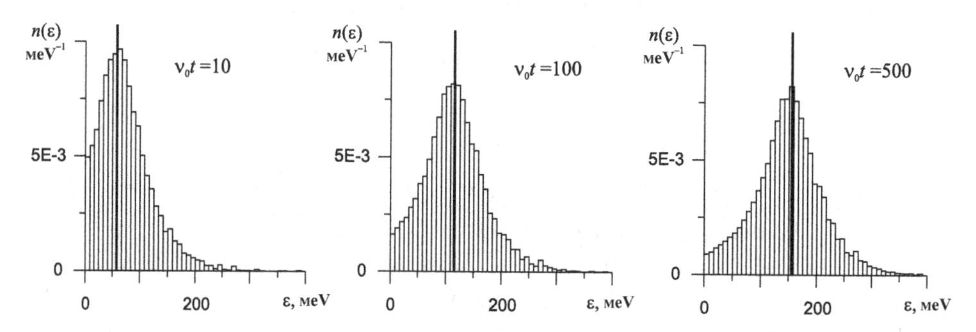

Figure 2.13 Evolution of the energetic distribution of carriers in localized states, $\varepsilon_0 = 40$ meV, $kT = 25$ meV.

Figures 2.12 and 2.13 present histograms of carrier energy distribution obtained by numerical Monte Carlo simulation of the filling of localized states in a time-of-flight experiment at two temperatures: $T = 115$ K and $T = 290$ K. Exponential density of states $\rho(\varepsilon) = (N_t/\varepsilon_0)\exp(-\varepsilon/\varepsilon_0)$ of width $\varepsilon_0 = 40$ meV was used for the purpose of simulation.The solid line shows the position of demarcation level $\varepsilon_*(t) = kT\ln(\nu_0 t)$. It follows that τ-approximation is rough for the second case, i.e., for $T = 290$ K and $\alpha \approx 0.6$, τ-in other words, the fraction of carriers in shallow traps, $\varepsilon < \varepsilon_*$ is large enough to be taken into account in calculations.

The approximation based on Eqs. (2.45) and (2.46), sometimes called the *tau*-approximation [Tyutnev et al. (2006); Nikitenko & Tyutnev (2007)], allows to express all the results for the exponential density of localized states in elementary functions, but poorly describes transport at dispersion parameters $\alpha > 0.5$, because the main condition of this approximation is violated. For $\alpha > 0.5$ the fraction of shallow-trap carriers cannot be neglected. Doubtless, Arkhipov and Rudenko (1982), understood this when they derived the τ–approximation for strongly nonequilibrium transport. Not surprisingly, equation (2.46) does not go over into the equation of normal transport for $\alpha > 1$, where $\alpha = kT/\varepsilon_0$ for the exponential density of states.

By way of example, the following equation for exponential density of states is obtained from Eq. (2.43):

$$\frac{1}{\theta(t)} = \frac{\alpha N_t}{(1-\alpha)N_c}[(\nu_0 t)^{1-\alpha} - 1].$$

For $\nu_0 t \gg 1$, one finds

$$\alpha < 1, \qquad \frac{1}{\theta(t)} \approx \frac{\alpha N_t}{(1-\alpha)N_c}(\nu_0 t)^{1-\alpha}, \qquad \mu_1(t) \propto t^{\alpha-1}, \quad D_1(t) \propto t^{\alpha-1};$$

$$\alpha \geq 1, \qquad \frac{1}{\theta(t)} \approx \frac{\alpha N_t}{(\alpha-1)N_c}, \qquad \mu_1(t) \approx \text{const}, \quad D_1(t) \approx \text{const}.$$

2.3.3 *Fractional equations for delocalized carriers*

Let $\Psi(t)$ be an asymptotically power-law distribution function

$$1 - \Psi(t) = \text{Prob}(\tau > t) = \int_t^\infty \psi(t')\, dt' \sim \frac{(ct)^{-\alpha}}{\Gamma(1-\alpha)}, \qquad \alpha \leq 1$$

then, relation (2.38) can be rewritten as

$$\frac{\partial n_t(\mathbf{r},t)}{\partial t} = (\tau_0 c^\alpha)^{-1} \frac{1}{\Gamma(1-\alpha)} \frac{\partial}{\partial t} \int_0^t \frac{n_d(\mathbf{r},t')}{(t-t')^\alpha} dt'. \tag{2.50}$$

In this expression, the integro-differential operator is the Riemann-Liouville fractional derivative of order $\alpha \leq 1$. Let us rewrite expression (2.50) in the form

$$\frac{\partial n_t(\mathbf{r},t)}{\partial t} = \frac{1}{\tau_0 c^\alpha}\, {}_0\mathsf{D}_t^\alpha n_d(\mathbf{r},t), \qquad K = c^\alpha l. \tag{2.51}$$

The last formula relates concentrations of quasi-free (delocalized) and localized carriers. This relation was written in terms of the Laplace transform for the multiple trapping model in Refs. [Noolandi (1977); Tiedje (1984)] and interpreted in terms of fractional derivatives in Ref. [Uchaikin & Sibatov (2007)]. Expression (2.50) coincides up to coefficients with the one obtained by Arkhipov, Popova, and Rudenko [Arkhipov et al. (1983)].

The flux is expressed through the concentration of delocalized carriers by the formula

$$\mathbf{q}(\mathbf{r},t) = \mathbf{v} n_d(\mathbf{r},t) - D\nabla n_d(\mathbf{r},t). \tag{2.52}$$

Here, $\mathbf{v}$ is the average velocity of directional motion, and D is the diffusion coefficient.

Substituting relation (2.52) into continuity equation

$$\frac{\partial n(\mathbf{r},t)}{\partial t} + \text{div}\ \mathbf{q}(\mathbf{r},t) = n(\mathbf{r},0)\ \delta(t). \tag{2.53}$$

and taking account of expression(2.51) yields the *fractional drift-diffusion equation* for a concentration of delocalized carriers:

$$\frac{\partial n_d(\mathbf{r},t)}{\partial t} + \frac{l}{\tau_0 K} \frac{\partial^\alpha n_d(\mathbf{r},t)}{\partial t^\alpha}$$
$$+\ \text{div}\Big(\mu\mathbf{E}\ n_d(\mathbf{r},t) - D\nabla n_d(\mathbf{r},t)\Big) = n(\mathbf{r},0)\ \delta(t). \tag{2.54}$$

2.3.4 *Fractional equation for the total concentration*

The equation for the total concentration is obtained taking into account that the overwhelming number of carriers are captured in traps for times $t \gg c^{-1}$, so that

$$n(\mathbf{r},t) \approx n_t(\mathbf{r},t).$$

This expression stems from Eq. (2.51). Then, using Eqs. (2.51–2.53), we obtain the fractional drift-diffusion equation for the total concentration:

$$\frac{\partial n(\mathbf{r},t)}{\partial t} + {}_0\mathsf{D}_t^{1-\alpha}\,\mathrm{div}\,\Big(\mathbf{K}n(\mathbf{r},t) - C\nabla n(\mathbf{r},t)\Big) = n(\mathbf{r},0)\,\delta(t), \tag{2.55}$$

where $\mathbf{K} = \tau_0 K\mathbf{v}/l$ and $C = \tau_0 KD/l$.

Application of the Riemann-Liouville fractional integral of order $(1-\alpha)$ to this equation leads to

$${}_0\mathsf{D}_t^{\alpha} n(\mathbf{r},t) + \mathrm{div}\,\Big(\mathbf{K}n(\mathbf{r},t) - C\nabla n(\mathbf{r},t)\Big) = \frac{t^{-\alpha}}{\Gamma(1-\alpha)} n(\mathbf{r},0). \tag{2.56}$$

Concentration $n(\mathbf{r},t)$ satisfies the normalization condition

$$\int n(\mathbf{r},t)\,dV = N,$$

where N is the number of charge carriers participating in the transport. In the time-of-flight experiment, N is the number of photoinjected carriers.

The fractional subdiffusion equation without drift,

$${}_0\mathsf{D}_t^{\alpha} n(\mathbf{r},t) = C\nabla^2 n(\mathbf{r},t) + \frac{t^{-\alpha}}{\Gamma(1-\alpha)}\delta(\mathbf{r}) \tag{2.57}$$

was derived by Nigmatullin (1984ab) for diffusion on fractal structures simulating porous and disordered media, and, independently, by Balakrishnan (1985) for the one-dimensional generalization of the Brownian motion. The Fokker-Planck fractional equation describing random walks of a particle in the external field was investigated in Refs. [Metzler et al. (1998); Sokolov (2001a); Barkai (2001a)]. Barkai (2001) was the first to show that the Fokker-Planck fractional equation written for the total carrier distribution density normalized to unity describes the anomalous transient current behavior (2.2).

Replacement of the first time derivative by the fractional one of order α in the normal diffusion equation yields the equation for the carrier density with non-conserved normalization:

$${}_0\mathsf{D}_t^{\alpha} n(x,t) = C\frac{\partial^2 n(x,t)}{\partial x^2},$$

which was investigated by Wyss (1986), Mainardi (1994) and Hilfer (1995, 2000). Nigmatullin (1986) linked this equation with percolation processes; Bisquert (2003,2005), Uchaikin and Sibatov (2005) interpreted it as a macroscopic diffusion equation for free carriers in the multiple trapping regime, i.e., $n(x,t) \equiv n_d(x,t)$. The expression for transient photocurrent obtained in Ref. [Uchaikin & Sibatov (2007)] by solving an analogous fractional dispersive-drift equation

$${}_0\mathsf{D}_t^{\alpha} n_d(x,t) = K\frac{\partial n_d(x,t)}{\partial x},$$

coincides with the formula derived in [Gusarov et al. (1999)] and used in [Uchaikin & Korobko (2004)] for description of multiple scattering on a fractal dust.

2.3.5 *Two-state dynamics*

A carrier in the multiple trapping model can reside in two dynamic states: rest and motion. The total walking time of a carrier consists of the total trapping time t_R and the motion time t_M:

$$t = t_R + t_M.$$

For the overwhelming majority of particles involved in dispersive transport, one has $t_R \gg t_M$. The displacement of a carrier from its initial position is directly related to motion time. Equation (2.56) is equivalent to a system of two equations, one of which relates carrier coordinates to motion time, and the other links the motion time and the total walking time [Sibatov & Uchaikin (2009)].

Conditional distribution density $p(t_M|t)$ can be found using the random walk model. Let us assume time t_M to be the process coordinate. The density $p(t_M|t)$ satisfies the equation of one-sided fractal walks (see Ref. [Sibatov & Uchaikin (2010)] for details):

$$ {}_0\mathsf{D}_t^\alpha p(t_M|t) + \frac{\tau_0 K}{l}\frac{\partial p(t_M|t)}{\partial t_M} = \frac{t^{-\alpha}}{\Gamma(1-\alpha)}\,\delta(t_M), \tag{2.58}$$

the solution of which is expressed through one-sided stable density with the characteristic exponent α:

$$p(t_M|t) = \frac{ct}{\alpha t_M}\left(\frac{t_M}{\tau_0}\right)^{-1/\alpha} g_+\left(ct\left(\frac{t_M}{\tau_0}\right)^{-1/\alpha};\alpha\right). \tag{2.59}$$

Conditional distribution density of coordinate $\mathbf{r}$ provided that the motion time is t_M, satisfies the ordinary Fokker–Planck equation:

$$\frac{\partial p(\mathbf{r}|t_M)}{\partial t_M} + \operatorname{div}\Big(\mu\mathbf{E}\; p(\mathbf{r}|t_M) - D\nabla p(\mathbf{r}|t_M)\Big) = \delta(t)p(\mathbf{r}|0). \tag{2.60}$$

The sought-for distribution $n(\mathbf{r},t)$ is found using the formula

$$n(\mathbf{r},t) = \int_0^t p(\mathbf{r}|t_M)p(t_M|t)dt_M. \tag{2.61}$$

Multiplying equation (2.58) by $p(\mathbf{r}|t_M)dt_M$ and integrating it over $(0,t)$ and taking Eq. (2.60) into account, we arrive at equation (2.56). This means that fractional equation (2.56) can be obtained by substituting the solution of equation (2.60) and expression (2.59) into formula (2.61).

2.3.6 *Delocalized carrier concentration*

In order to analyze the solution to fractional equation (2.54) for a delocalized carrier concentration, we use relationship (2.51) rewritten as

$$n_d(\mathbf{r},t) = \tau_0 c^\alpha\; {}_0\mathsf{D}_t^{1-\alpha} n(\mathbf{r},t).$$

Replacing $n(\mathbf{r},t)$ by Eq. (2.61) yields

$$n_d(\mathbf{r},t) = \tau_0 c^\alpha \int_0^t p(\mathbf{r}|t_M)\, {}_0\mathsf{D}_t^{1-\alpha} p(t_M|t)dt_M$$

The function $\tau_0 c^\alpha \, {}_0\mathsf{D}_t^{1-\alpha} p(t_M|t)$ represents conditional pdf $p(t|t_M)$. Indeed, it follows from the probability conservation law,

$$\frac{\partial p(t_M|t)}{\partial t} + \frac{\partial p(t|t_M)}{\partial t_M} = \delta(t)\delta(t_M),$$

and from equation (2.58), rewritten in the form

$$\frac{\partial p(t_M|t)}{\partial t} + \tau_0 c^\alpha \frac{\partial}{\partial t_M} \, {}_0\mathsf{D}_t^{1-\alpha} p(t_M|t) = \delta(t)\delta(t_M),$$

that

$$p(t|t_M) = \tau_0 c^\alpha \, {}_0\mathsf{D}_t^{1-\alpha} p(t_M|t). \tag{2.62}$$

Conditional distribution density $p(t|t_M)$ satisfies the equation

$$\frac{\partial p(t|t_M)}{\partial t_M} + (\tau_0 c^\alpha)^{-1} \, {}_0\mathsf{D}_t^{\alpha} p(t|t_M) = \delta(t)\delta(t_M).$$

By expressing the solution of the last equation through stable densities (see Ref. [Sibatov & Uchaikin (2010)] for details), we have:

$$p(t|t_M) = c\left(\frac{t_M}{\tau_0}\right)^{-1/\alpha} g_+\left(ct\left(\frac{t_M}{\tau_0}\right)^{-1/\alpha}; \alpha\right). \tag{2.63}$$

Therefore, the concentration of delocalized carriers is given by

$$n_d(\mathbf{r},t) = \int_0^t p(\mathbf{r}|t_M) p(t|t_M) dt_M, \tag{2.64}$$

where function $p(t|t_M)$ is defined by Eq. (2.63), and density $p(\mathbf{r}|t_M)$ satisfies the Fokker-Planck equation (2.60).

2.3.7 *Percolation and fractional kinetics*

The classical *percolation* problem for hopping conduction in disordered semiconductors can be formulated as follows [Bardou et al. (2002)]. We select a rate of transitions from one localized state to another, $\nu = \nu_0 \exp(-\xi)$, and assume the existence of a link between two nodes if the corresponding transition rate ν_{ij} exceed a given number ν. When a threshold value ξ is equal to critical one, $\xi = \xi_c$, an infinite percolation cluster arises piercing through the disordered grid. The critical value ξ_c defines parameters of temperature and concentration dependencies of transport coefficients [Baranovskii et al. (2002)].

The infinite percolation cluster above the percolation threshold has a fractal structure on scales that do not exceed the correlation length ζ. On larger scales, the infinite cluster becomes statistically homogeneous with dimension equal to the Euclidean dimension space in which it is embedded. To determine the characteristics of transfer, it is necessary to know the waiting time distribution for carriers in fractal fragments that make up the infinite cluster. At smaller distances, the transport occurs in the subdiffusion regime [Dykhne et al. (2004)].

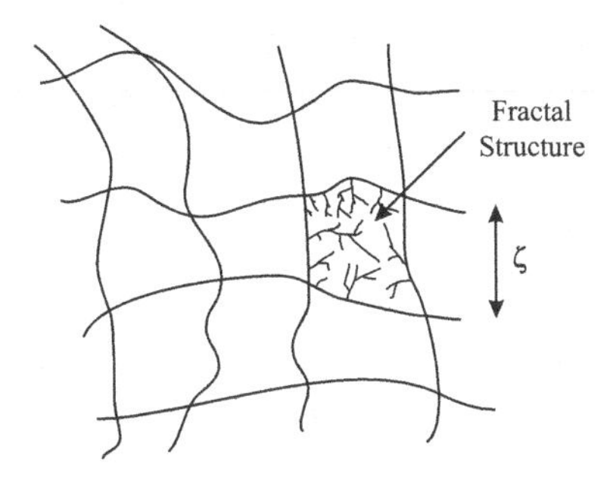

Figure 2.14 The structure of percolation cluster above percolation threshold (Bouchaud & Georges, 1990).

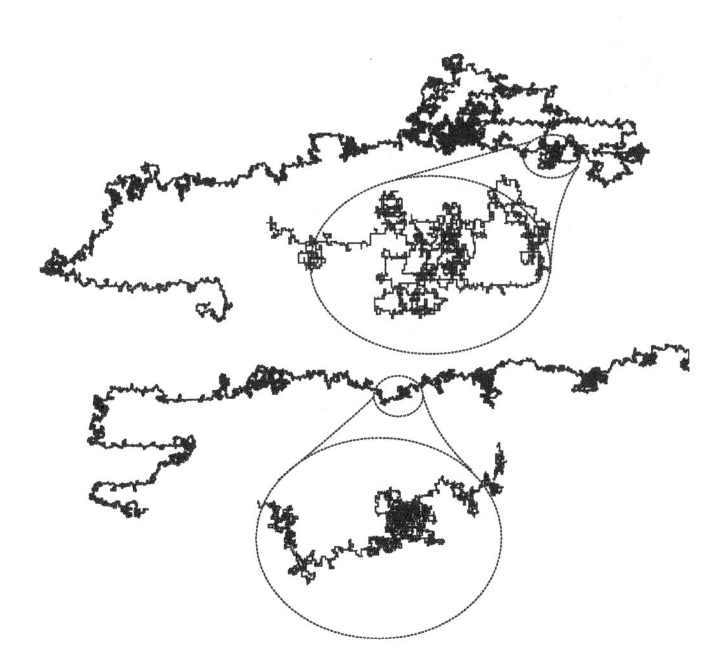

Figure 2.15 Two realizations of random walk of an electron in case of Gaussian disorder ($\sigma = 15\ kT$), localization radius $a = 0.5$ nm, distance between localized states $l = 5$ nm.

The simplest model of an infinite percolation cluster of dead ends is the so-called *comb model* (Fig. 2.16). At each node of a conductive bond, the walker can go into a dead branch of a cluster, represented by a tooth of the comb, where it undergoes diffusion (in the simplest case, the normal Gaussian diffusion). The carrier return from the dead loop in the main channel occurs through a random time distributed according to the Levy-Smirnov stable law,

$$q(t) \propto t^{-3/2}, \quad t \to \infty.$$

This involves the fractional time derivative of order 1/2 in differential equation describing the transport along the percolation channel. Under the assumption on subdiffusive motion in the dead branches, one can get other orders of the time derivative.

As indicared above, the fractional equation of diffusion was first associated with a comb model by Nigmatullin ([Nigmatullin (1986)]). A more detailed analysis of this model can be found in Arkhincheev's and coauthors works ([Arkhincheev (1990); Arkhincheev & Baskin (1991a); Arkhincheev (1991b)]).

Figure 2.16 The examples of simulated percolation clusters near the percolation threshold in the site problem from [Feder (1991)] (left panel). The size of quadratic lattice 160 × 160. Right panel: comb and bush models [Bouchaud & Georges (1990); Isichenko (1992)].

Thus, Arkhincheev and Baskin (1991a) considered diffusion on a comb with teeth directed along y-axis using the equation

$$\frac{\partial p}{\partial t} - D_x \delta(y) \frac{\partial^2 p}{\partial x^2} - D_y \frac{\partial^2 p}{\partial y^2} = \delta(x)\delta(y)\delta(t),$$

where D_x and D_y are diffusivities along $x-$ and $y-$axes respectively. After Fourier-Laplace transformation, the authors got the equation

$$\left(s + D_x k^2 \delta(y) - D_y \frac{\partial^2}{\partial y^2} \right) \tilde{p}(k, y, s) = \delta(y),$$

solution of which is of the form

$$\tilde{p}(k, y, s) = \frac{\exp\left[-\sqrt{s/D_y}|y|\right]}{2\sqrt{D_y s} + D_x k^2}.$$

Assuming $y = 0$ leads to the equation describing diffusion along the x-axis expressed in terms of Fourier-Laplace variables:

$$\left(s^{1/2} + \frac{D_x}{2\sqrt{D_y}}k^2\right)\tilde{p}(k,0,s) = 1,$$

Its inversion yields to the fractional equation for the dispersive diffusion along the x-axis:

$${}_0\mathsf{D}_t^{1/2}p(x,0,t) - \frac{D_x}{2\sqrt{D_y}}\frac{\partial^2 p(x,0,t)}{\partial x^2} = \delta(x)\delta(t).$$

Observe that the normalizing condition is not valid in this case. There is observed an appropriate analogy with the mobility edge in multiple trapping model and transport level in the hopping conductivity. The number of particles on the x-axis decreases with time (by a power law, see [Uchaikin (2008)]). The package carrier deepens along the y-axis.

The comb-model is developed and modified, for example, by truncation of teeth length, using random lengths or random positions of teeth on the x-axis. Some results in this way have obtained in [Arkhincheev (1998, 1999a,b)]

2.3.8 *The case of Gaussian disorder*

In case of Gaussian density of localized states, the carrier concentration n_{eff} near the transport level obeys the following equation

$$\frac{a_\beta}{c^\beta}\,{}_0\mathsf{D}_t^\beta n_{\text{eff}}(x,t) + a_\alpha A\frac{\partial}{\partial t}\int_0^t \frac{n_{\text{eff}}(x,t')}{(t-t')^{kT/\sigma}}\exp\left\{-\left(\frac{\sigma(t-t')}{4kT}\right)^{kT/2\sigma}\right\}dt'$$

$$= \langle l\rangle\,\frac{\partial n_{\text{eff}}(x,t)}{\partial x} - D^*\,\frac{\partial^2 n_{\text{eff}}(x,t)}{\partial x^2} = 0.$$

The first term containing the fractional derivative of order β takes into account the percolation nature of carrier trajectories, parameter β is usually close to the value 0.5.

A few words should be said about the trajectories of the carriers participating in the hopping conduction. Fig. 2.11 shows that the trajectory is not always a kind of Brownian motion even in the absence of a structural disorder. Especially clear difference is seen in the case of strong disorder and small radius of localization. The approach developed by Nikitenko [Nikitenko (2006)] does not take into account the topology of these trajectories in the derivation of transport equations. In principle, this is a rather complicated task for analytical consideration. Evidently, the structure of the trajectory has a significant influence on the spatial distribution of nonequilibrium carriers, on the values of transient current in the time-of-flight experiments, and on the kinetics of twin recombination. For analysis of such problems, the percolation model turns out to be suitable [Zvyagin (1984)]. In the next section, we consider the relation of percolation problem to fractional differential kinetics.

2.4 Some applications

2.4.1 *Dispersive diffusion*

The diffusion equation taking into account the quasi-free electron trapping by own localized states for times significantly exceeding the single trapping time w^{-1}, is written as follows:

$$ {}_0\mathsf{D}_t^\alpha n = C^2 \frac{\partial^2 n}{\partial x^2}, \qquad C = \sqrt{D_n/Aw^{1-\alpha}}. $$

To obtain the solution of this equation for different initial and boundary conditions, it is sufficient to know the Green's function, which satisfies the equation

$$ {}_0\mathsf{D}_t^\alpha G_n(x-\xi,t) = C^2 \frac{\partial^2 G_n(x-\xi,t)}{\partial x^2} + \delta(x-\xi)\delta(t). $$

Applying the Laplace transform yields

$$ G_n(x-\xi,t) = (C|x-\xi|)^{-2/\alpha} \quad g_+\left(t(C|x-\xi|)^{-2/\alpha};\alpha/2\right). $$

The total concentration obeys the equation

$$ \frac{\partial n_c}{\partial t} = C^2 \ {}_0\mathsf{D}_t^{1-\alpha} \frac{\partial^2 n_c}{\partial x^2}. $$

The corresponding Green's function is

$$ G_{n_c}(x-\xi,t) = (Ct)^{-\alpha/2} q\left((Ct)^{-\alpha/2}(x-\xi);2,\alpha,0\right), $$

where

$$ q(x;2,\alpha,0) = \frac{1}{\alpha|x|^{1+2/\alpha}} g_+\left(|x|^{-2/\alpha};\alpha/2\right) $$

is the fractional stable density.

2.4.2 *Photoluminescence decay*

The photoluminescence spectrum of amorphous semiconductors is usually composed of two bands [Seki et al. (2006)]. The main band arises due to radiative recombination of electrons and holes localized in the band tails. The less intense band is usually associated with radiative recombination via defect states. Seki et al. (2005, 2006) interested in the main band excited by relatively weak pulses of light at the absorption edge. Above a certain temperature, the photoluminescence decays according to asymptotical power law.

For moderate excitation intensities, the model of diffusion-controlled recombination assumes that luminescence decay is caused by *geminate recombination*, which predicts a power law with exponent -1.5: $I(t) \propto t^{-3/2}$. The model implies normal diffusion of carriers of geminate pairs. The power-law decay with exponent -3.2 is confirmed by various experiments. At low temperatures, the diffusion can be neglected. At higher temperatures, the model implies that one type of carrier is transferred by hopping via localized states in the band tail. Carriers of opposite sign are trapped in deep localized states in a short time after pulsed excitation. Many authors noted that the values of the decay exponent is less than predicted, $\delta \sim 1.1 \div 1.3 < 1.5$. It is also indicated that the rate increases with increasing temperature. These results can be interpreted as a deviation of the carrier diffusion from the Gaussian statistics [Seki et al. (2003, 2006)].

Recombination of geminate pairs is usually described in terms of *Onsager's theory*, which considers the walk of a pair of carriers in the Coulomb and external electric fields. Onsager's dynamic problem is solved by using the Fokker-Planck equation with the boundary condition of pair recombination on a sphere of radius R_a. It is assumed that a pair of charges is generated with uniform probability distribution on a sphere. The *survival probability* of the pair $\Omega(t, E_0)$ at time t in an external field E_0 and then the polarization current can be calculated.

The initial distance between carriers in a geminate pair may have a spatial distribution, while the asymptotic decay of photoluminescence is not sensitive to this distribution [Seki et al. (2005)]. This distribution is important for the analysis of the recombination kinetics at short times. We are interested only in the long-term kinetics as the authors [Seki et al. (2005)].

Onsager's approach considers the random walk of mobile carrier of geminate pairs, less mobile carrier is fixed in space. It is assumed that the movable twin performs diffusive motion in the Coulomb and external electric fields. This approach is applicable in the case of a low average length between two acts of localization. Recombination occurs instantaneously when the mobile carrier reaches the origin, which is as it should be fixed twin. That is, the rate of recombination is nothing but the density distribution of the first achievements in the problem of random walks with given initial conditions. Suppose now that the external electric field is zero.

Slightly modify the condition of recombination according to [Seki et al. (2006)]. Let the waiting time before recombination in the absence of random walk is given by the distribution function $\Phi(\mathbf{r}, t) = \int_0^t \phi(\mathbf{r}, t)dt$. Consider the case of the exponential distribution

$$\phi(\mathbf{r}, t) = \gamma_{\mathrm{rc}}(\mathbf{r}) \exp[-\gamma_{\mathrm{rc}}(\mathbf{r})t], \tag{2.65}$$

where $\gamma_{\mathrm{rc}}(\mathbf{r})$ is the recombination rate. Dependence of $\gamma_{\mathrm{rc}}(\mathbf{r})$ on distance $\mathbf{r}$ dividing carriers of geminate pairs is determined by type of recombination. In the case of tunneling mechanism, we have $\gamma_{\mathrm{rc}}(\mathbf{r}) \propto e^{-2\beta r}$, in the case of energetic transfers $\gamma_{\mathrm{rc}}(\mathbf{r}) \propto r^{-6}$ (see details in Ref. [Seki et al. (2006)]).

The equation for the total concentration has the form

$$\exp\left[-\gamma_{\mathrm{rc}}(\mathbf{r})t\right] \, {}_0\mathsf{D}_t^{\alpha} n(\mathbf{r}, t) \exp\left[\gamma_{\mathrm{rc}}(\mathbf{r})t\right] + (c^{\alpha}\tau_0)\mathrm{div}\left[\mu\mathbf{E}\, n(\mathbf{r}, t) - D\nabla n(\mathbf{r}, t)\right]$$

$$= \frac{t^{-\alpha}}{\Gamma(1-\alpha)} \exp\left[-\gamma_{\mathrm{rc}}(\mathbf{r})tS\right] n(\mathbf{r}, 0). \tag{2.66}$$

Instead of the standard Fokker-Planck equation, we used its fractional generalization and obtained the formula for the survival function (the generalized Mozumder formula) geminate pairs ($E_0 = 0$):

$$\Omega(t) = \exp\left(-\frac{r_c}{r_0}\,\mathrm{erfc}\left(\frac{r_0}{\sqrt{4Dt}}\right)\right)$$

for the case of dispersive transport of carriers:

$$\Omega(t) = \exp\left(-\frac{r_c}{r_0}\int_0^t d\tau \left\{\frac{ct}{\alpha\tau}\left(\frac{\tau}{\tau_0}\right)^{-1/\alpha} g_+\left(ct\left(\frac{\tau}{\tau_0}\right)^{-1/\alpha};\alpha\right)\mathrm{erfc}\left(\frac{r_0}{\sqrt{4D\tau}}\right)\right\}\right).$$

In the large times asymptotic $[\Omega(t) - \Omega_\infty]/\Omega_\infty \propto t^{-\alpha/2}$. The generalized Mozumder formula is in a good agreement with the observed asymptotics of luminescence intensity decay $\propto t^{-1-\alpha/2}$.

As noted in Ref. [Tyutnev et al. (2005)], for the analysis of geminate conductivity it is of great importance the question on the proportion of electrons that have been trapped at least once. This fraction can be determined by the formula (see the monograph [Tyutnev et al. (2005)] and references therein):

$$f_{tr} = \int_0^\infty \Omega(t)\tau_0^{-1}\exp(-t/\tau_0)dt,$$

where τ_0 is the lifetime of delocalized states before capture into traps. Substituting the generalized Mozumder formula into the last equation, we arrive at

$$f_{tr} = \frac{1}{\tau_0}\int_0^\infty \exp\left(-\frac{t}{\tau_0} - \frac{r_c}{r_0}\int_0^t d\tau \left\{\frac{ct}{\alpha\tau}\left(\frac{\tau}{\tau_0}\right)^{-1/\alpha} g_+\left(ct\left(\frac{\tau}{\tau_0}\right)^{-1/\alpha};\alpha\right)\mathrm{erfc}\left(\frac{r_0}{\sqrt{4D\tau}}\right)\right\}\right)dt.$$

The integral in the expression is easily calculated by the Monte-Carlo method.

2.4.3 *Including recombination*

Investigations of the photoconductivity, photoluminescence, and electron paramagnetic resonance provide a rich source of information about processes of electron-hole *recombination* in disordered solids. Thus, studies of photoluminescence of amorphous hydrogenated silicon (a-Si:H) have allowed to form a consensus model due to the fact that the Fermi level in this material can be controlled by doping or changing defect concentration. At sufficiently low temperatures, the excited carriers are rapidly trapped in band tail states, and then they radiatively recombine with charge carriers of an opposite sign in a tail of another band, or nonradiatively through the defective states. The radiative process can go in two ways [Fus & Yaan (1991)]: through the *twin recombination* when the photoexcited electrons and holes (the twin pair) are not separated by thermalization diffusion, and through the distant recombination, when each of the carriers diffuses, becomes localized and then recombines independently of another. In both cases, the recombination occurs by radiative tunneling process, the rate of which depends strongly on the distance between the carriers. The distribution of these distances causes a broad distribution of carrier lifetimes [Fus & Yaan (1991)]. The *geminate pair* does not contribute to the photoconductivity. Thus, the significant influence of temperature on photoconductivity and photoluminescence at high electric fields reveals the geminate nature of recombination in amorphous semiconductors.

Let us take into account recombination in fractional equations of dispersive transport. We start from monomolecular recombination of free carriers [Orenstein et al. (1982)] and the recombination of localized carriers by tunneling transitions [Seki et al. (2006)]. In the presence of monomolecular recombination, basic balance equations for the model of transport controlled by capture in localized states have the form [Main et al. (1992)]-[Nagase et al. (1999)]:

$$\frac{dn_d(\mathbf{r},t)}{dt} = -\sum_i \frac{dn_{ti}(\mathbf{r},t)}{dt} - \frac{n_d(\mathbf{r},t)}{\tau} + N\delta(t)$$

$$\frac{dn_{ti}(\mathbf{r},t)}{dt} = -r_i n_{ti}(\mathbf{r},t) + w_i n_d(\mathbf{r},t).$$

It follows from these equations that liner recombination gives rise to the corresponding term in the continuity equation:

$$\frac{\partial n(\mathbf{r},t)}{\partial t} + \text{div}\ \mathbf{q}(\mathbf{r},t) = -\frac{n_d(\mathbf{r},t)}{\tau} + n(\mathbf{r},0)\ \delta(t),$$

where τ is the lifetime of delocalized carriers prior to monomolecular recombination. The use of this equation instead of the continuity equation (2.53) leads to the generalized diffusion equation for nonequilibrium electron concentration $n_d(x,t)$, taking into consideration linear recombination:

$$\frac{\partial n_d(\mathbf{r},t)}{\partial t} + \frac{1}{\tau_{0n}c_n^\alpha}\frac{\partial^\alpha n_d(\mathbf{r},t)}{\partial t^\alpha}$$

$$+ \text{div}\Big(- \mu_n\mathbf{E}\ n_d(\mathbf{r},t) - D_n\nabla n_d(\mathbf{r},t)\Big) + \frac{n_d(\mathbf{r},t)}{\tau_n} = 0. \tag{2.67}$$

Then, the equation for total concentration $n(\mathbf{r},t)$, which is the sum of concentrations of free carriers and carriers localized in band tail states,

$$n(\mathbf{r},t) = n_t(\mathbf{r},t) + n_d(\mathbf{r},t),$$

can be written out, taking into account relationship (2.51), as

$$\frac{\partial n(\mathbf{r},t)}{\partial t} + \tau_{0n}c_n^\alpha\ {}_0\mathsf{D}_t^{1-\alpha}\left[\text{div}\Big(- \mu_n\mathbf{E}\ n(\mathbf{r},t) - D_n\nabla n(\mathbf{r},t)\Big) + \frac{n(\mathbf{r},t)}{\tau_n}\right]$$

$$= n(\mathbf{r},0)\ \delta(t), \tag{2.68}$$

or

$${}_0\mathsf{D}_t^\alpha n(\mathbf{r},t) + \text{div}\Big(- \mu_n\tau_{0n}c_n^\alpha\mathbf{E}\ n(\mathbf{r},t) - D_n\tau_{0n}c_n^\alpha\nabla n(\mathbf{r},t)\Big)$$

$$+\tau_{0n}K_n\frac{n(\mathbf{r},t)}{\tau_n} = \frac{t^{-\alpha}}{\Gamma(1-\alpha)}\,n(\mathbf{r},0). \tag{2.69}$$

The dispersive transport equation for tunneling transi- tions of localized carriers can be obtained based on the model of Seki, Wojcik, and Tachiya [Seki et al. (2003, 2006)]. These authors deduced a fractional equation for the first passage time distribution density. We derived equations for carrier con-centrations and found their solutions [Uchaikin & Sibatov (2007)].

Recombination waiting time in the absence of random walks is given by the distribution function

$$\Phi(\mathbf{r},t) = \int_0^t \phi(\mathbf{r},t)dt.$$

Seki et al. (2006) considered a case of exponential distribution

$$\phi(\mathbf{r},t) = \gamma_{\rm rc}(\mathbf{r})\exp[-\gamma_{\rm rc}(\mathbf{r})t], \tag{2.70}$$

where $\gamma_{\rm rc}(\mathbf{r})$ is the distance-dependent recombination rate.

The trap release time distribution density for the hopping process with recombination has the form

$$\psi_{\rm out}(\mathbf{r},t) = \psi(t)\ [1-\Phi(\mathbf{r},t)], \tag{2.71}$$

and recombination time distribution density is described by the expression

$$\psi_{\rm rc}(\mathbf{r},t) = \phi(\mathbf{r},t)\ [1-\Psi(t)]. \tag{2.72}$$

It is easy to show that the normalization condition

$$\int_0^\infty dt\ [\psi_{\rm out}(\mathbf{r},t)+\psi_{\rm rc}(\mathbf{r},t)] = 1$$

is fulfilled.

In the presence of recombination, the following relation between concentrations of free and localized carriers is obtained instead of relationship (2.38):

$$n_t(\mathbf{r},t) = \int_0^t \tau_0^{-1}n_d(\mathbf{r},t')\ [1-\Psi_{\rm out}(\mathbf{r},t-t')-\Psi_{\rm rc}(\mathbf{r},t-t')]\ dt', \tag{2.73}$$

where

$$\Psi_{\rm out}(\mathbf{r},t) = \int_0^t dt\ \psi_{\rm out}(\mathbf{r},t), \qquad \Psi_{\rm rc}(\mathbf{r},t) = \int_0^t dt\ \psi_{\rm rc}(\mathbf{r},t).$$

Taking into consideration that

$$\Psi_{\rm out}(\mathbf{r},t)+\Psi_{\rm rc}(\mathbf{r},t) = \Psi(t)+\Phi(\mathbf{r},t)-\Phi(\mathbf{r},t)\Psi(t),$$

relation (2.73) can be rewritten as

$$n_t(\mathbf{r},t) = \int_0^t dt'\ \tau_0^{-1} n_d(\mathbf{r},t')\ [1-\Psi(t-t')][1-\Phi(\mathbf{r},t-t')], \tag{2.74}$$

For exponential distribution $\phi(t)$ one obtains

$$\frac{\partial n_t(\mathbf{r},t)}{\partial t} = (\tau_0 c^\alpha)^{-1} \frac{1}{\Gamma(1-\alpha)} \frac{\partial}{\partial t} \int_0^t dt'\ \frac{n_d(\mathbf{r},t')}{(t-t')^\alpha} \exp[-\gamma_{\mathrm{rc}}(\mathbf{r})(t-t')]. \tag{2.75}$$

Substituting relation (2.74) into the continuity equation taking into account recombination, i.e.

$$\frac{\partial n(\mathbf{r},t)}{\partial t} + \mathrm{div}\ \mathbf{q}(\mathbf{r},t) + \phi(\mathbf{r},0) n_t(\mathbf{r},t) = n(\mathbf{r},0)\delta(t). \tag{2.76}$$

we arrive at the equation for the concentration of delocalized carriers that, in the case of exponential recombination time distribution, can be written out as

$$\frac{\partial n_d(\mathbf{r},t)}{\partial t} + (c^\alpha \tau_0)^{-1} \exp\left[-\gamma_{\mathrm{rc}}(\mathbf{r})t'\right]\ {}_0\mathsf{D}_t^\alpha n_d(\mathbf{r},t) \exp\left[\gamma_{\mathrm{rc}}(\mathbf{r})t\right]$$

$$+\mathrm{div}\left[\mu \mathbf{E} n_d(\mathbf{r},t) - D\nabla n_d(\mathbf{r},t)\right] = \delta(t) n(\mathbf{r},0). \tag{2.77}$$

The equation for the total concentration takes the form

$$\exp\left[-\gamma_{\mathrm{rc}}(\mathbf{r})t\right]\ {}_0\mathsf{D}_t^\alpha n(\mathbf{r},t) \exp\left[\gamma_{\mathrm{rc}}(\mathbf{r})t\right] + (c^\alpha \tau_0)\mathrm{div}\left[\mu \mathbf{E}\ n(\mathbf{r},t) - D\nabla n(\mathbf{r},t)\right]$$

$$= \frac{t^{-\alpha}}{\Gamma(1-\alpha)} \exp\left[-\gamma_{\mathrm{rc}}(\mathbf{r})tS\right] n(\mathbf{r},0). \tag{2.78}$$

The solution to the last equation is expressed through stable density [Uchaikin & Sibatov (2007)]:

$$n(\mathbf{r},t) = \exp\left[-\gamma_{\mathrm{rc}}(\mathbf{r})t\right]$$

$$\times \int_0^\infty \frac{1}{(4\pi D\tau)^{3/2}} \exp\left[-\frac{(\mathbf{r}-\mu\mathbf{E}\tau)^2}{4D\tau}\right] \frac{ct}{\alpha\tau_0(\tau/\tau_0)^{1+1/\alpha}}\ g_+\left(ct\,(\tau/\tau_0)^{-1/\alpha};\alpha\right)\ d\tau.$$

The spatial dependence of the recombination rate $\gamma_{\mathrm{rc}}(\mathbf{r})$ is defined by the distribution of recombination centers and the recombination type. In the case of the tunneling mechanism, one has $\gamma_{\mathrm{rc}}(\mathbf{r}) \propto e^{-2\beta r}$, and in the case of the energy transition model, $\gamma_{\mathrm{rc}}(\mathbf{r}) \propto r^{-6}$ ((see Ref. [Seki et al. (2006)] for details).

2.4.4 *Including generation*

As noted above, the Arkhipov-Rudenko equations were obtained for a pulse carrier injection. An arbitrary generation process can be included by inserting the corresponding source function in the right side of the transport equations.[4] In this section, we will not discuss all the features of the generation processes in disordered semiconductors but restrict ourselves to a particular case. Consider the equation

$$ {}_0\mathsf{D}_t^\alpha c(x,t) = D_\alpha \frac{\partial^2 c(x,t)}{\partial x^2} - v\frac{\partial c(x,t)}{\partial x} + \beta\left[c_1 - c(x,t)\right] + \frac{t^{-\alpha}}{\Gamma(1-\alpha)}c(x,0), $$

having the Laplace representation

$$ D_\alpha \frac{\partial^2 \widehat{c}(x,\lambda)}{\partial x^2} - v\frac{\partial \widehat{c}(x,\lambda)}{\partial x} - (\beta+\lambda^\alpha)\,\widehat{c}(x,\lambda) = -\frac{\beta c_1}{\lambda} - \lambda^{\alpha-1}c(x,0). \tag{2.79} $$

This is an ordinary non-homogeneous linear differential equation of the second order. The solution $\widehat{c}_0(x,\lambda)$ of the corresponding homogeneous equation obeying the requirement

$$ \lim_{|x|\to\infty} \widehat{c}_0(x,\lambda) \neq \infty $$

is of the form

$$ \widehat{c}_0(x,\lambda) = \widehat{c}_+(x,\lambda) + \widehat{c}_-(x,\lambda), $$

where

$$ \widehat{c}_+(x,\lambda) = A_+ \exp\left(\frac{v - \sqrt{v^2+4D_\alpha(\beta+\lambda^\alpha)}}{2D_\alpha}x\right), \tag{2.80} $$

and

$$ \widehat{c}_-(x,\lambda) = A_- \exp\left(\frac{v + \sqrt{v^2+4D_\alpha(\beta+\lambda^\alpha)}}{2D_\alpha}x\right). $$

The inhomogeneous equation can be solved by variation of arbitrary constants (Lagrangian method) in the solution of the corresponding homogeneous equation, assuming that $A_+ = A_+(x)$ and $A_- = A_-(x)$. As a result, we have

$$ \widehat{c}(x,\lambda) = B_2 \exp\left(\frac{v - \sqrt{v^2+4D_\alpha(\beta+\lambda^\alpha)}}{2D_\alpha}x\right) + \frac{\beta c_1}{\lambda(\beta+\lambda^\alpha)} $$

$$ + \frac{\lambda^{\alpha-1}}{\sqrt{v^2+4D_\alpha(\beta+\lambda^\alpha)}} \int_0^x d\xi\; c(\xi,0)\left[\exp\left(\frac{v - \sqrt{v^2+4D_\alpha(\beta+\lambda^\alpha)}}{2D_\alpha}(x-\xi)\right)\right. $$

$$ \left. - \exp\left(\frac{v + \sqrt{v^2+4D_\alpha(\beta+\lambda^\alpha)}}{2D_\alpha}(x-\xi)\right)\right]. $$

[4] This way is valid only in the linear case, that is, in the small-signal approximation.

For initial condition $c(x,0) = \delta(x)$, we get

$$\widehat{c}_\delta(x,\lambda) = B_2 \exp\left(\frac{v - \sqrt{v^2 + 4D_\alpha(\beta + \lambda^\alpha)}}{2D_\alpha} x\right) + \frac{\beta c_1}{\lambda(\beta + \lambda^\alpha)} +$$

$$+ \frac{\lambda^{\alpha-1}}{\sqrt{v^2 + 4D_\alpha(\beta + \lambda^\alpha)}} \left[\exp\left(\frac{v - \sqrt{v^2 + 4D_\alpha(\beta + \lambda^\alpha)}}{2D_\alpha} x\right) - \right.$$

$$\left. - \exp\left(\frac{v + \sqrt{v^2 + 4D_\alpha(\beta + \lambda^\alpha)}}{2D_\alpha} x\right)\right].$$

or, in some different form,

$$\widehat{c}_\delta(x,\lambda) = B_2 \exp\left(\frac{v - \sqrt{v^2 + 4D_\alpha(\beta + \lambda^\alpha)}}{2D_\alpha} x\right) + \frac{\beta c_1}{\lambda(\beta + \lambda^\alpha)} -$$

$$- \frac{2\lambda^{\alpha-1}}{\sqrt{v^2 + 4D_\alpha(\beta + \lambda^\alpha)}} \exp\left(\frac{vx}{2D_\alpha}\right) \sinh\left(\frac{\sqrt{v^2 + 4D_\alpha(\beta + \lambda^\alpha)}}{2D_\alpha} x\right).$$

This is the Laplace transform of the Green's function. To find its original, we analyze the asymptotic behaviour of $c(x,t)$ as $t \to \infty$. According to Tauberian theorem, it is connected with the Laplace transform at small λ. Assuming $\lambda^\alpha \ll \beta$ and introducing notation $V = \sqrt{v^2 + 4D_\alpha \beta}$, we represent the solution as

$$\widehat{c}_\delta(x,\lambda) \sim \frac{\beta c_1}{\lambda(\beta + \lambda^\alpha)} +$$

$$+ \frac{\lambda^{\alpha-1}}{V + \dfrac{\lambda^\alpha}{2V}} \left[\exp\left(\frac{(v-V)x}{2D_\alpha}\right) \exp\left(-\frac{\lambda^\alpha}{4VD_\alpha} x\right) - \exp\left(\frac{(v+V)x}{2D_\alpha}\right) \exp\left(\frac{\lambda^\alpha}{4VD_\alpha} x\right)\right].$$

The inversion transform leads to the convolution of a fractional stable density with the two-parameter Mittag-Leffler function. Thus, at point $x = 0$

$$\widehat{c}_\delta^{\,\mathrm{as}}(0,\lambda) = \frac{\beta c_1}{\lambda(\beta + \lambda^\alpha)},$$

and

$$c_\delta^{\mathrm{as}}(0,t) = c_1 \left[1 - E_\alpha(-\beta t^\alpha)\right],$$

where $E_\alpha(z)$ is the one-parameter Mittag-Leffler function.

2.4.5 *Bipolar dispersive transport*

Considering experiments with the double injection in disordered semiconductors, they usually assume that one type of carriers is much less mobile than the other, and often the first type of carrier mobility is neglected. This greatly simplifies the theoretical description of experiments on the time-dependent radiative conductivity, the study of diffusion-controlled recombination processes, etc. We derive here the bipolar diffusion equation in the framework of the fractional approach considered in our book.

The normal diffusion and normal drift of charge carriers in the case of double injection (holes and electrons) are characterized by the bipolar diffusion coefficient and bipolar mobility. In normal transport with $D_n > D_p$, the diffusion electron packet expands faster than the hole packet. The arising electric field (bipolar diffusion field) slows down electrons and accelerates holes. In the steady-state regime, hole and electron fluxes are identical. If there is a drift, the inequality of carrier mobilities μ_n and μ_p in an external electric field implies that electrons and holes have a common (bipolar) drift mobility. Time-of-flight experiments demonstrated that drift mobility in dispersive transport depends on the sample thickness and the electric field strength; therefore, drift mobility cannot be used as a characteristic of bipolar transport. The diffusion equation in the Arkhipov-Rudenko model contains time-dependent diffusion coefficients and mobility. Moreover, these quantities depend on the initial conditions and therefore are equally unsuitable for the description of the bipolar dispersive transport.

Let us assume that the semiconductor is homogeneous and the quasineutrality condition $\delta p = \delta n$ is fulfilled. Electron and hole lifetimes are identical: $\tau_n = \tau_p$. The dispersive transport equation (2.69) with monomolecular recombination for electrons and holes takes the form

$$ {}_0\mathsf{D}_t^{\alpha_n} n - \mu_n^* n \,\mathrm{div}\mathbf{E} - \mu_n^* \mathbf{E}\, \nabla n \;- D_n^*\, \nabla^2 n + \frac{\delta n}{\tau^*} = \frac{t^{-\alpha_n}}{\Gamma(1-\alpha_n)}\, \delta n(\mathbf{r},0), \qquad (2.81) $$

$$ {}_0\mathsf{D}_t^{\alpha_p} p + \mu_p^* p \,\mathrm{div}\mathbf{E} + \mu_p^* \mathbf{E}\, \nabla p \;- D_p^*\, \nabla^2 p + \frac{\delta p}{\tau^*} = \frac{t^{-\alpha_p}}{\Gamma(1-\alpha_p)}\, \delta p(\mathbf{r},0), \qquad (2.82) $$

where $D_n^* = D_n \tau_{0n} c_n^\alpha$ and $\mu_n^* = \mu_n \tau_{0n} c_n^\alpha$. Eliminating terms with $\mathrm{div}\mathbf{E}$ from these equations yields

$$ \frac{1}{\mu_n^* n + \mu_p^* p}\, \left(\mu_n^* n \; {}_0\mathsf{D}_t^{\alpha_p} + \mu_p^* p \; {}_0\mathsf{D}_t^{\alpha_n}\right) p + \frac{\mu_p^* \mu_n^* (n-p)}{\mu_n^* n + \mu_p^* p}\, \mathbf{E}\, \nabla p $$

$$ - \frac{\mu_n^* n D_p^* + \mu_p^* p D_n}{\mu_n^* n + \mu_p^* p}\, \nabla^2 p + \frac{\delta p}{\tau^*} $$

$$ = \frac{1}{\mu_n^* n + \mu_p^* p} \left[\mu_n^* n \frac{t^{-\alpha_p}}{\Gamma(1-\alpha_p)} + \mu_p^* p \frac{t^{-\alpha_n}}{\Gamma(1-\alpha_n)} \right] \delta p(\mathbf{r},0). \qquad (2.83) $$

Denoting by:

$\sigma_n = \mu_n n, \quad \sigma_p = \mu_p p, \quad \sigma = \sigma_n + \sigma_p$ electron/hole conductivity and total conductivity, respectively,

$$\mu_{\mathrm{amb}} = \frac{\mu_p^* \mu_n^* (n-p)}{\mu_n^* n + \mu_p^* p}$$

$$D_{\mathrm{amb}} = \frac{\mu_n^* n D_p^* + \mu_p^* p D_n}{\mu_n^* n + \mu_p^* p},$$

we rewrite the equation for bipolar dispersive transport with monomolecular recombination in the form

$$\frac{\sigma_n}{\sigma}\, {}_0\mathsf{D}_t^{\alpha_p} p(\mathbf{r},t) + \frac{\sigma_p}{\sigma}\, {}_0\mathsf{D}_t^{\alpha_n} p(\mathbf{r},t) + \mu_{\mathrm{amb}}\, \mathbf{E}\, \nabla p(\mathbf{r},t) \; - D_{\mathrm{amb}} \nabla^2 p(\mathbf{r},t) + \frac{\delta p(\mathbf{r},t)}{\tau^*}$$

$$= \left[\frac{\sigma_n}{\sigma} \frac{t^{-\alpha_p}}{\Gamma(1-\alpha_p)} + \frac{\sigma_p}{\sigma} \frac{t^{-\alpha_n}}{\Gamma(1-\alpha_n)} \right] \delta p(\mathbf{r},0). \tag{2.84}$$

The last bipolar transport equation contains two fractional derivatives of different orders in the general case.

Using the *Caputo derivative*

$${}_0^\alpha\mathsf{D}_t f(\mathbf{r},t) = \frac{1}{\Gamma(1-\alpha)} \int_0^t \frac{\partial f(\mathbf{r},t')/\partial t'}{(t-t')^\alpha} dt', \quad 0 < \alpha < 1.$$

we can rewrite the equation in the form

$$\frac{\sigma_n}{\sigma}\, {}_0^{\alpha_p}\mathsf{D}_t p(\mathbf{r},t) + \frac{\sigma_p}{\sigma}\, {}_0^{\alpha_n}\mathsf{D}_t p(\mathbf{r},t) + \mu_{\mathrm{amb}}\, \mathbf{E}\, \nabla p(\mathbf{r},t) \; - D_{\mathrm{amb}} \nabla^2 p(\mathbf{r},t) + \frac{\delta p(\mathbf{r},t)}{\tau^*} = 0.$$

In the long-time asymptotics, a term with lower order derivative predominates. This order coincides with the dispersion parameter of the corresponding type of carriers (holes or electrons). Thus, for preservation of the quasi-neutrality, the process is limited by transfer of carriers with a smaller dispersion parameter. Therefore, one can consider the motion of less mobile carriers amending results for the mobility and the diffusion coefficient.

2.4.6 *The family of fractional dispersive transport equations*

We conclude this chapter by listing some modifications of the FDT equation.

- The fractional Fokker-Planck equation for the total concentration of nonequilibrium carriers $n(\mathbf{r},t)$:

$${}_0^\alpha\mathsf{D}_t n(\mathbf{r},t) + \mathrm{div}\Big(\mathbf{K}\; n(\mathbf{r},t) - C\nabla n(\mathbf{r},t) \Big) = 0 \tag{2.85}$$

can be used for hopping conduction over the Poisson ensemble of traps and for multiple trapping into states on the band tail with the exponential energy distribution.

• The equation for the density of delocalized carriers in case of multiple trapping is of the form $n_d(\mathbf{r},t)$ is obtained

$$\frac{\partial n_d(\mathbf{r},t)}{\partial t} + \frac{l}{\tau_0 K}\ {}_0\mathsf{D}_t^\alpha\ n_d(\mathbf{r},t) + \mathrm{div}\Big(\mu\mathbf{E}\ n_d(\mathbf{r},t) - D\nabla n_d(\mathbf{r},t)\Big) = 0. \qquad (2.86)$$

Here $\mathbf{K}$ and C are coefficients of anomalous advection and diffusion. Parameter τ_0 is the mean sojourn time in delocalized state, l is the average length of delocalization, $K = c^\alpha l$, $c = w_0[\sin(\pi\alpha)/\pi\alpha]^{1/\alpha}$, w_0 is the capture rate of carriers into localized states, μ and D are mobility and diffusion coefficient of delocalized carriers, $\mathbf{K} = \tau_0 c^\alpha \mu\mathbf{E}$ is a dispersive advection, $C = \tau_0 c^\alpha D$ is an anomalous diffusion coefficient. For hopping conductivity with variable hopping length, $c = \nu_0[\sin(\pi\alpha)/\pi\alpha]^{1/\alpha}$, where ν_0 is the characteristic rate of jumps between the traps.

Solutions of fractional equation (2.85) are expressed through the solutions $n_1(\mathbf{r},t)$ of the ordinary Fokker-Planck equation by the relation:

$$n_\alpha(\mathbf{r},t) = \int_0^t d\tau\ \left\{\frac{ct}{\alpha\tau}\left(\frac{\tau}{\tau_0}\right)^{-1/\alpha} g_+\left(ct\left(\frac{\tau}{\tau_0}\right)^{-1/\alpha};\alpha\right) n_1(\mathbf{r},\tau)\right\}, \qquad (2.87)$$

where $g_+(t;\alpha)$ is the one-sided Lévy stable density. This relation allows to find analytical solutions in simple cases and to derive the general Monte-Carlo algorithm.

• The equation for the delocalized carrier concentration in case of multiple trapping and arbitrary density of states, which takes into account the percolative nature of conduction ways:

$$\frac{\partial n_d(\mathbf{r},t)}{\partial t} + \tau_0^{-1}\tau_\beta^\beta\ {}_0\mathsf{D}_t^\beta n_d(\mathbf{r},t) +$$

$$+\tau_0^{-1}\frac{\partial}{\partial t}\int_0^t n_d(\mathbf{r},t-t')\int_0^{\varepsilon_g}\rho(\varepsilon)\exp\left(-w_\varepsilon t'\ \mathrm{e}^{-\varepsilon/kT}\right)d\varepsilon\ dt' +$$

$$+\mathrm{div}[\ \mu\mathbf{E}\ n_d(\mathbf{r},t) - D\nabla n_d(\mathbf{r},t)] = 0. \qquad (2.88)$$

In case of hopping in a medium with the Gaussian energetic density of states, the equation for the carrier concentration near the transport layer is as follows

$$\frac{a_\beta}{c^\beta}\ {}_0\mathsf{D}_t^\beta n_{\mathrm{eff}}(x,t) + a_\alpha A\frac{\partial}{\partial t}\int\limits_0^t \frac{n_{\mathrm{eff}}(x,t')}{(t-t')^{kT/\sigma}}\exp\left\{-\left(\frac{\sigma(t-t')}{4kT}\right)^{kT/2\sigma}\right\}dt' =$$

$$= \langle l\rangle\ \frac{\partial n_{\mathrm{eff}}(x,t)}{\partial x} - D^*\ \frac{\partial^2 n_{\mathrm{eff}}(x,t)}{\partial x^2} = 0. \qquad (2.89)$$

The first term takes into account the percolation nature of carrier trajectories. The parameter β is usually close to 0.5.

• The transport equation taking into account the recombination of localized carriers has the form:

$$\exp\left[-\gamma_{\mathrm{r}}(\mathbf{r})t\right]\ {}_0^\alpha\mathsf{D}_t\ \exp\left[\gamma_{\mathrm{r}}(\mathbf{r})t\right]n(\mathbf{r},t) + (c^\alpha\tau_0)\mathrm{div}\left[\mu\mathbf{E}\ n - D\nabla n\right] = 0.$$

For tunneling and energy transfers, respectively,

$$\gamma_{\rm r}(\mathbf{r}) = k_0 e^{-2\beta \mathbf{r}}, \quad \gamma_{\rm r}(\mathbf{r}) = k_0 (R_F/r)^6,$$

where R_F is the Förster radius. Its solution for $n(\mathbf{r}, 0) = \delta(\mathbf{r})$ has the form

$$n(\mathbf{r}, t) = \exp\left[-\gamma_{\rm rc}(\mathbf{r})t\right] \int_0^\infty \frac{1}{(4\pi D\tau)^{3/2}} \exp\left[-\frac{(\mathbf{r} - \mu \mathbf{E}\tau)^2}{4D\tau}\right] \times$$

$$\times \frac{ct}{\alpha\tau_0(\tau/\tau_0)^{1+1/\alpha}}\, g_+\left(ct\,(\tau/\tau_0)^{-1/\alpha}; \alpha\right)\, d\tau. \tag{2.90}$$

Solution of the initial-value problem $n(\mathbf{r}, 0) = \delta(\mathbf{r})$ has the form

$$n(\mathbf{r}, t) = \exp\left[-\gamma_{\rm rc}(\mathbf{r})t\right] \int_0^\infty \frac{1}{(4\pi D\tau)^{3/2}} \exp\left[-\frac{(\mathbf{r} - \mu \mathbf{E}\tau)^2}{4D\tau}\right] \times$$

$$\times \frac{ct}{\alpha\tau_0(\tau/\tau_0)^{1+1/\alpha}}\, g_+\left(ct\,(\tau/\tau_0)^{-1/\alpha}; \alpha\right)\, d\tau. \tag{2.91}$$

• The transport equation taking into account the monomolecular recombination of delocalized carriers has the form:

$$\frac{\partial n(\mathbf{r}, t)}{\partial t} + \frac{\tau_0 K}{l}\, {}_0\mathsf{D}_t^{1-\alpha}\left[\operatorname{div}\left(-\mu \mathbf{E}\, \delta n(\mathbf{r}, t) - D\nabla n(\mathbf{r}, t)\right) + \frac{\delta n(\mathbf{r}, t)}{\tau_{\rm mr}}\right] = 0. \tag{2.92}$$

• The equation of bipolar dispersive diffusion

$$\left[\frac{\sigma_n}{\sigma}\, {}_0\mathsf{D}_t^{\alpha_p} + \frac{\sigma_p}{\sigma}\, {}_0\mathsf{D}_t^{\alpha_n}\right] p_f + \mu_{\rm amb}\, \mathbf{E}\, \nabla p_f \; - D_{\rm amb}\nabla^2 p_f + \frac{\delta p_f}{\tau^*} = 0. \tag{2.93}$$

Here, $\sigma_n = \mu_n n_f$, $\sigma_p = \mu_p p_f$ are conductivities of delocalized electrons and holes, $\sigma = \sigma_n + \sigma_p$; $\mu_{\rm amb} = \mu_p^* \mu_n^* (n_f - p_f)[\mu_n^* n_f + \mu_p^* p_f]^{-1}$ is a bipolar drift mobility, $D_{\rm amb} = (\mu_n^* n D_p^* + \mu_p^* p D_n)(\mu_n^* n + \mu_p^* p)^{-1}$ is a bipolar diffusion coefficient.

• In case of distributed dispersion parameter, the transport equation is of the form

$$\frac{\partial n(\mathbf{r}, t)}{\partial t} + \int_0^1 d\alpha\; \rho(\alpha)\; {}_0\mathsf{D}_t^{1-\alpha} \operatorname{div}\left(-\mathbf{K}_\alpha\, \delta n - C_\alpha \nabla n\right) = 0. \tag{2.94}$$

• For exponentially truncated power law distributions of localization times in the generalized Scher-Montroll model,

$$\frac{\partial n(\mathbf{r}, t)}{\partial t} + \operatorname{div}\left[e^{-\gamma t}\; {}_0\mathsf{D}_t^{1-\alpha} e^{\gamma t}\; \left(\mathbf{K}\; n(\mathbf{r}, t) - C\; \nabla n(\mathbf{r}, t)\right)\right] = 0. \tag{2.95}$$

Localization times have a finite variance, the CLT is applicable in this case, transport at large times is normal.

Chapter 3

Transient processes in disordered semiconductor structures

In this chapter the system of fractional kinetic equations obtained above, analytical and numeric methods of their solution are applied to the interpretation of

(1) time-of-flight experiments in thin layers of disordered semiconductors;
(2) measurements of radiation induced conductivity in polymers;
(3) measurements of frequency dependence of conductivity;
(4) transients in structures based on disordered semiconductors.

Comparison of the results with experimental data and the results of other approaches will evaluate the effectiveness of the fractional differential approach.

3.1 Time-of-flight method

3.1.1 *Transient current in disordered semiconductors*

The transient photocurrent $I(t)$ in a sample of the length L is determined through the conductivity current density as

$$I(t) = (1/L) \int_0^L j(x,t)dx. \tag{3.1}$$

The x-axis is assumed here to be directed along the electric field $\mathbf{E}$. Knowing that [Arkhipov et al. (1983); Tyutnev et al. (2006)]

$$j(x,t) = e\mu n_d E,$$

where n_d is the concentration of delocalized carriers, we find the density of transient current

$$I(t) = \frac{eKN\alpha}{L} t^{\alpha-1} \int_{\zeta_0}^{\infty} \zeta^{-\alpha} g_+(\zeta;\alpha)\, d\zeta, \quad \zeta_0 = t\,(L/K)^{-1/\alpha}. \tag{3.2}$$

Expressing the stable density in the form of a convergent series [Uchaikin & Zolotarev (1999)]

$$g_+(s;\alpha) = \sum_{n=1}^{\infty} \frac{(-1)^{n-1}}{n!} \frac{\Gamma(n\alpha+1)}{\Gamma(n\alpha)\Gamma(1-n\alpha)} s^{-n\alpha-1}$$

and substituting it into formula (3.2), we find after term by term integration

$$I(t) = eN\alpha\frac{L}{K}t^{-1-\alpha}\sum_{k=0}^{\infty}\frac{(-1)^k}{k!}\frac{t^{-\alpha k}(L/K)^k}{(k+2)\Gamma(1-\alpha-k\alpha)}. \tag{3.3}$$

For sufficiently small times, when $t < t_T$ and ζ_0 is small, using the formula for the moments of one-sided stable distributions

$$\int_0^{\infty} s^{-\alpha}g_+(s;\alpha)ds = \frac{1}{\Gamma(1+\alpha)},$$

we obtain

$$I(t) \cong \frac{eKN\alpha}{L\Gamma(1+\alpha)}t^{-1+\alpha}, \quad t < t_T. \tag{3.4}$$

At large times, when $\zeta_0 \gg 1$, the density $g_+(s;\alpha)$ can be presented as [Uchaikin & Zolotarev (1999)]:

$$g_+(s;\alpha) \sim \frac{\Gamma(1+\alpha)\sin\pi\alpha}{\pi s^{\alpha+1}}, \quad s \to \infty,$$

which leads to the asymptotics

$$I(t) \cong \frac{eNL\Gamma(1+\alpha)}{2K}\frac{\sin\pi\alpha}{\pi}t^{-1-\alpha}, \quad t > t_T. \tag{3.5}$$

Equating (3.4) and (3.5), we obtain for t_T:

$$t_T = \left(\frac{L}{K}\sqrt{\frac{\sin\pi\alpha}{2\pi\alpha}}\,\Gamma(1+\alpha)\right)^{1/\alpha}. \tag{3.6}$$

The last formula shows that in the model of multiple trapping at a fixed temperature, the transit time [Scher & Montroll (1975)]: $t_T \propto (L/E)^{1/\alpha}$. Drift mobility is given by

$$\mu_D = L/Et_T. \tag{3.7}$$

Comparison of photocurrent (3.2) with the experimental data for the organic complex trinitrofluorenone polyvinylcarbazole (TNF-PVK), $\alpha = 0.8$ and the result obtained by using the equation of the Arkhipov-Rudenko τ-formalism is shown in Fig. 3.1. Figure shows better agreement of results obtained on the basis of the fractional equation. Fig. 3.2 shows the theoretical dependence of various characteristics of the time-of-flight experiment in reduced coordinates for different values of the dispersion parameter α. In Fig. 3.3, the transient current curves found on the basis of the Arkhipov-Rudenko τ-approximation, the Nikitenko diffusion equation, and the fractional equation for the exponential density of localized states.

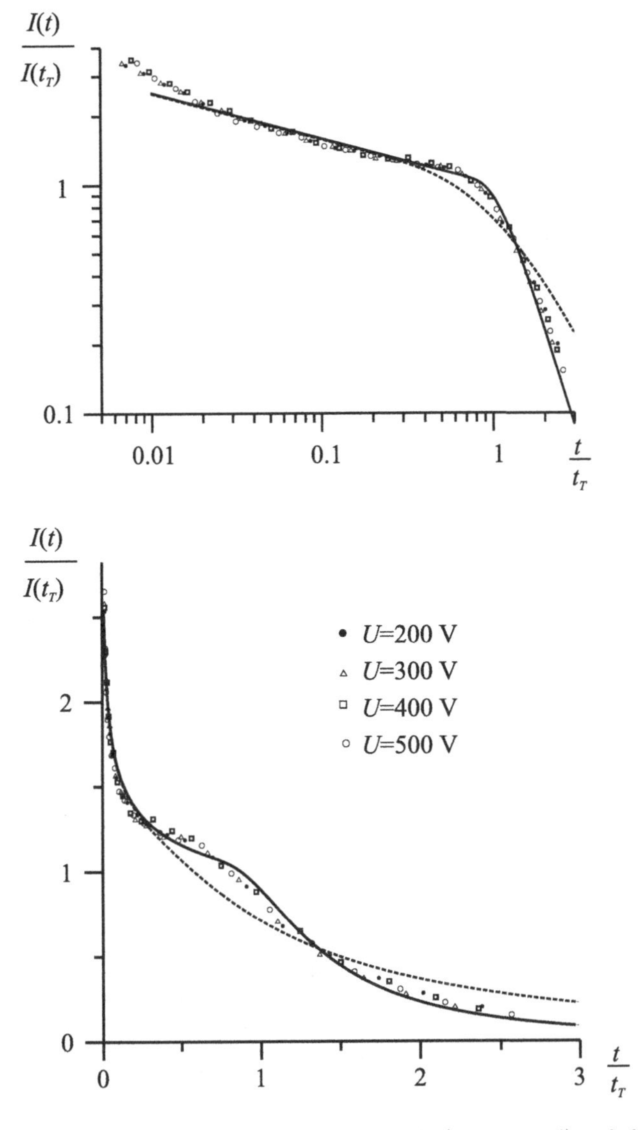

Figure 3.1 Transient current curves in the log-log scale (upper panel) and the linear scale (lower panel). Points are digitized from [Scher & Montroll, 1975], experimental data for organic complex of trinitrofluorenon and polyvinylcarbazole (TNF-PVK) obtained by Gill. Dashed line is the result obtained within the Arkhipov-Rudenko τ-approximation. Solid line is the transient current (2.11). Dispersive parameter $\alpha = 0.8$ [Scher & Montroll, 1975].

3.1.2 *Transient current in case of truncated waiting time distributions*

Koponen (1995) has shown that truncated Lévy flights can be analyzed easily by taking the distribution in the form of the power law multiplied by slowly decaying exponential function. Let waiting times be distributed according to function:

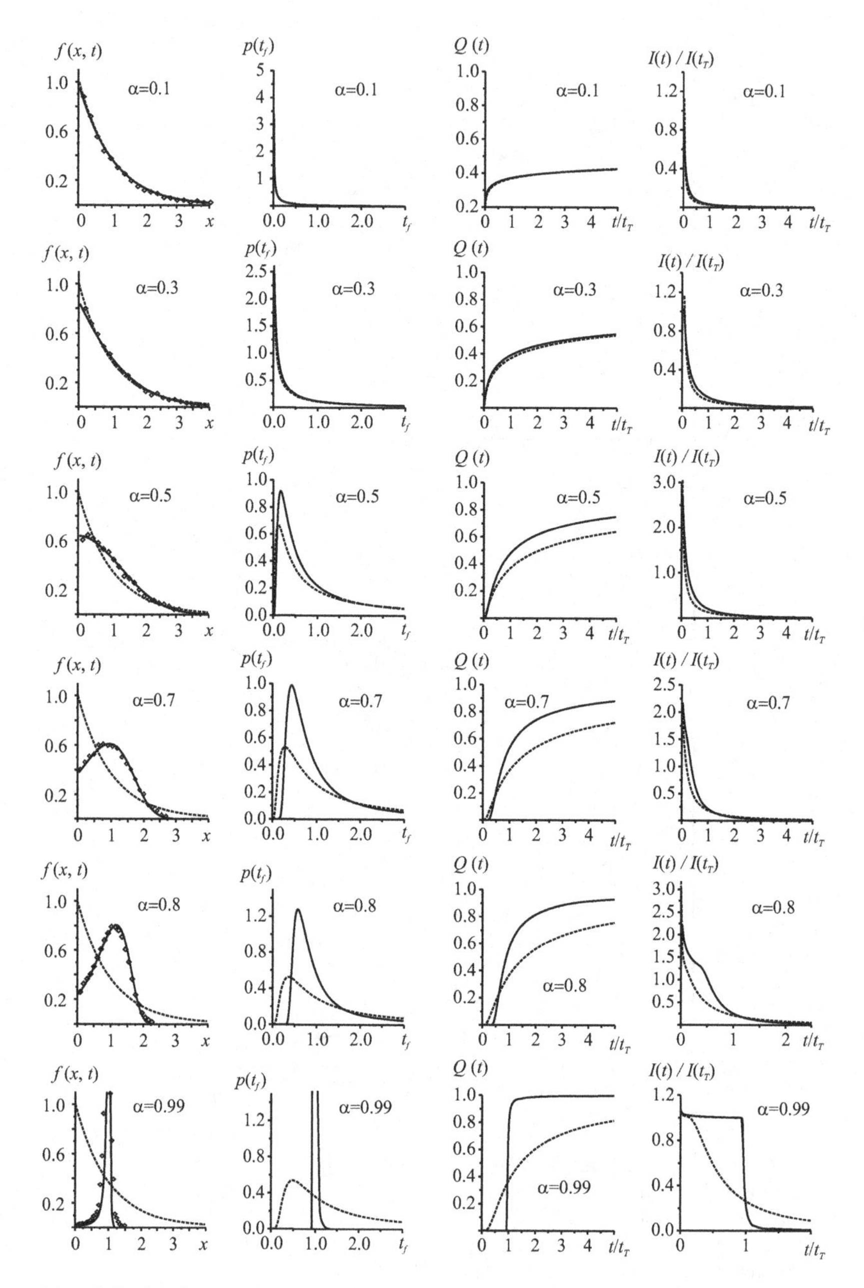

Figure 3.2 Calculated normalized total concentrations (1st column), pdfs of first passage times (2nd column), charge passed through a sample (3rd column) and transient current densities (4th column) for different α values in comparison with solutions of the Arkhipov-Rudenko τ-approximation (dashed lines).

$$Q(t) = 1 - \Psi(t) = \text{Prob}(T > t) \sim \frac{(ct)^{-\alpha}}{\Gamma(1-\alpha)} \exp(-\gamma t), \qquad t \gg c^{-1}. \tag{3.8}$$

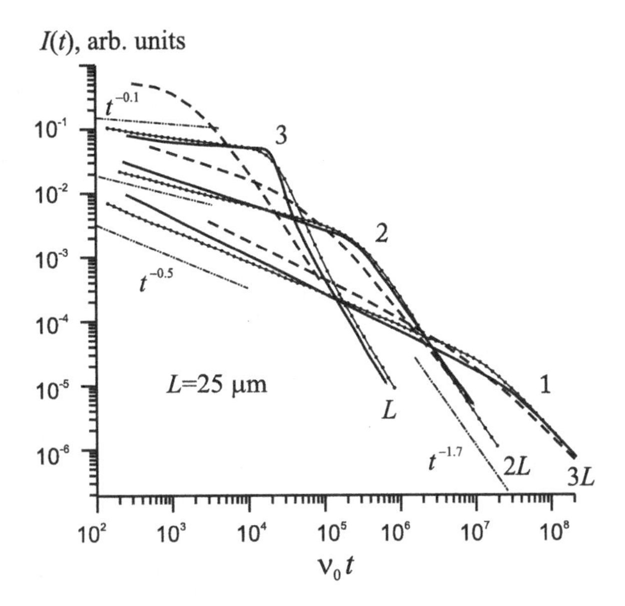

Figure 3.3 Transient current curves. Dashed lines are the solutions obtained within the Arkhipov-Rudenko τ-approximation, solid lines are the solutions of the diffusion Arkhipov-Rudenko-Nikitenko equation, dotted lines are the solutions of the fractional equation (2.85) for exponential density of localized states. $E = 5 \cdot 10^5$ V/cm, $w_0 = 10^6$ c^{-1}, $\mu_0\tau_0 = 2.5 \cdot 10^{-16}$ m^2/V, $l = 12,5$ nm. Parameters of fractional equations: 1) $\alpha = 0.5$, $K = 8$ μm/s$^{0.5}$, $L = 75$ μm; 2) $\alpha = 0.7$, $K = 73$ μm/s$^{0.7}$, $L = 50$ μm; 3) $\alpha = 0.9$, $K = 343$ μm/s$^{0.9}$, $L = 25$ μm.

The corresponding density is given by:

$$\psi(t) \sim \exp(-\gamma t)\,\frac{(ct)^{-\alpha}}{\Gamma(1-\alpha)}\left(\frac{\alpha}{t}+\gamma\right), \qquad t \gg c^{-1}. \tag{3.9}$$

The probability density function of the sum of N terms $T = T_1 + T_2 + \cdots + T_N$ is expressed through N-fold convolution: $q(t|N) = \psi^{*N}(t)$. Taking the Laplace transform $\widehat{q}(s|N) = [\widehat{\psi}(s)]^N$. The Laplace image of the density (3.9) has the form

$$\widehat{\psi}(s) \sim 1 - c^{-\alpha}[(s+\gamma)^\alpha - \gamma^\alpha], \quad s \ll c.$$

The Laplace transform of the density of the sum of N independent random times

$$\widehat{p}(s|N) \sim \exp\left[N\left(\frac{\gamma}{c}\right)^\alpha - N\left(\frac{s+\gamma}{c}\right)^\alpha\right], \quad s \ll c. \tag{3.10}$$

is the Laplace image of truncated stable distribution

$$p(t|N) \sim \exp\left[\left(\frac{\gamma}{c}\right)^\alpha - \gamma t\right]\, cN^{-1/\alpha} g_+(ctN^{-1/\alpha};\alpha). \tag{3.11}$$

In the case of large N, Eq.(3.10) satisfies the relation

$$\frac{\partial\widetilde{p}(s|N)}{\partial N} = \gamma^\alpha c^{-\alpha}\,\widetilde{p}(s|N) - c^{-\alpha}(s+\gamma)^\alpha \widetilde{p}(s|N) + \delta(N),$$

inverse Laplace transformation of which leads to the fractional equation

$$\frac{\partial p(t|N)}{\partial N} = \gamma^\alpha c^{-\alpha}\,p(t|N) - c^{-\alpha}\exp(-\gamma t)\ {}_0\mathrm{D}_t^\alpha\ \exp(\gamma t) p(t|N) + \delta(N)\delta(t). \tag{3.12}$$

All the time moments

$$\langle T^k \rangle = (-1)^k \left[\frac{d^k}{ds^k} \widetilde{p}(s|N) \right]_{s=0}$$

are finite now, and we write the mean value $\langle T \rangle_N$ and the variance V_N of the sum of N random waiting times as

$$\langle T \rangle_N = \frac{\alpha n}{c} \left(\frac{\gamma}{c} \right)^{\alpha-1}, \quad \text{and} \quad V_N = \frac{\alpha(1-\alpha)n}{c^2} \left(\frac{\gamma}{c} \right)^{\alpha-2},$$

respectively. Then according to the Central Limit Theorem for times $t \gg \gamma^{-1}$:

$$p_\mathcal{T}(t) \sim \frac{1}{\sqrt{2\pi V_N}} \exp\left((t - \langle T \rangle_N)^2 / 2V_N \right), \quad N \to \infty.$$

Thus, the stable distributions play here the role of intermediate asymptotics for sums of random variables under consideration. At sufficiently large number of terms the limit distribution is transformed from the stable one with exponent $\alpha < 1$ to the Gaussian one.

Obviously, the case of smoothly truncated power law distributions appears more suitable for analysis. The relation between concentrations of quasi-free and localized carriers has the form

$$n_t(\mathbf{r}, t) = \int_0^t \tau_0^{-1} n_d(\mathbf{r}, t') \left[1 - \Psi(t - t') \right] dt'.$$

Passing to Laplace transforms yields

$$(s + \gamma) \widetilde{n}_t(\mathbf{r}, s) = \tau_0^{-1} (s + \gamma)^\alpha \widetilde{n}_d(\mathbf{r}, s).$$

This is the Laplace transform of the equation

$$\frac{\partial n_t(\mathbf{r}, t)}{\partial t} + \gamma n_t(\mathbf{r}, t) = \tau_0^{-\alpha} e^{-\gamma t} \, {}_0\mathsf{D}_t^\alpha e^{\gamma t} n_d(\mathbf{r}, t).$$

Rewriting it in the form

$$n_d(\mathbf{r}, t) = \tau_0^\alpha e^{-\gamma t} \, {}_0\mathsf{D}_t^{1-\alpha} e^{\gamma t} n_t(\mathbf{r}, t),$$

and substituting the result in the continuity equation, we arrive at

$$\frac{\partial n(\mathbf{r}, t)}{\partial t} + \operatorname{div} \Big(\mu \mathbf{E} \, n_d(\mathbf{r}, t) - D \, \nabla n_d(\mathbf{r}, t) \Big) = 0.$$

In the dispersive transport case, most carriers are captured in traps, i.e., $n(\mathbf{r}, t) \approx n_t(\mathbf{r}, t)$; hence, it follows that

$$\frac{\partial n(\mathbf{r}, t)}{\partial t} + \operatorname{div} \left[e^{-\gamma t} \, {}_0\mathsf{D}_t^{1-\alpha} e^{\gamma t} \left(\mathbf{K} \, n(\mathbf{r}, t) - C \, \nabla n(\mathbf{r}, t) \right) \right] = 0. \quad (3.13)$$

Constants $\mathbf{K}$ and C are defined in the same way as in equation (2.85). By turning α to unity, the above equation goes over into the ordinary Fokker-Planck equation. At $\gamma = 0$ this equation coincides with equation (2.85) for the non-truncated power-law waiting time distribution.

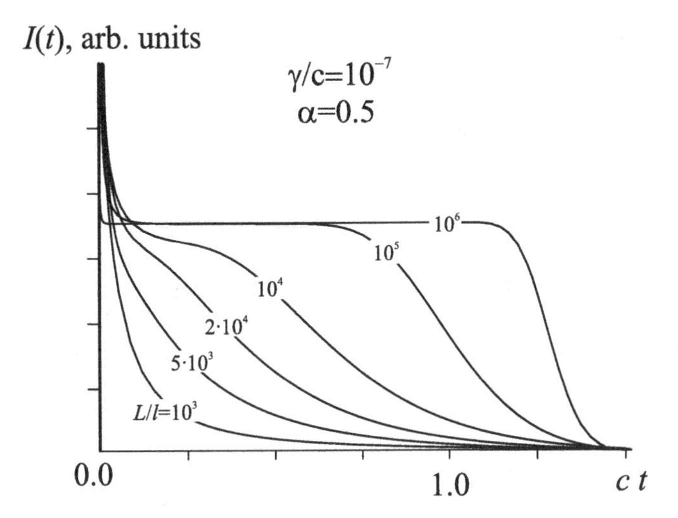

Figure 3.4 Transient photocurrent in case of truncated power law distributions of waiting times for different values of L/l-ratio.

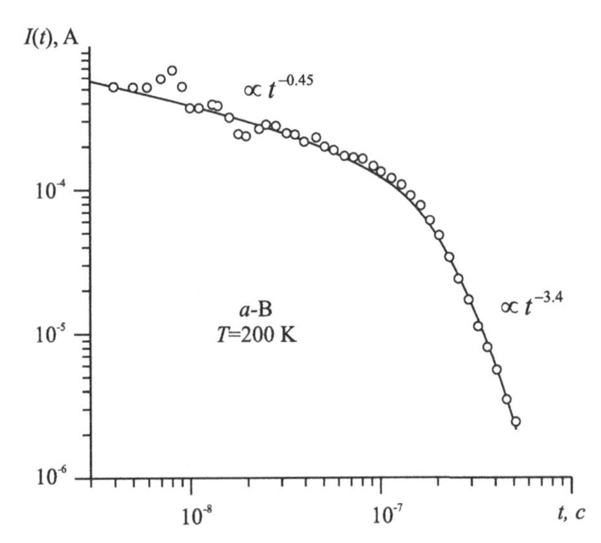

Figure 3.5 Transient photocurrent in amorphous boron. Points are the experimental data digitized from [Takeda et al. (1997)], lines are the calculations by formulas (3.1) and (3.14).

As follows from Eq. (3.11), conduction current density in pulsed injection for the case of truncated power-law waiting time distributions is expressed as

$$j(x,t) = eN \exp\left[\frac{x}{l}\left(\frac{\gamma}{c}\right)^{\alpha} - \gamma t\right] \; c\left(\frac{x}{l}\right)^{-1/\alpha} g_+\left(ct\left(\frac{x}{l}\right)^{-1/\alpha}; \alpha\right). \tag{3.14}$$

Transient current density is found by substituting this expression into formula (2.5).

If $\alpha = 1/2$, the expression for transient current takes the form

$$I(t) = \frac{eNl\sqrt{c}}{L}\left\{(\pi t)^{-1/2}\exp(-\gamma t) - (\pi t)^{-1/2}\exp\left(-\left(\sqrt{\gamma t} - \frac{1}{2\sqrt{\tau}}\right)^2\right)\right.$$

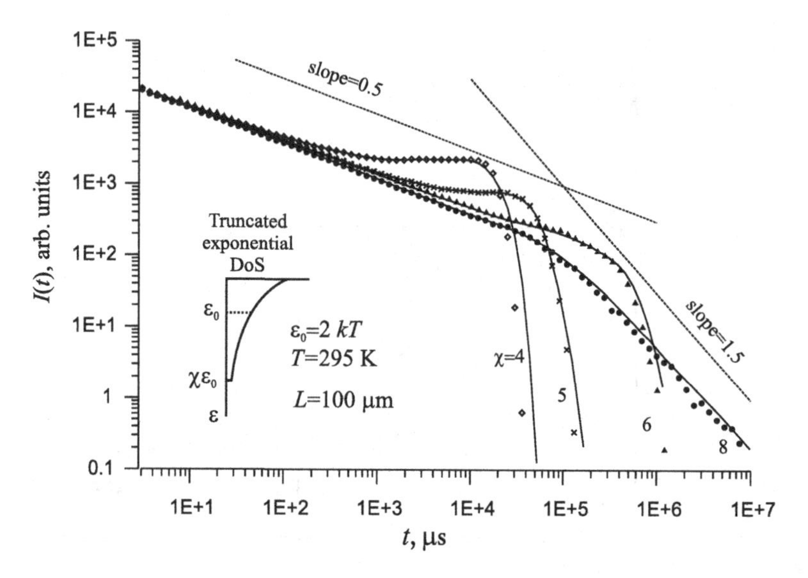

Figure 3.6 Transient current curves in case of the bounded exponential spectrum of localized states. Points are the results of Monte-Carlo simulation of multiple trapping, lines represents calculations according to Eqs. (3.1) and (3.14).

$$+\sqrt{\gamma}\left[\operatorname{erf}(\sqrt{\gamma t})-\operatorname{erf}\left(\sqrt{\gamma t}-\frac{1}{2\sqrt{T}}\right)\right]\Bigg\} \tag{3.15}$$

Fig. 3.4 illustrates transformation of transient current curves with an increasing L/l ratio. If the transient time is much smaller than the truncation time γ^{-1} the transport remains dispersive and does not pass to Gaussian asymptotics. When t_T becomes close to γ^{-1} the shape of the transient current curves undergoes modification, and they become inconsistent with the curves for normal and dispersive transport. For $t_T \gg \gamma^{-1}$, transport in the long-time asymptotic regime becomes normal.

In Ref. [Takeda et al. (1997)] properties of carrier transport in amorphous and β-rhombohedric boron are studied using the time-of-flight technique. The transient current decays as

$$I(t) \propto \begin{cases} t^{-1+\alpha_i}, \; t < t_T, \\ t^{-1-\alpha_f}, \; t > t_T, \end{cases}$$

where $\alpha_i = 0.55$ and $\alpha_f = 2.4$. Calculation according to (3.1) and (3.14) with parameters $\alpha = 0.7$, $\gamma = 0.45 \cdot 10^{7\ -1}$, $t_T = 2.2 \cdot 10^{-7}$ s provides quite good agreement with experimental data. Fig. 3.7 shows that analytical formulas (3.1) and (3.14) agree with results of numerical simulations for multiple trapping into localized states with truncated exponential density.

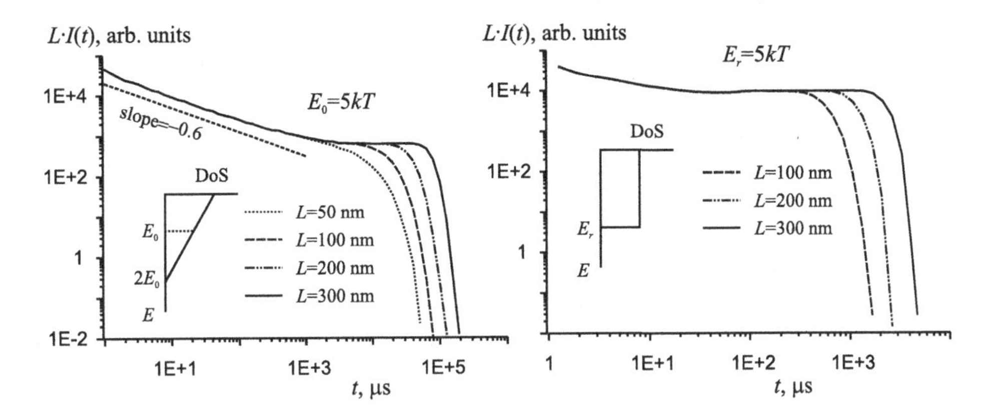

Figure 3.7 Transient current curves in case of multiple trapping into localized states with the triangular spectrum (left panel) and the rectangular one (right panel). Curves are found by integrating equation (3.16).

The advection-diffusion equation of multiple trapping in case of arbitrary density of states (for concentration of delocalized carriers) has the form

$$\frac{\partial n_d(\mathbf{r},t)}{\partial t} + \tau_0^{-1}\frac{\partial}{\partial t}\int_0^t n_d(\mathbf{r},t-t')\int_0^{\varepsilon_g}\rho(\varepsilon)\exp\left(-w_\varepsilon t'\,\mathrm{e}^{-\varepsilon/kT}\right)d\varepsilon\,dt'$$

$$+\mathrm{div}[\,\mu\mathbf{E}\,n_d(\mathbf{r},t) - D\nabla n_d(\mathbf{r},t)] = 0. \tag{3.16}$$

3.1.3 *Distributed dispersion parameter*

Transient current relaxation in certain disordered semiconductors, for example, porous silicon [Bisi et al. (2000)], assumes the form

$$I(t) \propto \begin{cases} t^{-1+\alpha_i}, \; t < t_T, \\ t^{-1-\alpha_f}, \; t > t_T, \end{cases} \quad 0 < \alpha_i \neq \alpha_f < 1. \tag{3.17}$$

The Scher-Montroll model of charge transfer in disordered semiconductors leads to the current dependence (2.2), where $\alpha_i = \alpha_f = \alpha$. It has been shown in Ref. [Averkiev et al. (2002)] that the value of α found from the dependence of carrier flight time in porous silicon on the electric field strength differs from that determined from transient photocurrent curves. The authors explain this discrepancy on the assumption of additional dispersion in terms of carrier mobility in structurally inhomogeneous porous silicon samples. It is natural to further extend this idea by assuming dispersion of the parameter α. It will be shown below that this assumption is enough to substantiate dependence (3.17), at least for the discrete spectrum $\{\alpha_1, \alpha_2, \ldots, \alpha_m\}$.

Let k_j be a fraction of traps that capture carriers for random time T distributed according to an asymptotically power law with exponent α_j. The averaged distribution function of a carrier residence time in the localized state has the form

$$\Psi(t) \sim 1 - \sum_j k_j \frac{(b_j\, t)^{-\alpha_j}}{\Gamma(1-\alpha_j)},$$

where b_j are normalization constants. Substituting this function into relationship (2.38) leads to the equation

$$\frac{\partial n_t(\mathbf{r},t)}{\partial t} = \frac{1}{\tau_0}\sum_j c_j^{-\alpha_j}\ {}_0\mathsf{D}_t^{\alpha_j} n_d(\mathbf{r},t), \qquad c_j = b_j\ (k_j)^{-1/\alpha_j}. \tag{3.18}$$

Taken together with continuity equation (2.53), expression (3.18) gives the drift-diffusion equation for the concentration of delocalized carriers in the case of the discretely distributed dispersion parameter:

$$\frac{\partial n_d(\mathbf{r},t)}{\partial t} + \frac{1}{\tau_0}\sum_j c_j^{-\alpha_j}\ {}_0\mathsf{D}_t^{\alpha_j} n_d(\mathbf{r},t') + \mathrm{div}\left(\mu\mathbf{E}\ n_d(\mathbf{r},t) - D\nabla n_d(\mathbf{r},t)\right) = n(\mathbf{r},0)\ \delta(t). \tag{3.19}$$

To calculate the transient current governed by the latter equation, we neglect diffusion, regard the electric field as being uniform, and align the x-axis along field $\mathbf{E}$. Then, equation (3.19) can be rewritten as

$$\frac{\partial n_d(x,t)}{\partial t} + \frac{1}{\tau_0}\sum_j c_j^{-\alpha_j}\ {}_0\mathsf{D}_t^{\alpha_j} n_d(x,t) + \mu E\ \frac{\partial n_d(x,t)}{\partial x} = N\delta(x)\ \delta(t).$$

The Laplace transform

$$\widetilde{n}_d(x,s) = \int_0^\infty dt\ n(x,t)\exp(-st)$$

satisfies the equation

$$s\widetilde{n}_d(x,s) + \frac{1}{\tau_0}\sum_j \left(\frac{s}{c_j}\right)^{\alpha_j} \widetilde{n}_d(x,s) + \mu E\ \frac{\partial \widetilde{n}_d(x,s)}{\partial x} = N\delta(x),$$

solution of which (for the case $\alpha_j < 1$) has the form

$$\widetilde{n}_d(x,s) = \frac{N}{\mu E A}\ \exp\left[-\frac{x}{\mu E\tau_0}\sum_j\left(\frac{s}{c_j}\right)^{\alpha_j}\right], \qquad \mu E\tau_0 = l,$$

with A standing for the sample area transverse to the electric field.

On assumption that traps are uniformly distributed in the sample, the time transform of the charge carrier density for $x \gg l$ is written as

$$\widetilde{n}(x,s) = \frac{N}{ls}\sum_j (s/c_j)^{\alpha_j} \exp\left(-\frac{x}{l}\sum\nolimits_j (s/c_j)^{\alpha_j}\right). \tag{3.20}$$

For the Laplace image of the transient current, we have:

$$\widetilde{I}(s) = \frac{eNl}{L}\ \frac{1-\exp\left(-L\sum_j (s/c_j)^{\alpha_j}/l\right)}{\sum_j (s/c_j)^{\alpha_j}}. \tag{3.21}$$

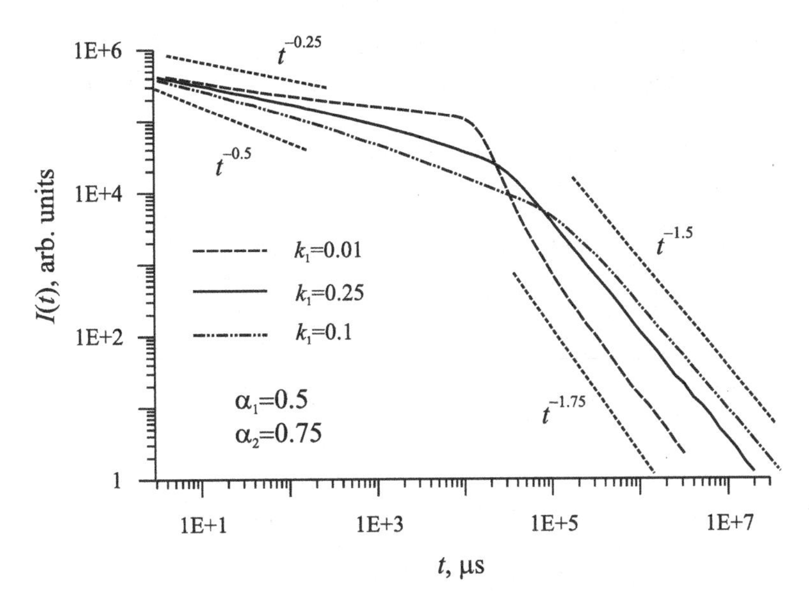

Figure 3.8 Transient current curves for dispersive transport characterized by two dispersion parameters $\alpha_1 = 0.5$ and $\alpha_2 = 0.75$. $\mu E \tau_0 = 10$ nm, $E = 10^6$ V/cm. Fractions k_1 of first kind traps ($\alpha_1 = 0.5$) are indicated in figure, $k_2 = 1 - k_1$. Dashed lines corresponds to power laws with exponents determined by dispersion parameters, $t^{-1\pm\alpha}$.

If the parameter α takes on a unique value, the inverse Laplace transform of formula (3.21) gives expression (2.11).

In order to see the long-time dependence of transient current, one should apply the Tauberian theorem, according to which the behavior of function $I(t)$ for $t \gg c_j^{-1}$ is determined by that of function (3.21) for $s \ll c_j$:

$$\widetilde{I}(s) \sim \frac{eNl}{L} \frac{2L\sum_j (s/c_j)^{\alpha_j}/l - \left(-L\sum_j (s/c_j)^{\alpha_j}/l\right)^2}{2\sum_j (s/c_j)^{\alpha_j}} \sim eN - \frac{eNL}{2l}(s/b_{\min})^{\alpha_{\min}}.$$

Here $\alpha_{\min}$ is the minimum value from the set $\{\alpha_1, \alpha_2, \ldots, \alpha_m\}$ and $b_{\min}$ is the corresponding value of the normalization constant. The inverse Laplace transformation leads to

$$I(t) \propto t^{-1-\alpha_{\min}}, \quad t \gg c_j^{-1}.$$

In the case of $s/c_j \gg (l/L)^{1/\alpha_j}$ for all j, it follows that

$$\widetilde{I}(s) \sim \frac{eNl}{L\sum_j (s/c_j)^{\alpha_j}} \sim \frac{eNl}{L(s/b_{\max})^{\alpha_{\max}}}, \quad s \gg c_j,$$

where $\alpha_{\max}$ is the maximum value from the set $\{\alpha_1, \alpha_2, \ldots, \alpha_m\}$ and $b_{\max}$ is the corresponding value of the normalization constant. Hence follows

$$I(t) \propto t^{-1+\alpha_{\max}}, \quad t \ll c_j^{-1}.$$

Thus, if the exponent in the carrier residence time distribution in traps takes on one of the values from an ordered set $\{\alpha_1, \alpha_2, \ldots, \alpha_m\}$ (discrete spectrum), the transient current behavior is determined by the maximum value of $\alpha_{\max} = \alpha_m$ in the initial time segment, and by the minimum value of $\alpha_{\min} = \alpha_1 \neq \alpha_m$ (Fig. 3.8) in the terminal one, in agreement with the results of the aforementioned experiments.

3.1.4 *Transient current curves in case of Gaussian disorder*

In frameworks of the hopping model of Gaussian disorder, the transport equation for effective concentration of carriers near the transport level is obtained in the form

$$\frac{a_\beta}{c^\beta}\,{}_0\mathsf{D}_t^\beta \eta(x,t) + a_\alpha A \frac{\partial}{\partial t}\int_0^t \frac{\eta(x,t-\tau)}{\tau^\alpha \exp(\gamma\tau^{\alpha/2})}d\tau + \tau_0\mu_h E\frac{\partial\eta(x,t)}{\partial x} - \tau_0 D_h \frac{\partial^2\eta(x,t)}{\partial^2 x} = 0. \tag{3.22}$$

According to the Comb model, the fractional derivative of order $\beta \in (0,1]$ in this equation takes into account percolative character of trajectories of hopping. The second term describes thermally activated hops between localized states with the Gaussian density $\rho(\varepsilon) \propto \exp(-\varepsilon^2/2\sigma^2)$. Comparison of the transient current curves calculated according to this equation with experimental data, Nikitenko and Tutnev's results, and Bassler's simulations for 1,1-bis(di-4-tolilaminophenyl)cyclogexane is presented in Fig 3.15.

As can be seen, Bassler's numerical results (1990) are consistent with experimental data, but calculations have sufficient statistical error. Observe that the curves $I(t)$ found from the Nikitenko diffusion equation are perfectly consistent with the results of our simulation of one-dimensional hopping. We associate this agreement with neglecting percolation nature of carrier trajectories in the diffusion equation with time-dependent coefficients and in the model of one-dimensional random walk. Including the fractional derivative into the diffusion equation, in agreement with the comb model of the percolation cluster, allows us to account the non-Brownian nature of the hopping trajectories (see inset in Fig. 3.15(a)) and better adjust the theoretical curves with the experimental data. Fig. 3.15(b) shows comparison of transient current curves found by equations of the Arkhipov-Rudenko τ-approximation, the Nikitenko diffusion equation, and the fractional equation for the case of the exponential density of localized states.

The conduction current density is expressed through the effective concentration of carriers near the transport level by the relation

$$j(x,t) = e\mu_h E\,\eta(x,t) - eD_h\frac{\partial\eta(x,t)}{\partial x}$$

where μ_h and D_h are effective mobility and diffusion coefficient at hopping.

3.1.5 *Percolation in porous semiconductors*

In Ref. [Aroutiounian et al. (2000)], the porous silicon is represented as a set of silicon atom clusters surrounded by complexes of SiO_x. Due to variation in size of

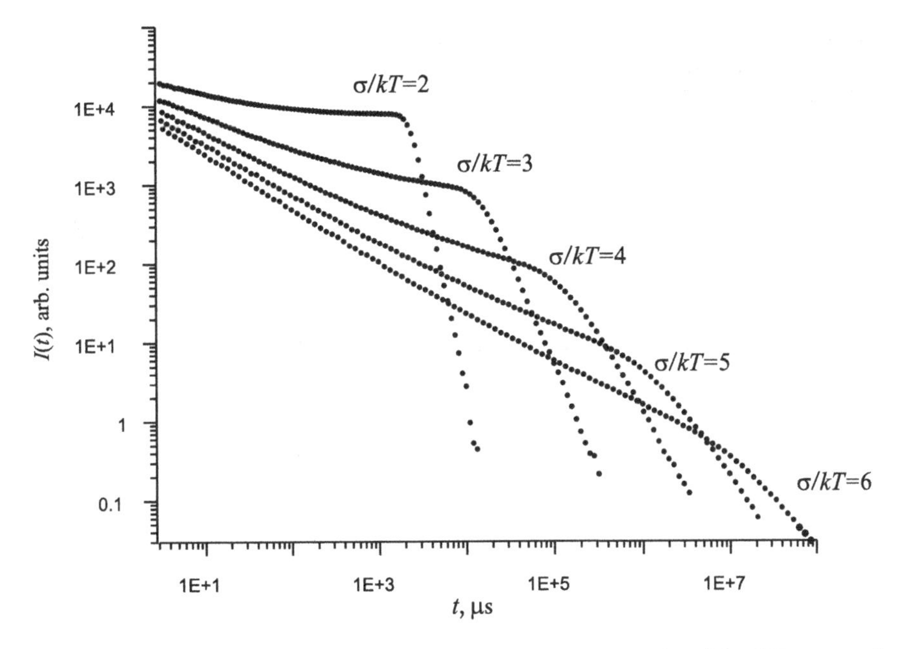

Figure 3.9 Transient current curves obtained by integrating equation (3.22) for different σ-values. $T = 295$ K, $E = 10^6$ V/cm.

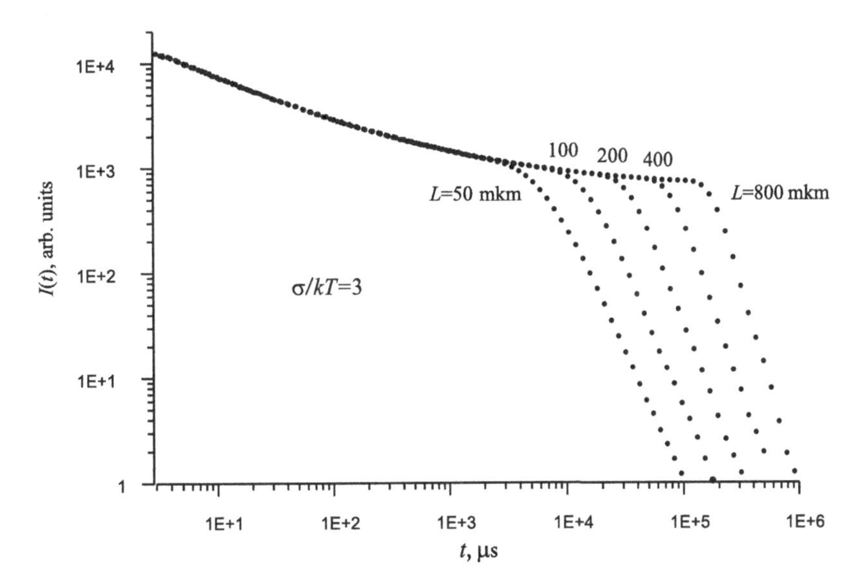

Figure 3.10 Theoretical transient current curves in case of Gaussian density of localized states for fixed $\sigma = 3kT$ and different values of sample thickness in the time-of-flight methodics.

nanocrystallites and distances between them, porous silicon is a disordered structure. Investigation of transient photocurrent in samples of porous silicon subjected to prolonged storage in air led the authors of Ref. [Averkiev et al. (2003)] to the

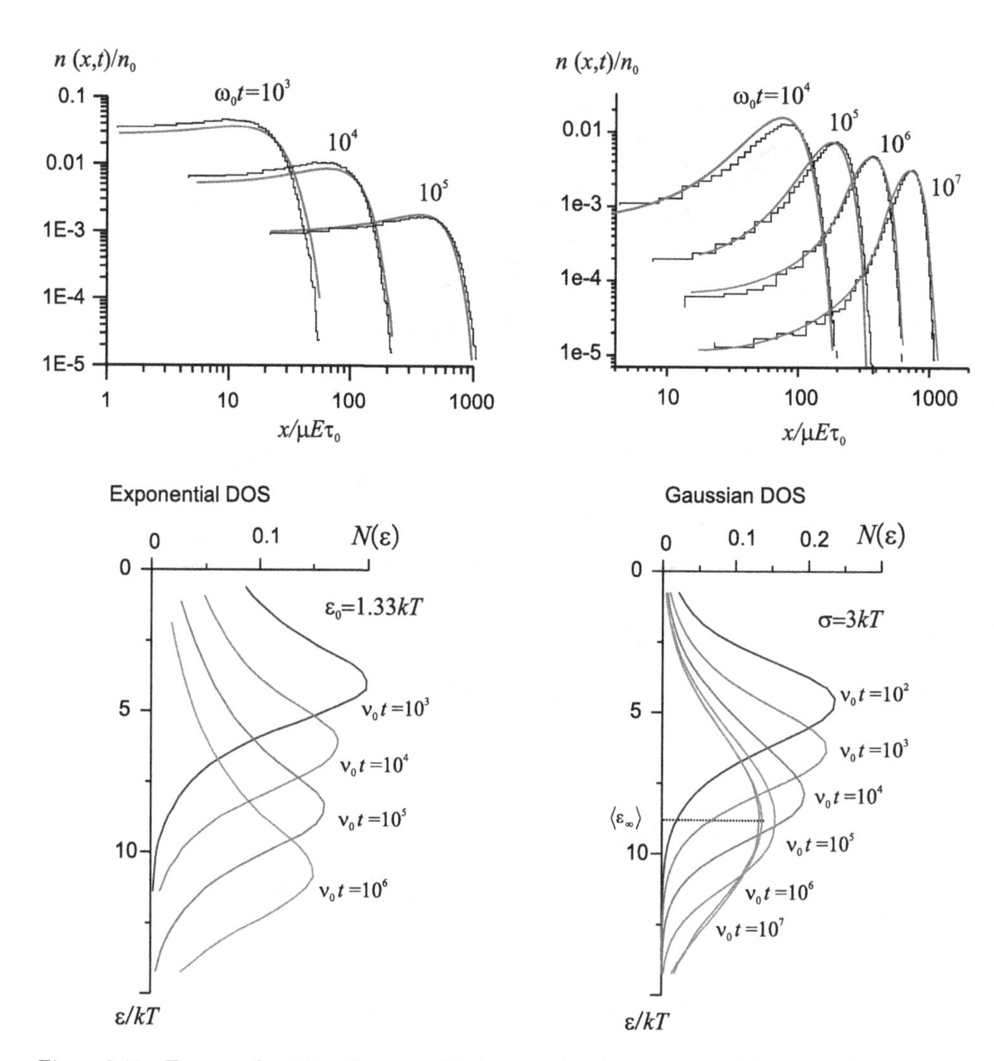

Figure 3.11 Energy relaxation of non-equilibrium carriers for exponential (left panels) and Gaussian (right panels) densities of localized states and corresponding profiles of charge carriers (upper panels). The histograms are the results of Monte-Carlo simulation.

conclusion about substantial effect of localized states on the charge transfer process. Experimental values of the dispersion parameter differ from those predicted by the multiple trapping model. In order to explain this discrepancy, the authors of Ref. [Averkiev et al. (2002)] assumed the existence of some additional diffusion dispersion of drifting packet due to the variations in charge carriers mobilities caused by heterogeneity of the structure of porous silicon layer. The nonlinear temperature dependence of the dispersion parameter in samples of porous silicon experimentally obtained in Ref. [Rao et al. (2002)] are contrary to the model of multiple trapping and thermally activated hopping, in which $\alpha \propto T$. How can the weak dependence

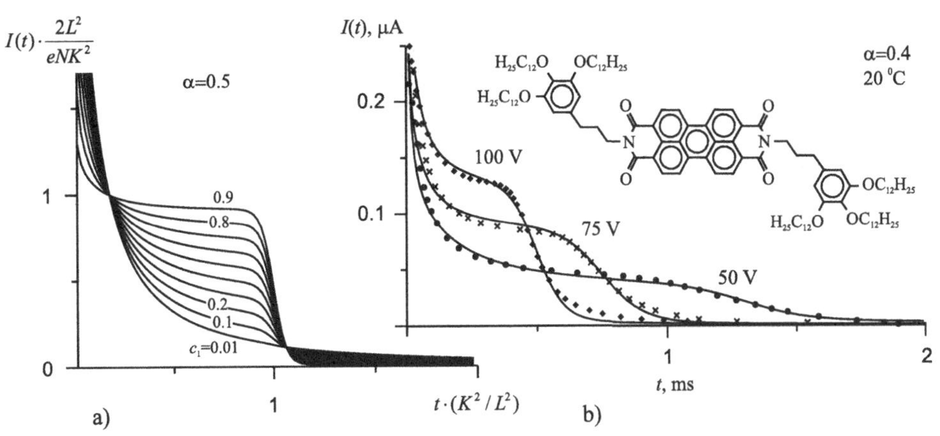

Figure 3.12 (a) Theoretical transient current curves in case of parallel channels with different statistics. First type channels are characterized by dispersive statistics ($\alpha = 0.5$), another ones by normal statistics. Fraction of carriers in channels of first type is $c_1 = 1 - c_\alpha$. (b) Comparison with experimental transient current curves for (3,4,5Pr)12G1-3-perylenetetracarboxyldiimide (data are digitized from [V. Duzhko et al. Applied Physics Letters 92, 113312 (2008)]). Adjustable parameters are $\alpha = 0.4$, $c_1 = 0.6$, $L/K = 0.8 \cdot 10^{-3}$ c$^{0.4}$.

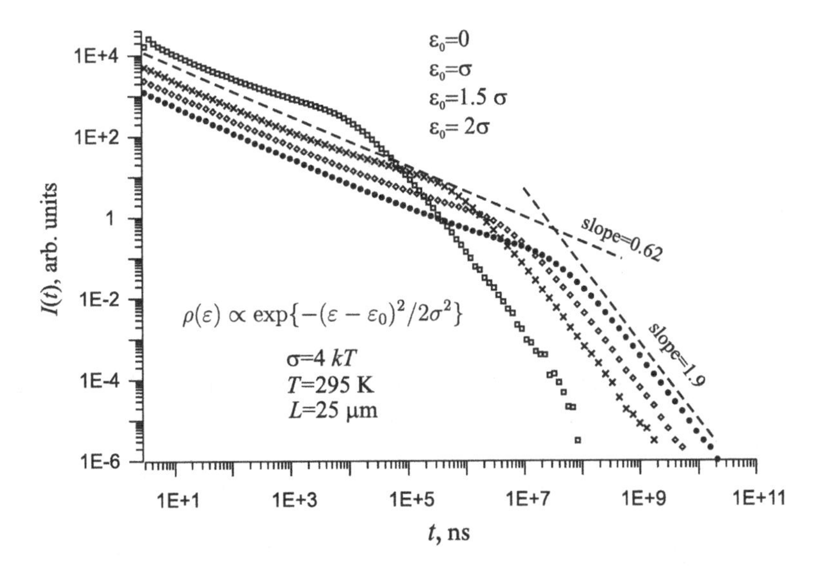

Figure 3.13 Transient current curves in case of multiple trapping into localized states with Gaussian density shifted below the mobility edge.

of dispersion parameter on temperature be explained, if tunneling mechanism does not dominate in the temperature range under consideration?

The problem of hopping conductivity can be directly reduced to percolation problems of bonds or sites [Zvyagin (1984)]. The infinite percolation cluster above the percolation threshold is a fractal structure on scales that do not exceed a certain

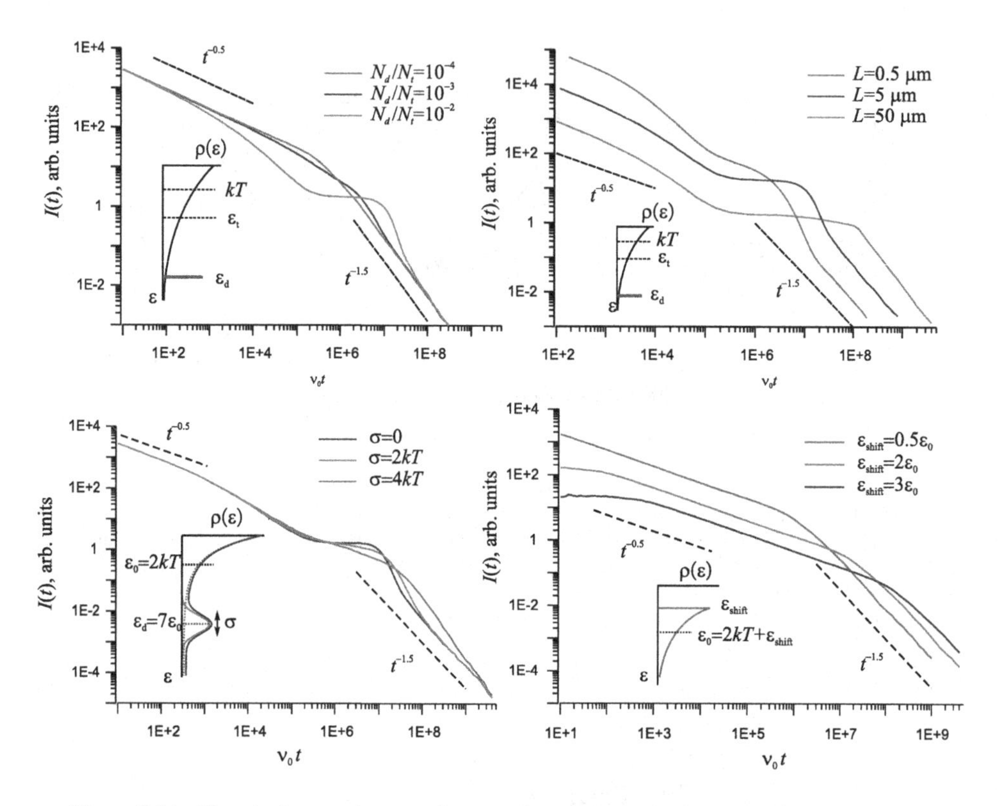

Figure 3.14 Transient current curves in case of nonmonotonic density of localized states.

correlation length ξ. At longer distances, an infinite cluster looks as statistically homogeneous and has a dimension equal to the Euclidean dimension of space in which it is located. To determine the transport characteristics, one has to know the waiting time distribution for carriers in fractal fragments that make up the infinite cluster. Authors of Ref. [Dykhne et al. (2004)] argues that the transport of particles occurs in the subdiffusion regime at distances shorter than the correlation length.

Within the above assumptions, the dispersion parameter of the percolation model depends only slightly on temperature and is controlled in many respects by the mediums porosity and structure. The features of the dispersive transport are retained, since the residence time in fractal fragments of the percolation cluster is distributed according to the asymptotic power law.

Let us apply the above assumptions to the experimental data [Rao et al. (2002)]. If we approximate the temperature dependence of the dispersion parameter by the linear equation $\alpha = c_1 T + c_2$, where c_1 is a relatively small coefficient, than the temperature dependence of the drift mobility, calculated by formulas (3.6–3.7), occurs in good agreement with the experimental data obtained in the same study (see Fig. 3.17).

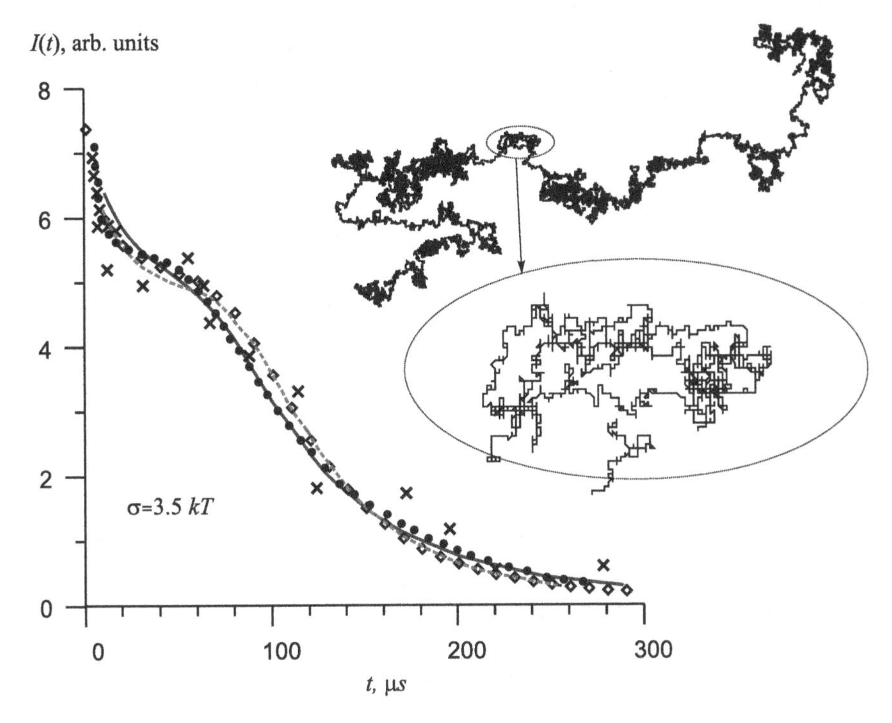

Figure 3.15 Transient current curves: circles and crosses are the results of the experiment and numerical simulations for 1,1-bis(di-4-tolilaminophenyl)cyclogexane, digitized from [Borsenberger P. M., Pautmeier L., Bassler H., J. Chem. Phys. 94 (1991)], dashed line is the solution from [Nikitenko V. R. & Tyutnev A. P. Semiconductors 41 (2007)], rhombuses are the Monte-Carlo simulation results of one-dimensional hopping conductivity, solid line is obtained with the help of equation (3.22): $\beta = 0.5$, $a_\beta = 1 - a_\alpha = 0.1$, $\alpha = kT/\sigma = 0.286$, $\gamma = 0.98$. The inset demonstrates the trajectory of the 2D-hopping diffusion with parameters $\sigma = 3.5kT = 95$ meV, T=312 K.

3.1.6 *Non-stationary radiation-induced conductivity*

Let us see how the transient current of uniform generation is connected with the transient current at pulsed injection of carriers. We will write $j_\delta(x,t)$, $f_\delta(x,t)$ and $I_{\mathrm{unif}}(t)$ for the conduction current density, concentration at pulsed injection, and the transient current at uniform generation, respectively. In the case of homogeneous generation of carriers without recombination, the conduction current density has the form:

$$I_{\mathrm{unif}}(t) = \frac{1}{L}\int_0^L dx\; j_{\mathrm{unif}}(x,t) = \frac{1}{L}\int_0^L dx \int_0^x d\xi\; G\, j_\delta(x-\xi) = \frac{eG}{L}\int_0^L dx\; (L-x) j_\delta(x,t).$$

Using the continuity equation, one can link the transient current to the concentration of carriers

$$I_{\mathrm{unif}}(t) = -\frac{eG}{L}\frac{\partial}{\partial t}\int_0^L dx\; (L-x)\int_0^x d\xi\; f(\xi,t) = -\frac{eG}{2L}\frac{\partial}{\partial t}\int_0^L dx\; (L-x)^2\, f_\delta(x,t).$$

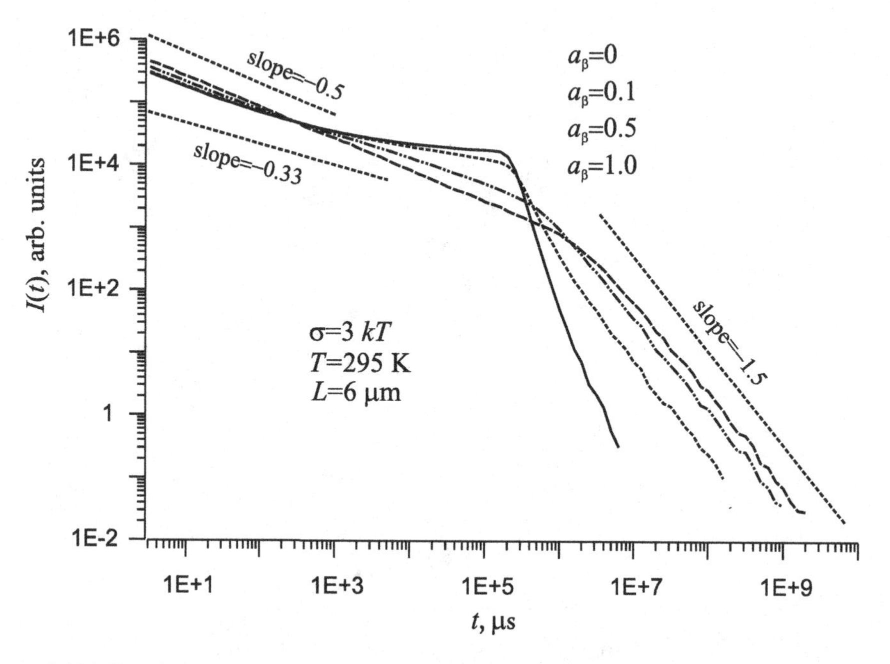

Figure 3.16 Transient current curves in case of Gaussian disorder ($\sigma = 3kT$) and percolation ($\beta = 0.5$), calculated via the solution of equation (3.22) for different a_β values, $E = 10^6$ V/cm, $\tau_0\mu_h E = 10$ nm, $\tau_0 D_h = 1$ nm^2, $c = 0.4 \cdot 10^6$ c^{-1}.

The Laplace transform of the transient current generated by the homogeneous irradiation of the sample

$$\widetilde{I}_{\text{unif}}(s) = \frac{eG}{L}\int\limits_0^L dx\;(L-x)\widetilde{j}_\delta(x,s) = \frac{eG}{L}\int\limits_0^L dx\;(L-x)\exp\left[-\frac{x}{l}\left(\frac{s}{c}\right)^\alpha\right].$$

After simple calculations we get

$$\widetilde{I}_{\text{unif}}(s) = eGl\,\frac{c^\alpha}{s^\alpha}\left\{1-\frac{lc^\alpha}{Ls^\alpha}\left[1-\exp\left(-\frac{Ls^\alpha}{lc^\alpha}\right)\right]\right\}. \tag{3.23}$$

This expression has the asymptotic

$$\widetilde{I}_{\text{unif}}(s) \sim \begin{cases} eGl\,\dfrac{c^\alpha}{s^\alpha}, & s\to\infty, \\ \dfrac{1}{2}\,eGL\left(1-\dfrac{Ls^\alpha}{3lc^\alpha}\right), & s\to 0, \end{cases} \qquad \alpha<1.$$

Consequently, the transient current for a uniform generation of carriers has the same asymptotic behavior as in the case of pulse injection (see Fig. 3.18):

$$I_{\text{unif}}(t) \sim \begin{cases} t^{-1+\alpha}, & t\to 0, \\ t^{-1-\alpha}, & t\to\infty, \end{cases} \qquad \alpha<1.$$

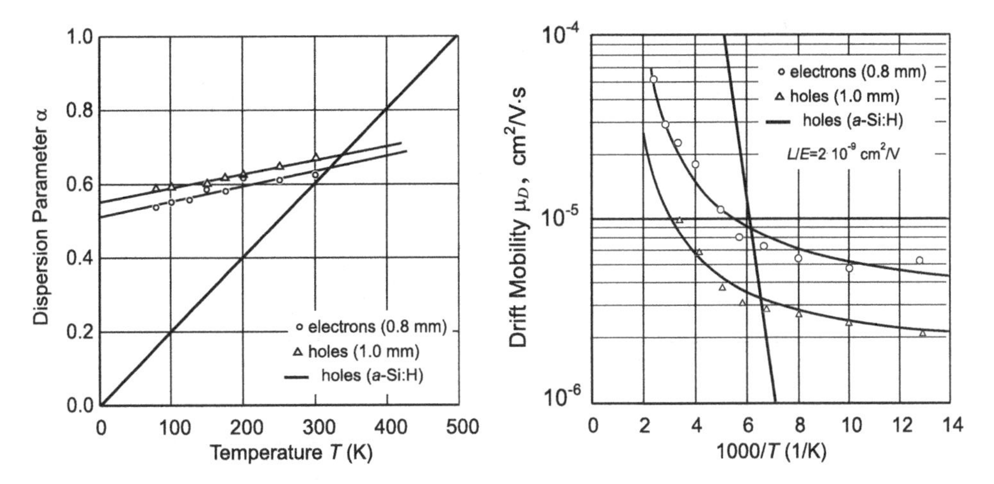

Figure 3.17 (a) Temperature dependence of the dispersion parameter in a porous silicon layer during dispersive transport (left panel): circles and triangles are the experimental data [Rao et al. (2002)] for electrons and holes, respectively; lines correspond to linear approximation. (b) Temperature dependence of the drift mobility (right panel): circles and triangles are the experimental data [Rao et al. (2002)] for electrons and holes, respectively; curves are calculated by formulas (3.6–3.7), The bold lines correspond to the experimental data for holes in hydrogenated amorphous silicon [Rao et al. (2002)]).

In the special case $\alpha = 1/2$ the inverse Laplace transform expression (3.23) leads to

$$I_{\text{unif}}(t) = \frac{eNl}{\sqrt{\pi ct}}\left[1 - \frac{l\sqrt{\pi ct}}{L}\,\text{erf}\left(\frac{L}{2l\sqrt{ct}}\right)\right].$$

In case of arbitrary $\alpha \in (0, 1]$, the current has the form

$$I_{\text{unif}}(t) = eGl\left[\frac{c^\alpha}{\Gamma(\alpha)}t^{\alpha-1} - \alpha c^\alpha \frac{l}{L}\,{}_0\mathsf{D}_t^{-\alpha}\left(t^{\alpha-1}\int\limits_{\zeta_0(t)}^{\infty}\zeta^{-\alpha}g_+(\zeta;\alpha)d\zeta\right)\right], \tag{3.24}$$

with

$$\zeta_0(t) = ct\left(\frac{L}{l}\right)^{-1/\alpha}.$$

The uniform generation is realized by fast electron irradiation in experiments on non-stationary radiation-induced electrical conductivity (see, for example, [Tyutnev et al. (2005)]). Tyutnev et al. (2005) write that the asymptotic behavior of the transient current in the dispersive transport regime in the region of small and large times is independent of the generation method. This result underlines the close relationship between the studies of charge carrier mobility by radiation-induced conductivity (uniform volumetric generation) or the time-of-flight method (near-surface generation of charge carriers).

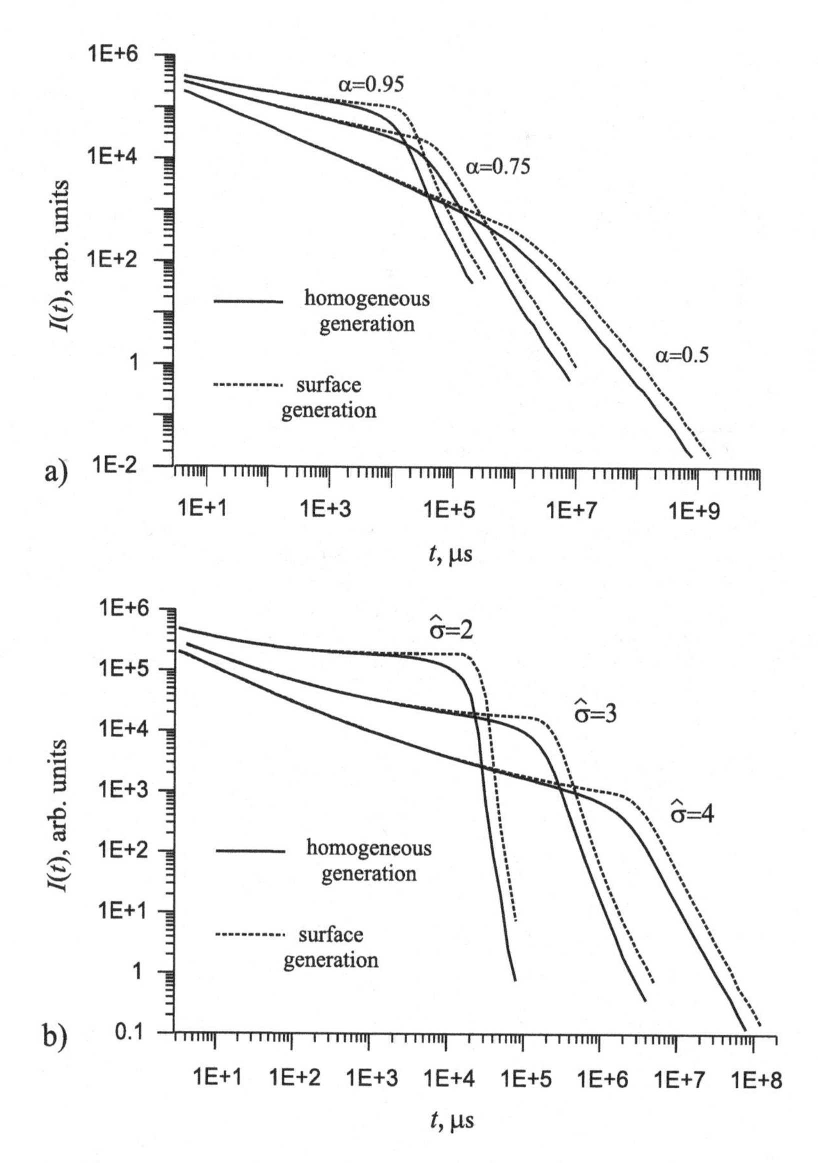

Figure 3.18 (a) Universal transient current density in case of surface injection (dashed line) and homogeneous pulsed generation of charge carriers (solid lines) for different values of dispersion parameters. $L = 25$ mkm, $E = 10^6$ V/cm. (b) The same for hopping via traps with Gaussian energetic disorder for different values of $\hat{\sigma} = \sigma/kT$. $L = 25$ mkm, $E = 10^6$ V/sm.

3.2 Non-homogeneous distribution of traps

3.2.1 *Non-uniform spatial distribution of localized states*

Let us calculate the transient current in the case of inhomogeneous spatial distribution of localized states. When traps are distributed over a sample with the density

$\rho(x)$, the average number of localizations of one carrier in a layer of thickness x is equal to

$$n = \int_0^x \rho(x)dx,$$

and the conduction current density is as follows

$$j(x,t) = eN \left[\int_0^x \rho(x)dx\right]^{-1/\alpha} g\left(ct\left[\int_0^x \rho(x)dx\right]^{-1/\alpha};\alpha\right). \tag{3.25}$$

From the continuity equation

$$\frac{\partial f(x,t)}{\partial t} + \frac{1}{eN}\frac{\partial j(x,t)}{\partial x} = \delta(x)\delta(t),$$

one can find the total concentration of carriers

$$f(x,t) = -\frac{1}{eN}\frac{\partial}{\partial x}\int_0^t j(x,t)dt = -\frac{\partial}{\partial x} G_+\left(ct\left[\int_0^x \rho(x)dx\right]^{-1/\alpha};\alpha\right)$$

$$= ct\ \alpha^{-1}\left[\int_0^x \rho(x)dx\right]^{-1/\alpha-1} \rho(x)\ g_+\left(ct\left[\int_0^x \rho(x)dx\right]^{-1/\alpha};\alpha\right). \tag{3.26}$$

Transient current can be calculated by substituting the expression for the conduction current density (3.25) into formula (2.5), or the expression for concentration (3.26) into formula (2.11). As a result, we obtain:

$$I(t) = \frac{eN}{L}\int_0^L dx\ \left[\int_0^x \rho(x)dx\right]^{-1/\alpha} g_+\left(ct\left[\int_0^x \rho(x)dx\right]^{-1/\alpha};\alpha\right),$$

We considered different types of spatial distribution of localized states, including those discussed in Ref. [Rybicki & Chybicki (1989)]. In particular, Fig. 3.19 shows the transient current curves in the case of surface layers depleted or enriched by traps. In the first case we observe the appearance of a maximum on the curves, in the second case we obtain a more diffuse characteristics than in the case of homogeneous distribution of traps in the sample. Analytical results are consistent with the simulation by the Monte-Carlo transport by multiple trapping.

3.2.2 *Multilayer structures*

Now, we consider the time-of-flight experiment for the three-layer structure, the outer layers are surface layers and the main bulk of the material is located between them. First, we obtain the results for the general case of three-dimensional structure, each layer is characterized by dispersive transport. Barrier effects are neglected, that is correct for large voltages applied to structures.

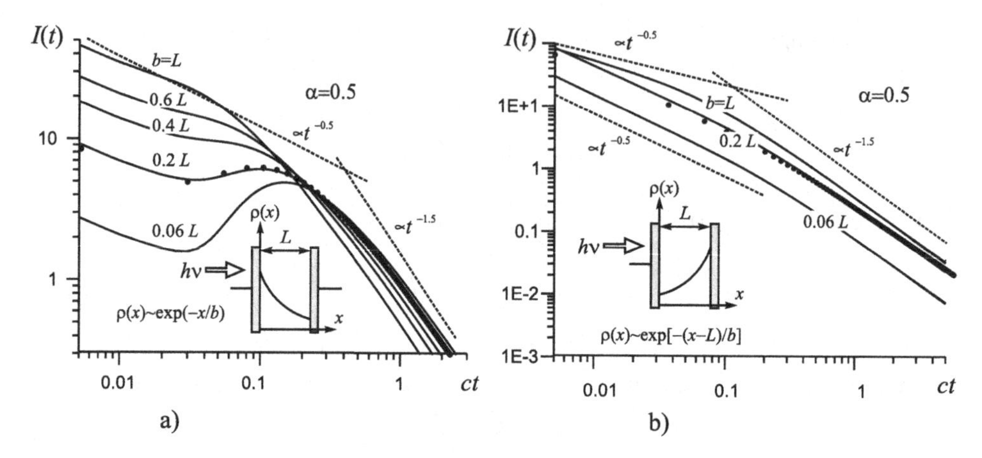

Figure 3.19 (a) Dispersive transient current curves ($\alpha = 0.5$) in case of non-uniform distribution of localized states $\rho(x) \propto \exp(-x/b)$ for different b values. (b) The same for $\rho(x) \propto \exp[-(x-L)/b]$. Points are the results of numerical simulation.

To write diffusion equations for each layer, we use the equation for concentration of quasi-free carriers:

$$\frac{l_1}{\tau_{01}K_1}\ {}_0\mathsf{D}_t^{\alpha_1}\eta_1(x,t) + \mu_1 E_1 \frac{\partial}{\partial x}\eta_1(x,t) = N\delta(x)\delta(t),$$

$$\frac{l_2}{\tau_{02}K_2}\ {}_0\mathsf{D}_t^{\alpha_2}\eta_2(x,t) + \mu_2 E_2 \frac{\partial}{\partial x}\eta_2(x,t) = \mu_1 E_1 \eta_1(L_1,t)\delta(x-L_1),$$

$$\frac{l_3}{\tau_{03}K_3}\ {}_0\mathsf{D}_t^{\alpha_3}\eta_3(x,t) + \mu_3 E_3 \frac{\partial}{\partial x}\eta_3(x,t) = \mu_2 E_2 \eta_2(L_1+L_2,t)\delta(x-L_1-L_2),$$

Transient current is related to solutions to these equations by the relation

$$I(t) = \frac{1}{L}\int_0^L j(x,t)dx$$

$$= \frac{\mu_1 E_1}{L_1+L_2+L_3}\int_0^{L_1}\eta_1(x,t)dx + \frac{\mu_2 E_2}{L_1+L_2+L_3}\int_{L_1}^{L_1+L_2}\eta_2(x,t)dx$$

$$+\frac{\mu_3 E_3}{L_1+L_2+L_3}\int_{L_1+L_2}^{L_1+L_2+L_3}\eta_3(x,t)dx.$$

Concentrations of quasi-free carriers are as follows

$$\eta_1(x,t) = \frac{N}{\mu_1 E_1}\ (x/K_1)^{-1/\alpha_1} g_+\left(t(x/K_1)^{-1/\alpha_1};\alpha_1\right),$$

$$\eta_2(x,t) = \frac{N\mu_1 E_1}{\mu_2 E_2}\left(\frac{x-L_1}{K_2}\right)^{-1/\alpha_2}\int\limits_0^t g_+\left((t-t')\left(\frac{x-L_1}{K_2}\right)^{-1/\alpha_2};\alpha_2\right)\eta_1(L_1,t')dt',$$

$$\eta_3(x,t) = \frac{N\mu_2 E_2}{\mu_3 E_3}\left(\frac{x-L_1-L_2}{K_3}\right)^{-1/\alpha_3}$$

$$\times\int\limits_0^t g_+\left((t-t')\left(\frac{x-L_1-L_2}{K_3}\right)^{-1/\alpha_3};\alpha_3\right)\eta_2(L_1+L_2,t')dt'.$$

Substituting these solutions into the previous relationship, we obtain

$$I(t) = I_1(t) + I_2(t) + I_3(t),$$

where

$$I_1(t) = \mathsf{L}^{-1}\left\{\frac{N}{L}\frac{K_1}{s^{\alpha_1}}\left[1-\exp\left(-\frac{L_1}{K_1}s^{\alpha_1}\right)\right]\right\},$$

$$I_2(t) = \mathsf{L}^{-1}\left\{\frac{N}{L}\frac{K_2}{s^{\alpha_2}}\exp\left(-\frac{L_1}{K_1}s^{\alpha_1}\right)\left[1-\exp\left(-\frac{L_2}{K_2}s^{\alpha_2}\right)\right]\right\},$$

$$I_3(t) = \mathsf{L}^{-1}\left\{\frac{N}{L}\frac{K_3}{s^{\alpha_3}}\exp\left(-\frac{L_1}{K_1}s^{\alpha_1}\right)\exp\left(-\frac{L_2}{K_2}s^{\alpha_2}\right)\left[1-\exp\left(-\frac{L_3}{K_3}s^{\alpha_3}\right)\right]\right\},$$

where L^{-1} is the operator of inverse Laplace transformation.

A similar calculation was performed in the case of hopping in a material with Gaussian energetic disorder. In Fig. 3.20 and 3.21, calculations for the time-of-flight method and the non-stationary radiation-induced electrical conduction for three-, four- and six-layer systems are presented.

3.2.3 *The "disordered – crystalline" semiconductor structure*

Now, we consider transport of charge carriers in the system disordered semiconductor (labelled 1) – crystalline semiconductor (labelled 2). In the layer of disordered semiconductor, the concentration of excess delocalized carriers $n_{f,1}(x,t)$ for times $t \gg w^{-1}$ obeys the equation

$$(c\tau_0^\alpha)^{-1}\ {}_0\mathsf{D}_t^\alpha n_{f,1}(x,t) + \mu_1 E_1 \frac{\partial n_{f,1}(x,t)}{\partial x} = 0, \tag{3.27}$$

where L_1, μ_1, w, α, E_1 are the layer thickness, mobility of delocalized carriers, capture rate into localized states, dispersive parameter and electric field in the first layer. In the crystalline semiconductor, drift of free carriers is described by the equation

$$\frac{\partial n_{f,2}(x,t)}{\partial t} + \mu_2 E_2 \frac{\partial n_{f,2}(x,t)}{\partial x} = 0. \tag{3.28}$$

Suppose, that the condition on the boundary between semiconductors is as follows (conduction currents are proportional to concentrations of delocalized carriers)

$$n_{f,1}(L_1,t) = n_{f,2}(L_1,t). \tag{3.29}$$

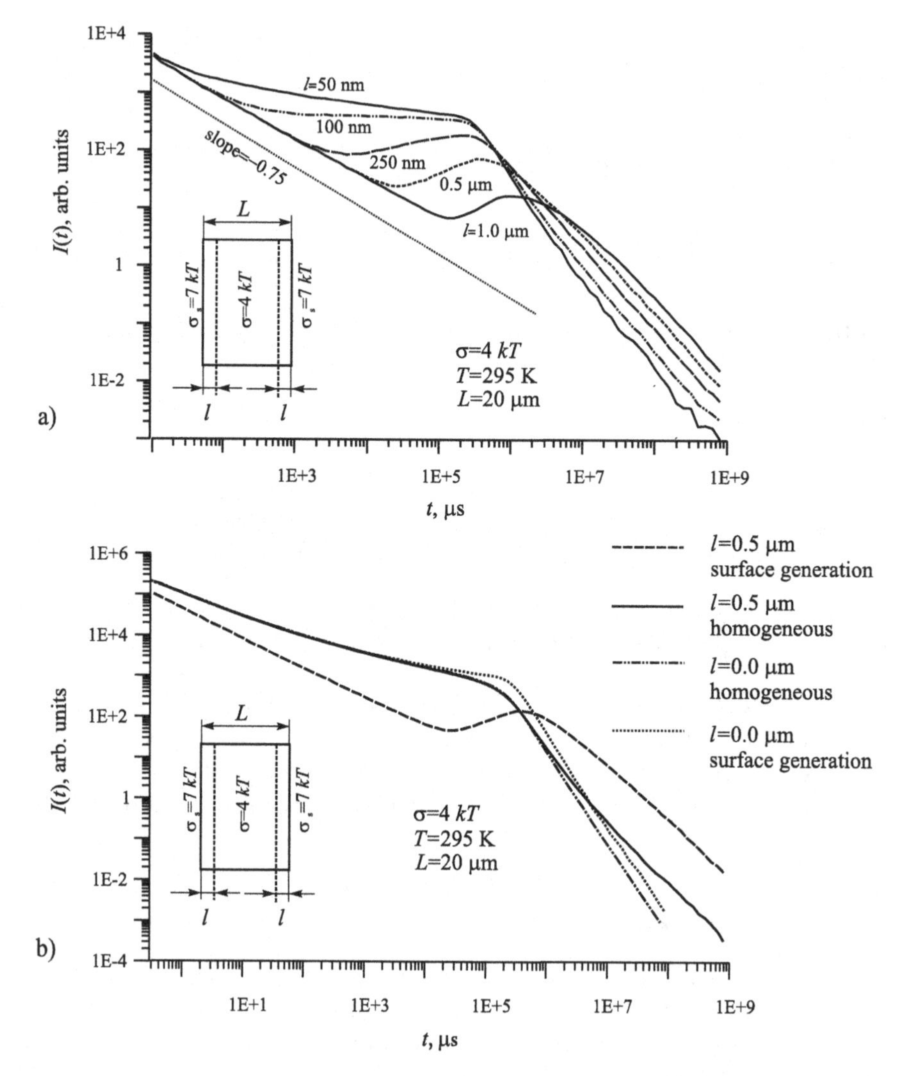

Figure 3.20 Transient currents with taking into account surface layers of different thickness (T = 295 K, L = 20 mkm, $E = 10^6$ V/cm, $\hat{\sigma} = 4$ is the parameter of Gaussian disorder of the bulk, and $\hat{\sigma} = 7$ of the surface layers. b) Influence of surface layers on transient current curves in time-of-flight method (near electrode generation of carriers), and in case of uniform generation of carriers.

Photoinjection of charge carriers is realized on the surface of disordered semiconductor,

$$n_{f,1}(x,0) = N\delta(x), \tag{3.30}$$

where N is a number of photoinjected carriers on a unit area of the surface.

The solution to equation (3.27) with initial condition (3.30) has the form

$$n_{f,1}(x,t) = (N/\mu_1 E_1)(x/K)^{-1/\alpha} g_+ \left(t(x/K)^{-1/\alpha};\alpha\right),$$

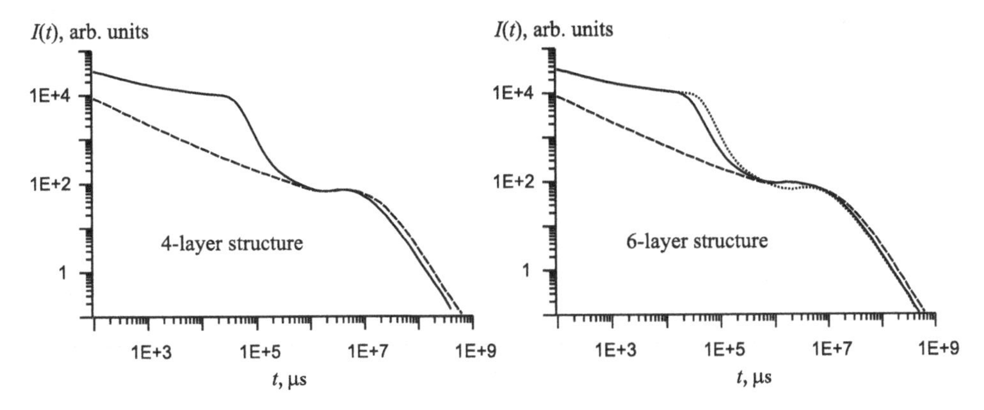

Figure 3.21 Transient current curves in multilayer structures. In both cases $T = 295$ K, $L =$ 20 mkm, $E = 10^6$ V/cm and layers are characterized by Gaussian disorder. Left panel: solid line corresponds to the 4-layer structure ($l = L/4$) $\{\hat{\sigma_1}, \hat{\sigma_2}, \hat{\sigma_3}, \hat{\sigma_4}\} = \{3, 5, 3, 5\}$, dashed line – $\{\hat{\sigma_1}, \hat{\sigma_2}, \hat{\sigma_3}, \hat{\sigma_4}\} = \{5, 3, 5, 3\}$. Right panel: solid line – $\{\hat{\sigma_1}, \hat{\sigma_2}, \hat{\sigma_3}, \hat{\sigma_4}, \hat{\sigma_5}, \hat{\sigma_6}\} = \{3, 5, 3, 5, 3, 5\}$, dashed line – $\{\hat{\sigma_1}, \hat{\sigma_2}, \hat{\sigma_3}, \hat{\sigma_4}, \hat{\sigma_5}, \hat{\sigma_6}\} = \{5, 3, 5, 3, 5, 5\}$; for comparison, dotted line demonstrates $I(t)$ in 4-layer structure. In calculations, barrier effects have been neglected.

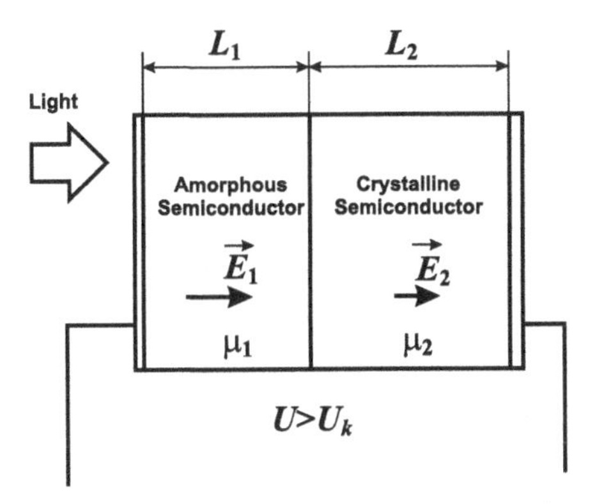

Figure 3.22 The structure amorphous semiconductor – crystalline semiconductor.

where $l = \mu_1 E_1/w_0$, $K = c_1^\alpha \mu_1 E_1/w_0$, $c_1 = w_0 A^{-1/\alpha}$, $A = \pi\alpha/\sin\pi\alpha$. The solution to equation (3.28) with boundary condition (3.29) has the form

$$n_{f,2}(x,t) = \int_0^t n_{f,1}(L_1, t-\tau)\delta\left(\tau - \frac{x - L_1}{\mu_2 E_2}\right) d\tau.$$

Thus, we have

$$n_f(x,t) = \begin{cases} \frac{N}{l}\left(\frac{xA}{l}\right)^{-1/\alpha} g_+\left(wt\left(\frac{xA}{l}\right)^{-1/\alpha};\alpha\right), & x < L_1, \\ \frac{N}{l}\left(\frac{L_1A}{l}\right)^{-1/\alpha} g_+\left(w\left(t - \frac{x-L_1}{\mu_2E_2}\right)\left(\frac{L_1A}{l}\right)^{-1/\alpha};\alpha\right), & x - L_1 < \mu_2E_2t, \\ 0, & x - L_1 > \mu_2E_2t. \end{cases}$$

The transient current in a double layer structure is composed of two components:

$$I(t) = I_1(t) + I_2(t),$$

where

$$I_1(t) = \frac{e\mu_1E_1}{L_1}\int\limits_0^{L_1} n_{f,1}(x,t)dx = \frac{e\mu_1E_1N\alpha}{AL_1}(wt)^{-1+\alpha}\int\limits_\tau^\infty s^{-\alpha}g_+(s;\alpha)ds,$$

$$\tau = wt(L_1A/l)^{-1/\alpha}$$

is the transient current in the layer of disordered semiconductor

$$I_2(t) = \frac{e\mu_2E_2}{L_2}\int\limits_{L_1}^{L_1+L_2} n_{f,2}(x,t)dx = \frac{N}{l}\frac{e(\mu_2E_2)^2}{L_2w}$$

$$\begin{cases} G_+\left(wt\left(\frac{L_1A}{l}\right)^{-1/\alpha};\alpha\right), & t < \frac{L_2}{\mu_2E_2}, \\ G_+\left(wt\left(\frac{L_1A}{l}\right)^{-1/\alpha};\alpha\right) - G_+\left(w\left(t - \frac{L_2}{\mu_2E_2}\right)\left(\frac{L_1A}{l}\right)^{-1/\alpha};\alpha\right), & t > \frac{L_2}{\mu_2E_2}, \end{cases}$$

is the transient current in the layer of crystalline semiconductor. The series expansion of the stable distribution is as follows

$$G_+(s;\alpha) = \int\limits_0^x g_+(s;\alpha)ds = 1 + \sum_{n=1}^\infty \frac{(-1)^n}{n!}\frac{s^{-n\alpha}}{\Gamma(1-n\alpha)}.$$

Introducing the notation

$$C_1 = (N/l)e\mu_1E_1\alpha(L_1A/l)^{-1/\alpha}, \quad \gamma = (\mu_2E_2)^2(L_1A/l)^{1/\alpha}/(L_2w\mu_1E_1\alpha), \quad (3.31)$$

$$\tau = wt(L_1A/l)^{-1/\alpha}, \quad \tau_0 = w(L_2/\mu_2E_2)(L_1A/l)^{-1/\alpha}, \quad (3.32)$$

we rewrite these expressions for currents in more compact form

$$I_1(t)/C_1 = \tau^{-1+\alpha}\int\limits_\tau^\infty s^{-\alpha}g_+(s;\alpha)ds,$$

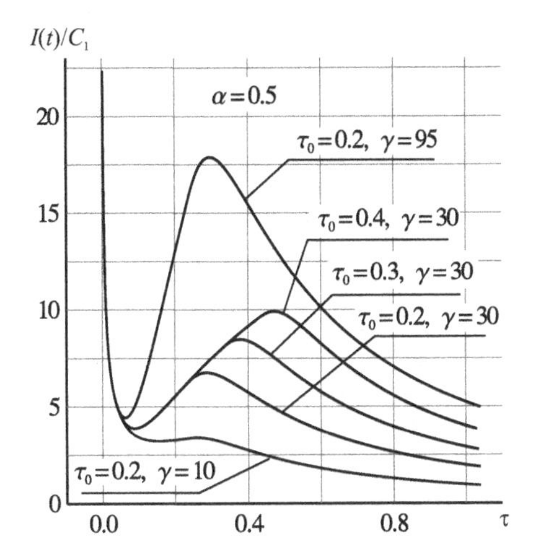

Figure 3.23 Theoretical curves of transient photocurrent in the structure disordered semiconductor – crystalline semiconductor.

$$I_2(t)/C_1 = \begin{cases} \gamma G_+(\tau;\alpha), & \tau < \tau_0, \\ \gamma\left[G_+(\tau;\alpha) - G_+(\tau-\tau_0;\alpha)\right], & \tau > \tau_0. \end{cases} \tag{3.33}$$

Quantities τ_0 and γ determining the shift and magnitude of a maximum on the transient current curves are expressed as

$$\tau_0 = \frac{t_{T2}}{t_{T1}}\left[\frac{\Gamma(1+\alpha)}{\sqrt{2A}}\right]^{1/\alpha}, \qquad \gamma = \frac{\mu_2 E_2}{\mu_1 E_1}\frac{1}{\tau_0},$$

where t_{T1} and t_{T2} are transient times for the layer of disordered semiconductor and the layer of crystalline semiconductor, respectively. If the crystal serves as a low-resistance contact, then $E_1 \gg E_2$ and $\gamma \ll 1$ and the transport characteristics for the two-layer structure is almost identical to the characteristics of transport in a monolayer of disordered semiconductor.

Fig. 3.24 displays the comparison of theoretical results obtained within the fractional model of charge transport with the time-of-flight experimental data for the amorphous semiconductor– crystalline semiconductor (solid lines are the experimental data for the structure a-$Se_{95}As_5$ – c-CdSe from [Kazakova & Lebedev (1998)], dashed lines are theoretical results). For the first curve ($U_1 = 6.4\ V$) we found $\alpha = 0.78$, $C_1 = 6.7\ \mu A$, $\gamma = 4400$, $w(L_1A/l)^{-1/\alpha} = 0.018$, $\tau_0 = 1.4 \cdot 10^{-5}$. For the second curve ($U_2 = 12.8\ V$), as follows from formulas (3.31) and (3.32), $C_1' = C_1(U_2/U_1)^{1/\alpha} = 16.29\ \mu A$, $\gamma' = \gamma(U_2/U_1)^{1-1/\alpha} = 3620$, $\tau_0' = \tau_0(U_2/U_1)^{1/\alpha-1} = 1.7 \cdot 10^{-5}$, $\alpha = 0.78$.

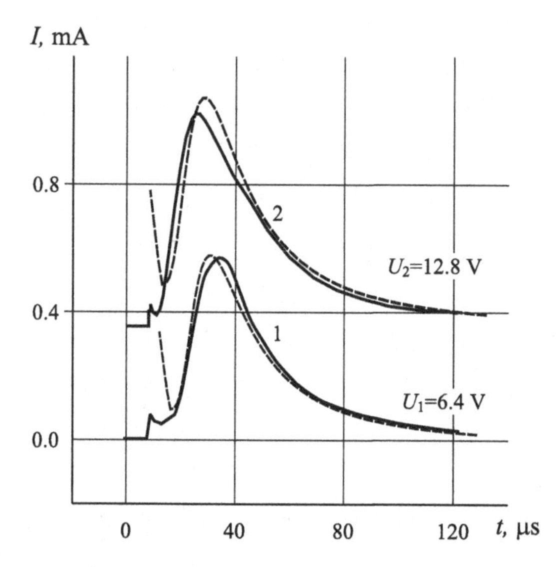

Figure 3.24 Transient current in the structure a-$Se_{95}As_5$ – c-CdSe (solid lines present the experimental data digitized from [Kazakova & Lebedev (1998)], dashed lines are the results of calculations by Eq. (3.33)).

3.3 Transient processes in a diode under dispersive transport conditions

3.3.1 *Turning on by the current step*

The fact that the fractional differential approach allows us to describe both normal and dispersive transport in terms of the unified formalism can be used for analysis of transients in structures based on disordered semiconductors, by analogy with similar structures based on crystalline semiconductors. We demonstrate this by calculating transition process in a semiconductor diode under conditions of dispersive transport. In this case, the current $I(t)$ and/or the voltage $U(t)$ play the role of time-dependent transient parameters. The diode performs the transition from the neutral state to the conducting one due to the current step, i.e. the load resistance R_l is substantially greater than the resistance of the diode R_d [Gaman (2000)]. On the assumption that low injection conditions are fulfilled, the author of Ref. [Sibatov (2012)] calculated the process for semi-infinite planar diode with n-type base. Recombination and generation in the space charge region are neglected. Holes are injected from the p-region into the n-region with a sharp turn on of current. Later, an equilibrium distribution of holes for a given current step I_s is established as a result of competition between the injection and recombination processes in the base.

The energetic diagram of a diode based on disordered semiconductors is presented in Fig. 3.25. This diagram takes into account only distributed traps in the mobility gap. As a consequence, one must use corresponding transport equations. The dispersive transport of non-equilibrium holes is described by the generalized

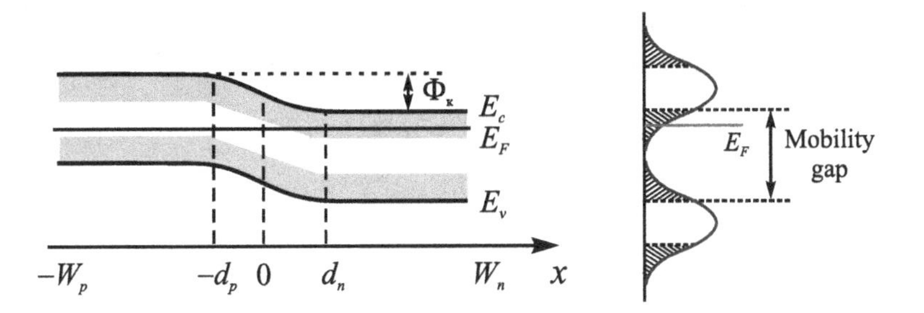

Figure 3.25 The simplest energetic diagram of a diode based on disordered semiconductors.

Fokker-Planck equation

$$\frac{\partial p_d(\mathbf{r},t)}{\partial t} + \frac{\tau_l^\alpha}{\tau_d} e^{-\gamma_l t} \, {}_0\mathsf{D}_t^\alpha \left[e^{\gamma_l t} \, p_d(\mathbf{r},t)\right]$$

$$+ \operatorname{div}\left[\mu \mathbf{E} \, p_d(\mathbf{r},t) - D_p \nabla p_d(\mathbf{r},t)\right] + \gamma_f p_d(\mathbf{r},t) = 0.$$

Here $p_d(\mathbf{r},t)$ is the concentration of non-equilibrium holes. In the case of one-dimensional diffusion (planar diode), it can be rewritten in the form:

$$\frac{\partial p_d(x,t)}{\partial t} + \frac{\tau_l^\alpha}{\tau_d} e^{-\gamma_l t} \, {}_0\mathsf{D}_t^\alpha \left[e^{\gamma_l t} \, p_d(x,t)\right] - D_p \frac{\partial^2 p_d(x,t)}{\partial x^2} + \gamma_f p_d(x,t) = 0.$$

Here γ_l and γ_f are parameters of recombination of localized and quasi-free carriers, respectively. This equation is written for concentration of mobile (quasi-free) carriers, which is applicable in the model of multiple trapping or percolation model "backbone – dead ends". Making the Laplace transformation on time yields

$$s\widehat{p}_d(x,s) + \frac{\tau_l^\alpha}{\tau_d} (s+\gamma_l)^\alpha \widehat{p}_d(x,s) - D_p \frac{\partial^2 \widehat{p}_d(x,t)}{\partial x^2} + \gamma_f \widehat{p}_d(x,s) = p_d(x,0).$$

Using the evident conditions

$$p_d(x,0) = 0, \quad \lim_{x\to\infty} p_d(x,t) = 0,$$

and neglecting by time of flight through the spatial charge region of the diode

$$p_d(0,t) = p_n \left[\exp\left(\frac{eU(t)}{kT}\right) - 1\right], \tag{3.34}$$

we obtain solution to this equation in the form $\tilde{p}_d(x,s) =$

$$= \sqrt{\gamma_f \tau_d + (\gamma_l \tau_l)^\alpha} \; p_n \left[\exp\left(\frac{eU_c}{kT}\right) - 1\right] \frac{\exp\left(-x\sqrt{s+\gamma_f+\tau_l^\alpha \tau_d^{-1}(s+\gamma_l)^\alpha}\right)}{s\sqrt{\tau_d s + \gamma_f \tau_d + \tau_l^\alpha (s+\gamma_l)^\alpha}}.$$

At point $x = 0$

$$\tilde{p}_d(0,s) = p_n \left[\exp\left(\frac{eU_c}{kT}\right) - 1\right] \frac{\sqrt{\gamma_f \tau_d + (\gamma_l \tau_l)^\alpha}}{s\sqrt{\tau_d s + \gamma_f \tau_d + \tau_l^\alpha (s+\gamma_l)^\alpha}}.$$

In the case of dispersive transport, carriers are localized in traps for vast time interval, and one can neglect of recombination of mobile (delocalized) carriers

$$\gamma_f \tau_d \ll (\gamma_l \tau_l)^\alpha, \quad \tau_d s + \gamma_f \tau_d \ll \tau_l^\alpha (s + \gamma_l)^\alpha.$$

As a result, we obtain the expression:

$$\tilde{p}_d(0, s) = p_n \left[\exp\left(\frac{eU_c}{kT}\right) - 1\right] \frac{\gamma_l^{\alpha/2}}{s(s + \gamma_l)^{\alpha/2}}.$$

Performing the inverse Laplace transformation, we find

$$p_d(0, t) = \gamma_l^{\alpha/2}\, p_n \left[\exp\left(\frac{eU_c}{kT}\right) - 1\right] \int_0^t \frac{\tau^{\alpha/2-1}}{\Gamma(\alpha/2)}\, e^{-\gamma_l \tau} d\tau$$

$$= p_n \left[\exp\left(\frac{eU_c}{kT}\right) - 1\right] \frac{1}{\Gamma(\alpha/2)} \int_0^{\gamma_l t} \xi^{\alpha/2-1}\, e^{-\xi} d\xi$$

$$= p_n \left[\exp\left(\frac{eU_c}{kT}\right) - 1\right] \frac{\Gamma(\alpha/2; \gamma_l t)}{\Gamma(\alpha/2)},$$

where $\Gamma(\nu; t)$ is the incomplete gamma-function [Abramowitz & Stegun (1972)]. Comparing this relation with Eq. (3.34), we obtain the equation

$$\left[\exp\left(\frac{eU(t)}{kT}\right) - 1\right] = \left[\exp\left(\frac{eU_c}{kT}\right) - 1\right] \frac{\Gamma(\alpha/2; \gamma_l t)}{\Gamma(\alpha/2)}.$$

Solving this equation with respect to $U(t)$ yields

$$U(t) = \frac{kT}{e} \ln\left\{1 + \left[\exp\left(\frac{eU_c}{kT}\right) - 1\right] \frac{\Gamma(\alpha/2; \gamma_l t)}{\Gamma(\alpha/2)}\right\}.$$

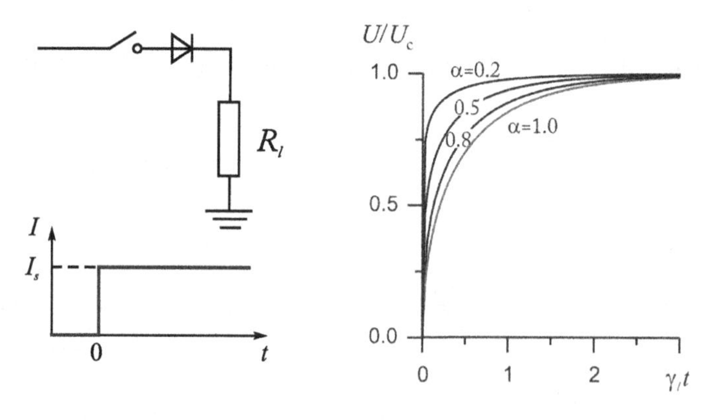

Figure 3.26 The kinetics of the diode voltage at switching on by the current step under dispersive transport conditions.

It is easy to obtain approximate formulas for the two cases

$$U(t) \approx U_c \frac{\Gamma(\alpha/2;\gamma_l t)}{\Gamma(\alpha/2)}, \quad \text{for} \quad U_c \ll kT/e,$$

and

$$U(t) \approx U_c + \frac{kT}{e} \ln\left(\frac{\Gamma(\alpha/2;\gamma_l t)}{\Gamma(\alpha/2)}\right), \quad \text{for} \quad U_c \gg kT/e.$$

In the case of normal transport $\alpha = 1$, and taking into account

$$\frac{\Gamma(1/2;\gamma_l t)}{\Gamma(1/2)} = \mathrm{erf}(\sqrt{\gamma_l t}),$$

we arrive at the expression for the diode based on crystalline semiconductors.

Fig. 3.26 shows the voltage kinetics for different values of dispersion parameter of holes in the n-region, when the diode is switching on by a current step.

3.3.2 *Turning off by interruption of circuit*

Now we are interested in the transition process when a diode is turned off by interruption of circuit in case of dispersive transport. The diode with a semi-infinite base of n-type is in the conductive state until the time moment $t = 0$, in which the circuit is opened. The voltage across the diode abruptly decreases by a voltage ohmic resistance. Concentration of excess holes in the base will dissipate during some time. Recombination determines the kinetics of the voltage at low injection levels, because current through the diode does not flow.

Stationary distribution of excess carriers in the base before turning off has an exponential form

$$p(x,0) = p_0 \exp(-x/L_p).$$

When the fractional process we are going to investigate at $t > 0$ has been existed at $t \leqslant 0$ initial conditions at $t = 0$ are not enough. We have to specify behavior of the function on the all time interval $t < 0$, i.e. to specify "prehistory" [Samko et al. (1987)]. Assume that the equilibrium distribution of quasi-free carriers takes place on the wide time interval in the past. In case of quasi-continuous distribution of localized states in the mobility gap, in equilibrium all carriers are localized.

Write an equation of dispersive diffusion for the concentration of mobile excess holes in the base of the planar diode in the form:

$$\frac{\partial p_d(x,t)}{\partial t} + \frac{\tau_l^\alpha}{\tau_d} e^{-\gamma_l t} {}_0\mathsf{D}_t^\alpha e^{\gamma_l t} p_d(x,t) - D_p \frac{\partial^2 p_d(x,t)}{\partial x^2} + \gamma_f p_d(x,t) = 0.$$

For given boundary conditions

$$\left.\frac{\partial p_f}{\partial x}\right|_{x=0} = 0, \quad \lim_{x\to\infty} p_d(x,t) = 0, \quad p_d(0,t) = p_n \left[\exp\left(\frac{eU(t)}{kT}\right) - 1\right],$$

after Laplace transformation on time we arrive at the following equation

$$s\tilde{p}_d(x,s) + \frac{\tau_l^\alpha}{\tau_d}\,(s+\gamma_l)^\alpha \tilde{p}_d(x,s) - D_p \frac{\partial^2 \tilde{p}_d(x,s)}{\partial x^2} + \gamma_f \tilde{p}_d(x,s) = p_0\, e^{-x/L_p}.$$

Representing it in the form

$$\frac{\partial^2 \tilde{p}_d(x,s)}{\partial x^2} - a(s)\tilde{p}_d(x,s) = p_0 \exp(-x/L_p),$$

where function $a(s)$ is as follows

$$a(s) = \frac{(s+\gamma_f)\tau_d + (s+\gamma_l)^\alpha \tau_l^\alpha}{D_p\, \tau_d},$$

we write the solution in the following form

$$\tilde{p}_d(x,s) = \frac{p_0}{L_p^{-2} - a}\left[\frac{e^{-x\sqrt{a(s)}}}{L_p\sqrt{a(s)}} - e^{-x/L_p}\right].$$

At $x = 0$ we have:

$$\tilde{p}_d(0,s) = \frac{p_0}{\sqrt{a(s)}(\sqrt{a(s)} + L_p^{-1})}.$$

Neglecting of recombination of quasi-free carriers, i.e. assuming

$$\tau_d s + \gamma_f \tau_d \ll \tau_l^\alpha (s+\gamma_l)^\alpha,$$

we obtain

$$\tilde{p}_d(0,s) = \frac{D_p \tau_d \tau_l^{-\alpha} p_0}{(s+\gamma_l)^{\alpha/2}\left[\sqrt{D_p\tau_d/L_p^2\tau_l^\alpha} + (s+\gamma_l)^{\alpha/2}\right]}.$$

The inverse Laplace transformation of this equation gives

$$p_d(0,t) = p_0 L_p^2\, \frac{e^{-\gamma_l t}}{t}\left(\frac{t}{\tau_l}\right)^{\alpha/2}\left[\frac{1}{\Gamma(\alpha/2)} - E_{\alpha/2,\alpha/2}\left(-\left(\frac{t}{\tau_l}\right)^{\alpha/2}\right)\right],$$

where

$$E_{\alpha,\,\beta}(z) = \sum_{n=0}^{\infty} \frac{z^n}{\Gamma(\alpha n + \beta)}$$

is the two-parameter *Mittag-Leffler function*. Comparing this result with formula (3.34), we obtain for the kinetics of voltage

$$U(t) = \frac{kT}{e}\ln\left\{1 + \frac{p_0 L_p^2}{p_n}\,\frac{e^{-\gamma_l t}}{t}\left(\frac{t}{\tau_l}\right)^{\alpha/2}\left[\frac{1}{\Gamma(\alpha/2)} - E_{\alpha/2,\alpha/2}\left(-\left(\frac{t}{\tau_l}\right)^{\alpha/2}\right)\right]\right\}.$$

3.4 Frequency properties of disordered semiconductor structures

3.4.1 *Frequency dependence of conductivity*

The frequency dependence of the real component of conductivity in many disordered semiconductors is fairly well described by the power law:

$$\text{Re}\ \sigma(\omega) = A\omega^{\nu}, \tag{3.35}$$

where the exponent ν normally takes on values from 0.7 to 1 [Pollak & Geballe (1961); Zvyagin (1984); Jonscher (1986)]. Conductivity is related to mobility by the expression

$$\sigma(\omega) = e\eta\mu(\omega).$$

Here, η is the concentration of effective carriers. The Nyquist formula (generalized Einstein relation) linking mobility with the diffusion coefficient at nonzero frequencies has the form

$$\mu(\omega) = (e/kT)D(\omega),$$

where the noise spectrum according the Wiener-Khintchin theorem is expressed through the Fourier transform of the velocity autocorrelation function [Lax (1960)]

$$\text{Re}\ D(\omega) = \int\limits_0^\infty \cos(\omega t)\langle v(t)v(0)\rangle dt. \tag{3.36}$$

As emphasized by Scher and Lax [Scher & Lax (1973)], Eq. (3.36) is important, because "a knowledge of the fluctuations of the equilibrium ensemble in the absence of the electric field permits a calculation of the linear response of the system (mobility)." The authors of Ref. [Scher & Lax (1973)] showed that relation (3.36) can be represented as

$$D(\omega) = -\frac{1}{6}\,\omega^2 \int\limits_0^\infty dt\ e^{-i\omega t}\left\langle [\mathbf{r}(t) - \mathbf{r}(0)]^2 \right\rangle = -\omega^2 \int\limits_0^\infty dx\ x^2\,[\widetilde{n}(x,s)]_{s=i\omega}\,, \tag{3.37}$$

where $\widetilde{n}(x,s)$ is the Laplace image in time of the solution to the diffusion equation.

Let us return to Eq. (3.13), reducing it to the one-dimensional form with coordinate-independent coefficient C in the absence of a field

$$\frac{\partial n(x,t)}{\partial t} = C\ e^{-\gamma t}\ {}_0\mathsf{D}_t^{1-\alpha}\left[e^{\gamma t}\ \frac{\partial^2 n(x,t)}{\partial x^2}\right].$$

The Laplace transform of this equation yields

$$s\ \widetilde{n}(x,s) = C\ (s+\gamma)^{1-\alpha}\ \frac{\partial^2 \widetilde{n}(x,s)}{\partial x^2} + \delta(x).$$

Substituting the solution of this equation

$$\widetilde{n}(x,s) = \frac{s^{-1/2}(s+\gamma)^{(\alpha-1)/2}}{\sqrt{C}} \exp\left(-\frac{|x|}{\sqrt{C}}\ \sqrt{s(s+\gamma)^{\alpha-1}}\right)$$

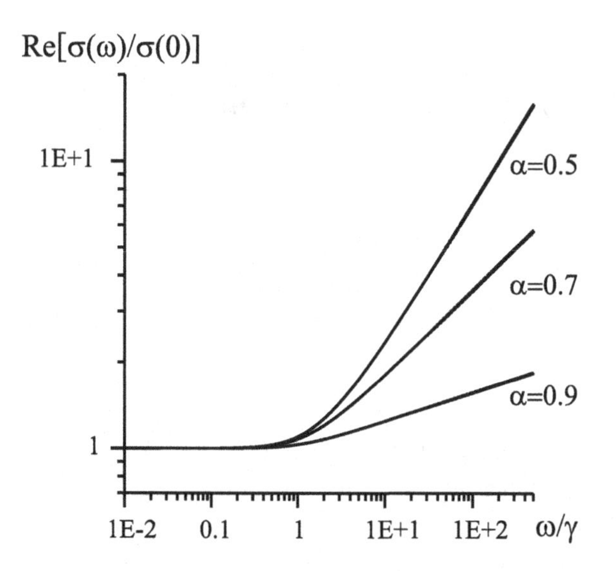

Figure 3.27 Frequency dependencies of the conductivity for different α values.

into relation (3.37), gives the expression

$$D(\omega) = 2C(\gamma + i\omega)^{1-\alpha}.$$

Finally, we get

$$\text{Re } \sigma(\omega) = (e^2\eta/kT) \text{ Re } D(\omega)$$

$$= 2C(e^2\eta/kT)\ (\gamma^2 + \omega^2)^{(1-\alpha)/2} \cos((1-\alpha)\arctan(\omega/\gamma)). \tag{3.38}$$

For frequencies $\omega \gg \gamma$, it becomes

$$\text{Re } \sigma(\omega) = 2C(e^2\eta/kT)\ \omega^{1-\alpha} \sin(\pi\alpha/2).$$

Figure 3.27 presents frequency-dependent conductivity curves calculated from Eq. (3.38), which predicts the power-law dependence of conductivity on o at high frequencies in the dispersive transport case. The exponent may acquire values from 0 to 1; in normal transport $\alpha = 1$, the conductivity is totally frequency independent. In transport driven by the multiple trapping mechanism, exponent a grows linearly with temperature. Consequently, in the case of alternating current and $\alpha \propto T$, exponent $\nu = 1 - \alpha$ in the frequency dependence of conductivity must linearly decrease with increasing temperature. Such temperature behavior has been reported for a variety of semiconductors (see, for instance, [Ghosh et al. (2006)]).

3.4.2 *A diode at dispersive transport conditions*

Now, we calculate the frequency dependence of semiconductor diode conductivity governed by dispersive transport. To this end, an expression is needed for the density of alternating current flowing through a diode to the n-type base of which the constant displacement U is applied in the direction of transmission and weak variable signal $u_1 \exp(i\omega t)$, with the amplitude $u_1 \ll U$ and $u_1 \ll kT/e$, i.e.

$$u(t) = U + u_1 \exp(i\omega t). \tag{3.39}$$

The total hole concentration in the diode base can be presented as the sum of constant $p_c^-(x)$ and variable $p_d^\sim(x,t)$ components, i.e., $p_d(x,t) = p_f^-(x) + p_d^\sim(x,t)$.

The solution to this problem for a crystalline semiconductor diode can be found in Refs. [Gaman (2000); Bonch-Bruevich & Kalashnikov (1977)]. The case of a low injection level is considered, in which alternating current density across the p-n-junction $J^\sim = -eD_p \partial p_f^\sim / \partial x|_{x=0}$. The expression for $p_d^\sim(x,t)$ is obtained by solving the generalized diffusion equation, taking into account linear monomolecular recombination (see Eq. (2.67))

$$\frac{\partial p_d(x,t)}{\partial t} + \frac{1}{\tau_{0p} c_p^\alpha} {}_0\mathsf{D}_t^\alpha p_d(x,t) = D_p \frac{\partial^2 p_d(x,t)}{\partial x^2} - \frac{\Delta p_d(x,t)}{\tau_p},$$

where $p_d(x,t)$ is the concentration of delocalized nonequilibrium carriers, and τ_0 is the mean time of a single capture on localized band tail states. Separating the total concentration of nonequilibrium holes into constant and variable constituents yields

$$\frac{\partial p_d^\sim(x,t)}{\partial t} + \frac{1}{\tau_{0p} c_p^\alpha} {}_0\mathsf{D}_t^\alpha p_d^\sim(x,t) = D_p \frac{\partial^2 p_f^-(x,t)}{\partial x^2} + D_p \frac{\partial^2 p_d^\sim(x,t)}{\partial x^2} - \frac{p_f^-(x) + p_f^\sim(x,t) - p_n}{\tau_p},$$

where p_n is the equilibrium concentration of delocalized holes in the n-region. Because the expression

$$D_p \frac{\partial^2 p_f^-(x)}{\partial x^2} = \frac{p_f^-(x) - p_n}{\tau_p},$$

holds for the constant constituent, the variable one satisfies the following equation

$$\frac{\partial p_d^\sim(x,t)}{\partial t} + \frac{1}{\tau_{0p} c_p^\alpha} {}_0\mathsf{D}_t^\alpha p_d^\sim(x,t) = D_p \frac{\partial^2 p_d^\sim(x,t)}{\partial x^2} - \frac{p_d^\sim(x,t)}{\tau_p} \tag{3.40}$$

Applying the Laplace transform

$$L_{p_f^\sim}(x,s) = \int_0^\infty p_d^\sim(x,t) \exp(-st) dt,$$

to diffusion equation (3.40), we obtain

$$D_p \frac{\partial^2 L_{p_f^\sim}(x,s)}{\partial x^2} = \left(s + \frac{1}{\tau_{0p} c_p^\alpha} s^\alpha + 1/\tau_p \right) L_{p_f^\sim}(x,s). \tag{3.41}$$

Let us formulate boundary conditions. We consider the variable signal with frequencies, at which the time of flight through a spatial charge region of the p-n junction is smaller than ω^{-1}:

$$p_f^-(0) + p_d^\sim(0,t) = p_n \exp\left(\frac{e}{kT} (U + u_1 \exp(i\omega t)) \right) \tag{3.42}$$

$$\approx p_n \exp\left(\frac{eU}{kT} \right) + \frac{ep_n}{kT} \exp\left(\frac{eU}{kT} \right) u_1 \exp(i\omega t). \tag{3.43}$$

Hence, one obtains

$$p_d^{\sim}(0,t) = \frac{ep_n}{kT}\exp\left(\frac{eU}{kT}\right)u_1\exp(i\omega t).$$

The second boundary condition is $p_d^{\sim}(\infty,t) = 0$. Taking account of boundary conditions for the solution of Eq. (3.41) gives

$$L_{p_f^{\sim}}(x,s) = L_{p_f^{\sim}}(0,s)\exp\left[-x\sqrt{D_p^{-1}\left(s+\frac{1}{\tau_{0p}c_p^{\alpha}}s^{\alpha}+1/\tau_p\right)}\right]. \tag{3.44}$$

Function $\exp\left(-x\sqrt{D_p^{-1}(s+1/(\tau_{0p}c_p^{\alpha})s^{\alpha}+1/\tau_p)}\right)$ is the Laplace image of a certain function $y(x,t)$. It follows from formulas (3.39, 3.44) that

$$p_f^{\sim}(x,t) = \int_0^{\infty} p_f^{\sim}(0,t-t')y(x,t')dt'$$

$$= \frac{ep_n}{kT}\exp\left(\frac{eU}{kT}\right)u_1\exp(i\omega t)\exp\left[-x\sqrt{D_p^{-1}\left(i\omega+\frac{1}{\tau_{0p}c_p^{\alpha}}(i\omega)^{\alpha}+1/\tau_p\right)}\right].$$

The density of alternating current flowing through the p-n junction is given by

$$J^{\sim} = -eD_p\frac{\partial p_f^{\sim}(x,t)}{\partial x}|_{x=0}$$

$$= \frac{e^2p_nD_p}{kT}\exp\left(\frac{eU}{kT}\right)\sqrt{D_p^{-1}\left(i\omega+\frac{1}{\tau_{0p}c_p^{\alpha}}(i\omega)^{\alpha}+1/\tau_p\right)}\quad u_1\exp(i\omega t).$$

The complex p-n-junction conductivity is

$$Y = S_{p-n}\frac{e^2p_nD_p}{kT}\exp\left(\frac{eU}{kT}\right)\sqrt{D_p^{-1}\left(i\omega+\frac{1}{\tau_{0p}c_p^{\alpha}}(i\omega)^{\alpha}+1/\tau_p\right)},$$

where S_{p-n} is the p-n junction area. Separating out the real and imaginary parts brings Y to the form

$$Y = G_{p-n} + i\omega C_D,$$

where G_{p-n} is the total conductivity, and C_D is the diffusion capacity of the p-n junction:

$$Y = S_{p-n}\sqrt{\frac{D_p}{2}}\,\frac{e^2p_n}{kT}\exp\left(\frac{eU}{kT}\right)$$

$$\times\sqrt{\frac{1}{\tau_p}+\frac{1}{\tau_{0p}c_p^{\alpha}}\cos\frac{\pi\alpha}{2}\omega^{\alpha}+\sqrt{\left(\frac{1}{\tau_p}+\frac{1}{\tau_{0p}c_p^{\alpha}}\cos\frac{\pi\alpha}{2}\omega^{\alpha}\right)^2+\left(\omega+\frac{1}{\tau_{0p}c_p^{\alpha}}\sin\frac{\pi\alpha}{2}\omega^{\alpha}\right)^2}}$$

$$+iS_{p-n}\sqrt{\frac{D_p}{2}}\,\frac{e^2p_n}{kT}\exp\left(\frac{eU}{kT}\right)$$

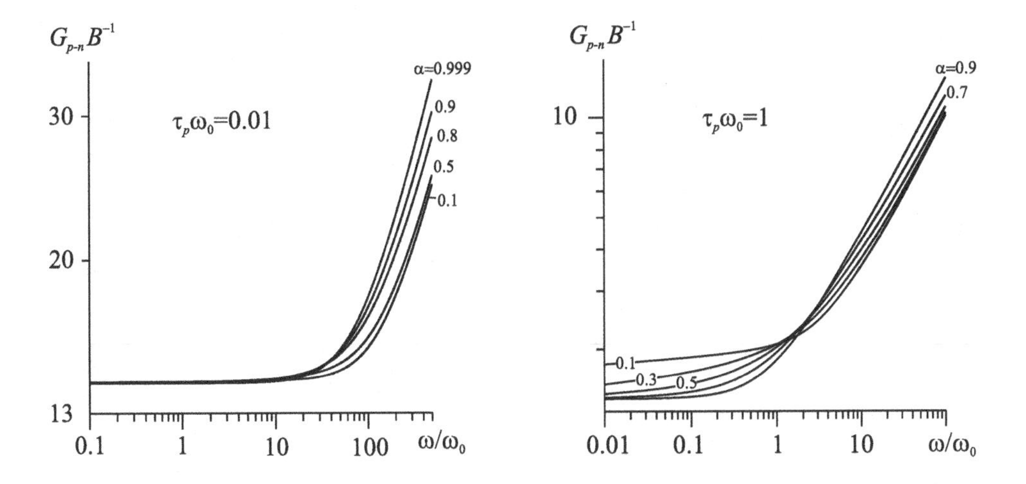

Figure 3.28 Frequency dependence of the diode conductivity for different α values.

$$\times \frac{\omega + \dfrac{1}{\tau_{0p} c_p^\alpha} \sin \dfrac{\pi\alpha}{2} \omega^\alpha}{\sqrt{\dfrac{1}{\tau_p} + \dfrac{1}{\tau_{0p} c_p^\alpha} \cos \dfrac{\pi\alpha}{2} \omega^\alpha + \sqrt{\left(\dfrac{1}{\tau_p} + \dfrac{1}{\tau_{0p} c_p^\alpha} \cos \dfrac{\pi\alpha}{2} \omega^\alpha\right)^2 + \left(\omega + \dfrac{1}{\tau_{0p} c_p^\alpha} \sin \dfrac{\pi\alpha}{2} \omega^\alpha\right)^2}}}, \tag{3.45}$$

For the total p-n-junction conductivity, one has

$$G_{p-n} = S_{p-n} \sqrt{\frac{D_p}{2}} \frac{e^2 p_n}{kT} \exp\left(\frac{eU}{kT}\right)$$

$$\times \sqrt{\frac{1}{\tau_p} + \frac{1}{\tau_{0p} c_p^\alpha} \cos \frac{\pi\alpha}{2} \omega^\alpha + \sqrt{\left(\frac{1}{\tau_p} + \frac{1}{\tau_{0p} c_p^\alpha} \cos \frac{\pi\alpha}{2} \omega^\alpha\right)^2 + \left(\omega + \frac{1}{\tau_{0p} c_p^\alpha} \sin \frac{\pi\alpha}{2} \omega^\alpha\right)^2}}.$$

Graphs of the conductivity frequency dependences, based on this formula, are plotted in Fig. 3.28. Here,

$$\omega_0 = (\tau_0 c^\alpha)^{-1/(1-\alpha)}, \quad A = S_{p-n} \sqrt{D_p/2} (e^2 p_n/kT) \exp(eU/kT) \omega_0^{-1}.$$

When $\alpha \to 1$, we come to the classical expression for a crystalline semiconductor-based diode (such as presented in Refs. [Gaman (2000); Bonch-Bruevich & Kalashnikov (1977)]):

$$G_{p-n}^{\text{cryst}} = S_{p-n} \sqrt{\frac{D_p}{2}} \frac{e^2 p_n}{kT} \exp\left(\frac{eU}{kT}\right) \sqrt{\tau_p^{-1} + \sqrt{\tau_p^{-2} + \omega^2}}.$$

Tending $\omega \to 0$ leads to a differential conductivity of the p-n junction in an alternating current:

$$G_{p-n}^0 = S_{p-n} \sqrt{\frac{D_p}{2\tau_p}} \frac{e^2 p_n}{kT} \exp\left(\frac{eU}{kT}\right).$$

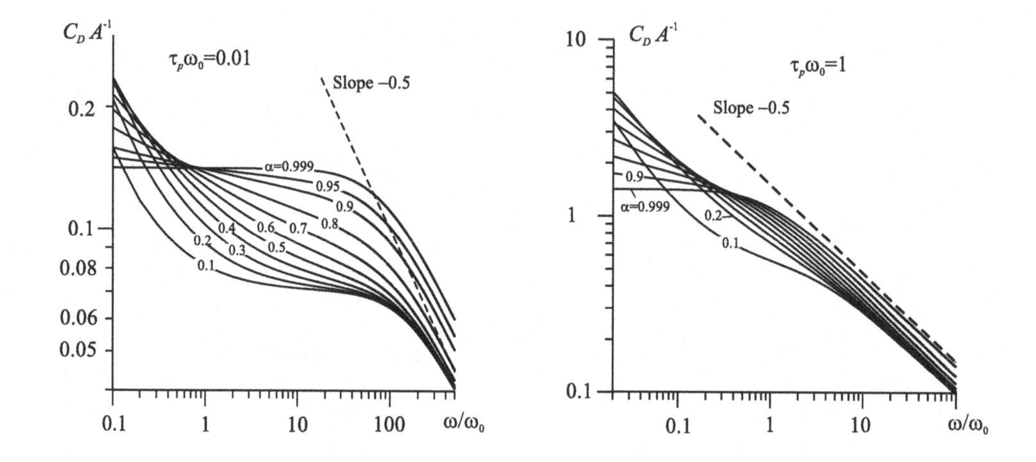

Figure 3.29 Frequency dependence of the diffusion capacity for different α values.

In the high-frequency region for which $\omega \gg \dfrac{1}{\tau_{0p}c_p^\alpha}\omega^\alpha$ and $\omega \gg 1/\tau_p$, we have

$$G_{p-n} = S_{p-n}\sqrt{\frac{D_p}{2}}\,\frac{e^2p_n}{kT}\exp\left(\frac{eU}{kT}\right)\sqrt{\omega}.$$

In other words, neither carrier capture in localized band tail states nor carrier recombination via deep centers affects conductivity in the high-frequency region. In the medium-frequency region, where $\omega^{1-\alpha} \ll \frac{1}{\tau_{0p}c_p^\alpha}$, but $\omega^{1-\alpha} \gg 1/\tau_p$, we have p-n-junction conductivity for dispersive transport in disordered semiconductors:

$$G_{p-n} = S_{p-n}\sqrt{D_p}\,\frac{e^2p_n}{kT}\exp\left(\frac{eU}{kT}\right)\frac{1}{\sqrt{\tau_{0p}c_p^\alpha}}\cos\left(\frac{\pi\alpha}{4}\right)\omega^{\alpha/2}.$$

A similar expression was obtained for diffusion capacity [Sibatov & Uchaikin (2006)]. By tending α to unity ($\alpha \to 1$), we come to the expression reported for crystalline semiconductor diodes in (see Refs. [Gaman (2000); Bonch-Bruevich & Kalashnikov (1977)]):

$$C_D = S_{p-n}\sqrt{\frac{D_p}{2}}\,\frac{e^2p_n}{kT}\exp\left(\frac{eU}{kT}\right)\left[\frac{1}{\tau_p}+\sqrt{\frac{1}{\tau_p^2}+\omega^2}\right]^{-1/2}.$$

In the high-frequency region for which $\omega \gg \frac{1}{\tau_{0p}c_p^\alpha}\omega^\alpha$ and $\omega \gg 1/\tau_p$, hold true, we have

$$C_D = S_{p-n}\sqrt{\frac{D_p}{2}}\,\frac{e^2p_n}{kT}\exp\left(\frac{eU}{kT}\right)\frac{1}{\omega}.$$

The frequency dependencies of diode diffusion capacity in the given coordinates are shown in Fig. 3.29.

Analogous relations can be obtained for tunneling recombination.

The aim of the above consideration was to illustrate the convenience of fractional equations for the analysis of frequency characteristics of devices based on disordered semiconductors, compared to the Arkhipov-Rudenko equation (2.42). The last equation incorporates the time-dependent diffusion coefficient, and the Fourier transform converts it into the equation containing an integral convolution of the Fourier image of time-dependence of the diffusion coefficient with the transformant of carrier concentration. In other words, we arrive at an integral equation which is much more difficult to solve than the algebraic equation obtained by Fourier transformation of a fractional equation.

Chapter 4

Fractional kinetics in quantum dots and wires

Some features of the transport in disordered semiconductors can be clarified when studying electron transfer in arrays of colloidal quantum dots. Investigations of these artificial materials with controlled properties can additionally elucidate fundamental concepts of disordered solid physics such as localization, nonlinear effects associated with long-range Coulomb correlations, occupancy of traps and the Coulomb blockade.

As noted in [Novikov (2003)], understanding the dynamics of localization processes in individual nanocrystals can be made more deep by using data on their fluorescence blinking. Therefore, we start with the problem of blinking fluorescence statistics. On the base of analysis of the single-photon count statistics, a new mathematical model of the process is developed in our work [Uchaikin & Sibatov (2009b); Sibatov & Uchaikin (2010)]. It is based on fractional generalization of the telegraph process and gives a unified description of both the scintillation exponential and power distributions for the on-and off-intervals. Further, we study statistical properties of anomalous transport in ordered quantum dot ensembles. The last section is devoted to recent theoretical work [Falceto & Gopar (2010); Beenakker et al. (2009)] studied the conductivity of one-dimensional quantum system with a fractal-type disorder, characterized by asymptotically power law distribution of distances between scattering barriers. These studies were motivated by the experimental work [Kohno & Yoshida (2004)] demonstrating the scaling behavior of the diameter modulations in semiconductor SiC nanowires. The observed behavior corresponds to the non-Gaussian nature of fluctuations of thickness which is typical for some synthesized nanowires and polymer nanofibrills.

4.1 Fractional optics of quantum dots

4.1.1 *Off- and on-intervals statistics*

Colloidal quantum dots (QDs) have a broad absorption spectrum, a narrow emission spectrum, a high luminescence quantum yield, and moreover they are stable with respect to light degradation [Chakraborty (1999); Shimizu et al. (2001); Heiss

(2005); Osadko (2006)]. Due to these properties, QDs are promising as active media for lasers, single photon sources, and luminescent marks in chemistry and biology. However, the *blinking of QDs fluorescence* limits their application (on states, in which many photons are emitted alternate with off states, in which nanoparticles do not emit). Experimental studies of blinking fluorescence of QDs [Shimizu et al. (2001); Kuno et al. (2001)] show that the on and off times are distributed according to the power laws:

$$\Psi_{\text{on}}(t) = \text{Prob}\{T_{\text{on}} > t\} \propto t^{-\alpha}, \qquad \Psi_{\text{off}}(t) = \text{Prob}\{T_{\text{off}} > t\} \propto t^{-\beta}, \tag{4.1}$$

where $0 < \alpha,\ \beta < 1$ (see Fig. 4.1). The parameters α and β remain nearly the same upon variations in the measurement conditions, such as temperature, laser intensity, and size of QDs. Because of this, the determination of the blinking fluorescence mechanism is a rather complicated problem.

Because blinking process is stochastic by nature, the information experimentally obtained should inevitably be interpreted in terms of stochastic processes. Obviously, it should be a two state (binary) renewal process (see [Zumofen & Klafter (1993); Uchaikin & Zolotarev (1999); Jung et al. (2002)]). It is known that random variables distributed according to laws (4.1) have infinite average values; therefore, the Gaussian statistics is inapplicable to the description of the process and it is necessary to invoke the Lévy generalized limit theorem and fractional stable distributions [Uchaikin & Zolotarev (1999)]. In turn, these distributions satisfy equations with fractional derivatives; therefore, in the long term asymptotic limit, the power law blinking process will also be described by equations with fractional order derivatives.

The study of the photon counting statistics in the case $0 < \alpha = \beta < 1$ [Jung et al. (2002)] shows that relative fluctuations do not decrease with time, and the Mandel parameter increases proportionally to the time of observation. The correlation functions of a process with two states were calculated in [Margolin & Barkai (2004)] for equal on and off exponents. Many theoretical works are devoted to the search for a microscopic mechanism of blinking $\alpha = \beta$ [Tang & Marcus (2005); Jung et al. (2002)]. However, numerous experimental data show [Kuno et al. (2001)] that the exponent α is by no means always equal to β. The case $\alpha \neq \beta$ has been considered in [Sibatov & Uchaikin (2010)]. Involving the fractional stable statistics and fractional calculus has helped to surmount calculational problems that arise in description of the stochastic behavior of the considered binary process.

In the next subsection, we present a brief review of physical mechanisms that are proposed by various authors as responsible for power law blinking. In subsection 4.1.3, we calculate the distribution of the total fluorescence time; formulate the equation for the density of this distribution; consider the asymptotic behavior of this density for three cases $\alpha > \beta$, $\alpha = \beta$ and $\alpha < \beta$, and compare these calculations with the results of numerical simulation of the binary process by the Monte Carlo method. Using the obtained distribution, we will analyze fluctuations in the total fluorescence time and in the number of counted photons.

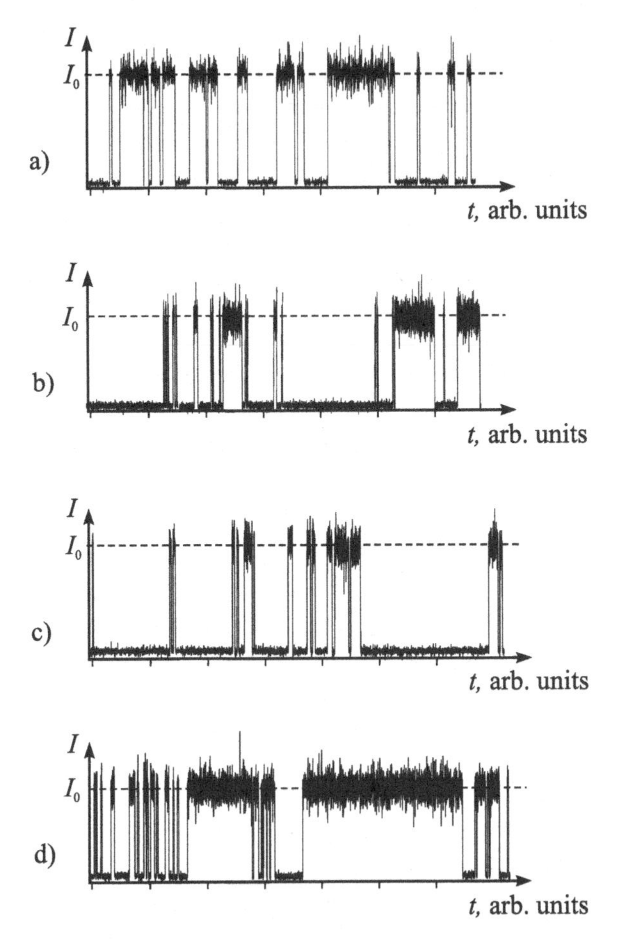

Figure 4.1 Monte Carlo simulation of blinking quantum dot fluorescence as a Poisson process embedded into an alternating renewal process: (a) on- and off-intervals are distributed according to exponential law; (b) according to power laws with exponent 0.5; (c) on-times have exponential pdf and off-times have asymptotically power law pdf ($\beta = 0.5$); (d) on-times are distributed according to power law ($\alpha = 0.5$), off-intervals are exponentially distributed.

4.1.2 *Physical mechanisms of power law blinking*

The majority of physical models of blinking fluorescence are based on the idea that the transition from the on to the off state occurs due to Auger ionization of QDs [Efros & Rosen (1997)]. In the case of formation of two electron-hole pairs in a QD during a short period of time, annihilation of one of them results in that the electron (hole) is knocked out of the other pair, i.e., the QD becomes ionized. Three-level concept used in spectroscopy of single molecules leads to exponential on- and off- distributions. The question arises: which structure of energy levels in QD should be in order to lead to power law distributions of the on- and off-times.

The model of activation kinetics [Kuno et al. (2001)] predicts that the exponent of the off time distribution depends on temperature. An electron (hole) captured in a trap of the QD environment can escape from the trap and then recombine with a hole (electron) in the QD core. The escape rate depends on the trap depth and temperature as follows:

$$\gamma(\varepsilon) = A\exp(-\varepsilon/kT).$$

If the energy distribution of traps is exponential,

$$\rho(\varepsilon) = \alpha\exp(-\alpha\varepsilon),$$

the distribution of off-times is asymptotically of the power law type

$$\psi_{\text{off}}(t) = \langle\gamma(\varepsilon)\exp(-\gamma(\varepsilon)t)\rangle = \int\limits_0^\infty \rho(\varepsilon)\gamma(\varepsilon)\exp(-\gamma(\varepsilon)t)d\varepsilon \propto t^{-1-\alpha kT},$$

$$t \gg [\gamma(\alpha^{-1})]^{-1}.$$

However, experiments [Shimizu et al. (2001)] show that, the power law kinetics is weakly sensitive to temperature changes in a wide region from low to room temperatures. For this reason, thermodynamic processes most likely do not play a significant role in the intermittent fluorescence of QDs.

A possible relation between the power law statistics and the theory of the first passage time upon walk in the coordinate or configurational (energy) space was discussed in [Empedocles & Bawendi (1999)] and [Jung et al. (2002)]. The simplest formal model considers a one-dimensional Brownian motion from the coordinate origin along the x-axis that contains an absorption point $x = a$. In this case, the first passage time T is distributed according to the stable Lévy-Smirnov law with the density

$$p_T(t) = \frac{1}{\sqrt{2\pi}}\, t^{-3/2}e^{-a^2/2t}.$$

This density has a power law tail with the exponent $-3/2$. Experimentally obtained exponents are different in studies by various groups of researchers (see [Shimizu et al. (2001); Kuno et al. (2001); Stefani et al. (2005); Osadko (2006); Jung et al. (2002)]), but they are often close to 3/2. In some theoretical works, models were developed exactly for this case, i.e., $\alpha+1 \approx \beta+1 \approx 3/2$ [Tang & Marcus (2005)]. To answer the question why the exponents of power law densities of distributions are close to $-3/2$, rather than to other values and why the power law of distributions of off times is universal for capped colloidal QDs for different intensities, temperatures, and sizes, many researchers address diffusion models.

One of these models considers the walk of a carrier in the coordinate space of a nanocrystal. The size of QDs is comparable to the electron wavelength; consequently, quantum effects should play a primary role, and the model of classical diffusion in space is unlikely to be applied in this case. In addition, the Monte Carlo simulation of the process showed that, for QDs of sizes considered, diffusion does not yield power law distributions off times within eight orders of time observed in experiment.

Margolin and Barkai (2004) proposed another model, which explains blinking by three-dimensional hopping transport of electrons knocked out of QDs in the surrounding space. A positively charged QD remains dark until the electron returns back. The wide range of on-times is explained by the occurrence of long lived states of holes located in the environment of QDs. As long as the hole remains captured and the electron diffuses, the QD core remains in the neutral state and is capable of absorbing and emitting photons. However, in this model, the probability that the itinerant electron does not return back to the QD core is nonzero contradicts experiments [Kuno et al. (2001); Shimizu et al. (2001)].

A remarkable review on intermittent radiation was made by Kuno et al. (2001). The authors consider physical mechanisms of blinking and propose a new model of tunneling charge through fluctuating barriers. In this model, the power law distribution of the on times arises because there are many on states in a QD, the transitions between which do not occur within the on period. A barrier that separates states on the QD surface and core fluctuates in energy or in width, which, in the quasiclassical approximation, yields a wide distribution of transition times. Calculations show that the described situation leads to a richer family of distributions of on and off times, which is not observed in reality (in many experiments $\alpha \approx \beta \approx 0.5$). Tang & Marcus (2005) note, it remains unclear how, for such small sized QDs, tunneling paths can be spread within five orders of magnitude to ensure eight orders of the power law distribution of off times observed in the experiment.

Shimizu et al. [Shimizu et al. (2001)] suggested a model of resonant transitions of an electron between an excited state of a QD and a long lived intermediate state. An event of electronic transition that switches the luminescence intensity only occurs if the excited state and trap are at resonance. The energy of the lowest excited state of the QD plays the role of coordinate. The notion of the occurrence of a disorder in the environment of QDs gives rise to doubt in the assumption that there exists a single electron trap level with a narrow energy width that is localized in the environment of each individual QD. To explain the occurrence of these energy levels, Tang and Marcus [Tang & Marcus (2005)] assume that the trap is crystalline surface state.

Frantsuzov and Marcus [Frantsuzov & Marcus (2005)] apply a different model of intermittent luminescence, which does not imply the occurrence of any long lived trap level. After the photoexcitation, the QD always returns to the ground neutral state either directly or via a surface state. The on and off switching of the QDs luminescence is ensured by a large spread in the rate of nonradiative relaxation of the excited electron state to the ground state via surface states of holes. It is assumed that holes are captured by an induced Auger process. The power law blinking of the QDs luminescence is a result of a rapid nonradiative relaxation of

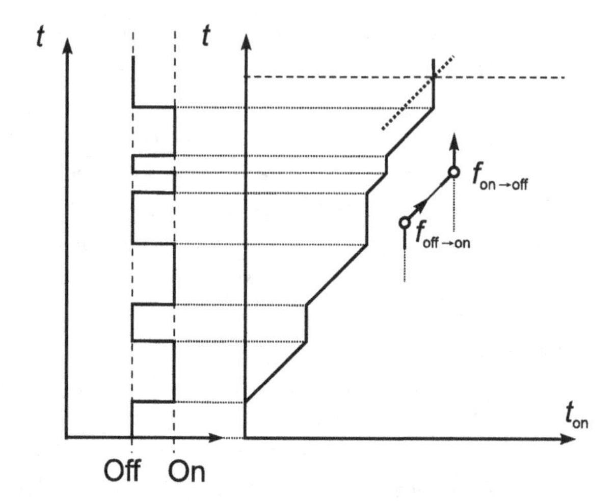

Figure 4.2 Schematic presentation of trajectories in the coordinates: state vs observation time (left panel) and effective radiation time vs observation time (right panel).

the excitation by the transition of the hole to deep lying surface states.

The power law character of the distributions gives grounds to relate the QDs blinking with the Lévy walk model [Jung et al. (2002)]. In $(t_{\rm on}-t_{\rm off})$-plane, blinking looks as a one-dimensional fractal walk with an infinite rate of free motion (Lévy flight), whereas, in $(t_{\rm on}-t)$-plane it looks like a random walk with a finite rate (Lévy walk) equal to unity (Fig. 4.2).

4.1.3 *Two-state renewal model*

Unlike Markov processes, renewal processes possess the memory but lose it at random renewal instants [Cox & Smith (1967)]. In the two-state model considered here, renewal times are associated with transitions between the states. Consider a process starting from the instant at which an off-to-on or on-to-off transition occurs with probability $\varepsilon_{\rm on}$ or $\varepsilon_{\rm off}=1-\varepsilon_{\rm on}$. The total on-time is a random variable conditioned on the observation time $(0,t)$ with probability density function denoted by $p_{\rm on}(t_{\rm on}|t)$. Defining $f_{\rm on\to off}(t_{\rm on},t)dt_{\rm on}dt$ as the average number of on-to-off transitions within the rectangle $(t_{\rm on},t_{\rm on}+dt_{\rm on})\times(t,t+dt)$ and $f_{\rm off\to on}(t_{\rm on},t)dt_{\rm on}dt$ as the average number of off-to-on transitions within $(t_{\rm on},t_{\rm on}+dt_{\rm on})\times(t,t+dt)$, we express the on-time probability distribution in terms of transition probability densities as

$$p_{\rm on}(t_{\rm on}|t)=\int\limits_0^{t_{\rm on}} f_{\rm off\to on}(t_{\rm on}-t',t-t')\Psi_{\rm on}(t')\,dt'+\int\limits_0^{t} f_{\rm on\to off}(t_{\rm on},t-t')\Psi_{\rm off}(t')\,dt', \tag{4.2}$$

where

$$\Psi_{\text{on,off}}(t) = \int_t^\infty \psi_{\text{on,off}}(t')dt'$$

is the probability that the waiting time in either state exceeds t. The transition probability densities satisfy the system of integral equations

$$f_{\text{on}\to\text{off}}(t_{\text{on}}, t) = \int_0^{t_{\text{on}}} f_{\text{off}\to\text{on}}(t_{\text{on}} - t', t - t')\psi_{\text{on}}(t')dt' + \varepsilon_{\text{off}}\delta(t_{\text{on}})\delta(t), \tag{4.3}$$

$$f_{\text{off}\to\text{on}}(t_{\text{on}}, t) = \int_0^t f_{\text{on}\to\text{off}}(t_{\text{on}}, t - t')\psi_{\text{off}}(t')\, dt' + \varepsilon_{\text{on}}\delta(t_{\text{on}})\delta(t), \tag{4.4}$$

which can readily be derived by considering sample paths in the $(t_{\text{on}},\ t)$ plane (Fig. 4.2). Taking the double Laplace transform (with respect to t and t_{on}):

$$\widetilde{p}(s_{\text{on}}|s) = \int_0^\infty dt_{\text{on}} \int_0^\infty dt\ \exp(-s_{\text{on}}t_{\text{on}} - st)\ p(t_{\text{on}}|t)$$

of Eqs. (4.2)-(4.4), we obtain

$$\widetilde{p}_{\text{on}}(s_{\text{on}}|s) = \widetilde{f}_{\text{off}\to\text{on}}(s_{\text{on}}, s)\ \widehat{\Psi}_{\text{on}}(s + s_{\text{on}}) + \widetilde{f}_{\text{on}\to\text{off}}(s_{\text{on}}, s)\ \widehat{\Psi}_{\text{off}}(s),$$

$$\widetilde{f}_{\text{on}\to\text{off}}(s_{\text{on}}, s) = \widetilde{f}_{\text{off}\to\text{on}}(s_{\text{on}}, s)\ \widehat{\psi}_{\text{on}}(s + s_{\text{on}}) + \varepsilon_{\text{off}},$$

$$\widetilde{f}_{\text{off}\to\text{on}}(s_{\text{on}}, s) = \widetilde{f}_{\text{on}\to\text{off}}(s_{\text{on}}, s)\ \widehat{\psi}_{\text{off}}(s) + \varepsilon_{\text{on}}.$$

Thus, the double Laplace transform reduces the system of integral equations to three algebraic ones. Solving them, we express the Laplace transforms of the transition rates as

$$\widetilde{f}_{\text{on}\to\text{off}}(s_{\text{on}}, s) = \frac{\varepsilon_{\text{on}}\widehat{\psi}_{\text{on}}(s + s_{\text{on}}) + \varepsilon_{\text{off}}}{1 - \widehat{\psi}_{\text{on}}(s + s_{\text{on}})\widehat{\psi}_{\text{off}}(s)},$$

$$\widetilde{f}_{\text{off}\to\text{on}}(s_{\text{on}}, s) = \frac{\varepsilon_{\text{off}}\widehat{\psi}_{\text{off}}(s) + \varepsilon_{\text{on}}}{1 - \widehat{\psi}_{\text{on}}(s + s_{\text{on}})\widehat{\psi}_{\text{off}}(s)},$$

and the Laplace transform of the desired probability density function as

$$\widetilde{p}_{\text{on}}(s_{\text{on}}|s) = \frac{\varepsilon_{\text{off}}\widehat{\psi}_{\text{off}}(s) + \varepsilon_{\text{on}}}{1 - \widehat{\psi}_{\text{on}}(s + s_{\text{on}})\widehat{\psi}_{\text{off}}(s)}\,\frac{1 - \widehat{\psi}_{\text{on}}(s + s_{\text{on}})}{s + s_{\text{on}}} + \frac{\varepsilon_{\text{on}}\widehat{\psi}_{\text{on}}(s + s_{\text{on}}) + \varepsilon_{\text{off}}}{1 - \widehat{\psi}_{\text{on}}(s + s_{\text{on}})\widehat{\psi}_{\text{off}}(s)}\,\frac{1 - \widehat{\psi}_{\text{off}}(s)}{s}. \tag{4.5}$$

The last expression is identical (up to notation) to that for the characteristic function of photon counts obtained in [Jung et al. (2002)], where it was used to analyze the asymptotic behavior of Mandel's Q parameter. Performing the Laplace inversion, one can obtain an asymptotic expression for the conditional probability density function $p_{\text{on}}(t_{\text{on}}|t)$ and demonstrate that this density satisfies a fractional equation.

4.1.4 *Fractional blinking process*

We will assume further that the waiting times T_{on} and T_{off} are distributed according densities

$$\psi_{\text{on}}(t) = c_{\text{on}}^{\alpha} t^{\alpha-1}\, E_{\alpha,\alpha}(-c_{\text{on}}^{\alpha} t^{\alpha}), \qquad \psi_{\text{off}}(t) = c_{\text{off}}^{\beta} t_{\text{off}}^{\beta-1}\, E_{\beta,\beta}(-c_{\text{off}}^{\beta} t^{\beta}). \tag{4.6}$$

Their Laplace transforms are:

$$\widehat{\psi}_{\text{on}}(s_{\text{on}}) = \frac{1}{1 + (s_{\text{on}}/c_{\text{on}})^{\alpha}}, \qquad \widehat{\psi}_{\text{off}}(s_{\text{off}}) = \frac{1}{1 + (s_{\text{off}}/c_{\text{off}})^{\beta}}, \qquad \alpha, \beta \le 1. \tag{4.7}$$

Substituting Laplace transforms (4.7) into (4.5), we obtain

$$\widetilde{p}_{\text{on}}(s_{\text{on}}|s) = \frac{c_{\text{on}}^{-\alpha}(s+s_{\text{on}})^{\alpha-1} + c_{\text{off}}^{-\beta} s^{\beta-1} + c_{\text{on}}^{-\alpha} c_{\text{off}}^{-\beta} s^{\beta}(s+s_{\text{on}})^{\alpha}\left[\varepsilon_{\text{on}}(s+s_{\text{on}})^{-1} + \varepsilon_{\text{off}} s^{-1}\right]}{c_{\text{on}}^{-\alpha}(s+s_{\text{on}})^{\alpha} + c_{\text{off}}^{-\beta} s^{\beta} + c_{\text{on}}^{-\alpha} c_{\text{off}}^{-\beta}(s+s_{\text{on}})^{\alpha} s^{\beta}}. \tag{4.8}$$

When $\alpha = 1$ and $\beta = 1$, we have the result corresponding to exponential on- and off-time distributions:

$$\widetilde{p}_{\text{on}}(s_{\text{on}}|s) = \frac{1 + C + c_{\text{on}}^{-1}(s + \varepsilon_{\text{off}} s_{\text{on}})}{s + C(s+s_{\text{on}}) + c_{\text{on}}^{-1}(s+s_{\text{on}})s}$$

Let us rewrite Eq. (4.8) as

$$s^{\beta}\, \widetilde{p}_{\text{on}}(s_{\text{on}}|s) + C(s+s_{\text{on}})^{\alpha}\, \widetilde{p}_{\text{on}}(s_{\text{on}}|s) + c_{\text{on}}^{-\alpha}(s+s_{\text{on}})^{\alpha} s^{\beta}\, \widetilde{p}_{\text{on}}(s_{\text{on}}|s)$$

$$= C(s+s_{\text{on}})^{\alpha-1} + s^{\beta-1} + c_{\text{on}}^{-\alpha} s^{\beta}(s+s_{\text{on}})^{\alpha}\left[\varepsilon_{\text{on}}(s+s_{\text{on}})^{-1} + \varepsilon_{\text{off}} s^{-1}\right],$$

where $C = c_{\text{off}}^{\beta}/c_{\text{on}}^{\alpha}$. Taking the double inverse Laplace transform and using the identity

$$\frac{1}{(2\pi i)^2}\int_{\Gamma} ds \int_{\Gamma'} ds_{\text{on}}\; \exp(st + s_{\text{on}} t_{\text{on}})(s+s_{\text{on}})^{\alpha}\widetilde{p}_{\text{on}}(s_{\text{on}}|s)$$

$$= \frac{1}{2\pi i}\int_{\Gamma} ds\; \exp(st)\left[\exp(-st_{\text{on}})\; {}_0\mathsf{D}_{t_{\text{on}}}^{\alpha} \exp(st_{\text{on}})\widehat{p}_{\text{on}}(t_{\text{on}}|s)\right] \tag{4.9}$$

$$= \mathsf{T}_t^{-t_{\text{on}}}\; {}_0\mathsf{D}_{t_{\text{on}}}^{\alpha}\; \mathsf{T}_t^{t_{\text{on}}}\; p_{\text{on}}(t_{\text{on}}|t), \tag{4.10}$$

where T_t^{a} is the shift operator:

$$\mathsf{T}_t^{a}\; p_{\text{on}}(t_{\text{on}}|t) = p_{\text{on}}(t_{\text{on}}|t+a),$$

we obtain a fractional equation for the probability density of the total on-time:

$${}_0\mathsf{D}_t^{\beta}\; p_{\text{on}}(t_{\text{on}}|t) + C\; \mathsf{T}_t^{-t_{\text{on}}}{}_0\mathsf{D}_{t_{\text{on}}}^{\alpha}\mathsf{T}_t^{t_{\text{on}}}\; p_{\text{on}}(t_{\text{on}}|t) + c_{\text{on}}^{-\alpha}\; {}_0\mathsf{D}_t^{\beta}\; \mathsf{T}_t^{-t_{\text{on}}}{}_0\mathsf{D}_{t_{\text{on}}}^{\alpha}\mathsf{T}_t^{t_{\text{on}}}\; p_{\text{on}}(t_{\text{on}}|t)$$

$$= C\,\delta(t - t_{\text{on}})\frac{t^{-\alpha}}{\Gamma(1-\alpha)} + \delta(t_{\text{on}})\frac{t^{-\beta}}{\Gamma(1-\beta)} - \frac{c_{\text{on}}^{-\alpha} t_{\text{on}}^{-\alpha}(t-t_{\text{on}})^{-\beta}}{\Gamma(1-\alpha)\Gamma(1-\beta)}\left(\frac{\alpha\varepsilon_{\text{off}}}{t_{\text{on}}} + \frac{\beta\varepsilon_{\text{on}}}{t-t_{\text{on}}}\right).$$

Now, we assume that the blinking process starts with the off→on transition; i.e., we set $\varepsilon_{\text{on}} = 1$ and $\varepsilon_{\text{off}} = 0$. By the Tauberian theorem (e.g., see [Feller (1967)]) expression (4.8) in the long-time asymptotic regime ($t \gg c_{\text{on}}^{-1},\ c_{\text{off}}^{-1}$) reduces to

$$\widetilde{p}_{\text{on}}(s_{\text{on}}|s) \sim \frac{C(s+s_{\text{on}})^{\alpha-1}+s^{\beta-1}}{C(s+s_{\text{on}})^{\alpha}+s^{\beta}}, \qquad C = c_{\text{off}}^{\beta}/c_{\text{on}}^{\alpha}. \tag{4.11}$$

Rewriting this expression as

$$\left[C(s+s_{\text{on}})^{\alpha}+s^{\beta}\right]\widetilde{p}_{\text{on}}(s_{\text{on}}|s) = C(s+s_{\text{on}})^{\alpha-1}+s^{\beta-1}.$$

and taking the double inverse Laplace transform, we obtain an asymptotic fractional equation for the probability density function of the total on-time:

$${}_0\mathsf{D}_t^{\beta}\, p_{\text{on}}(t_{\text{on}}|t) + C\ \mathsf{T}_t^{-t_{\text{on}}}\,{}_0\mathsf{D}_{t_{\text{on}}}^{\alpha}\,\mathsf{T}_t^{+t_{\text{on}}}\ p_{\text{on}}(t_{\text{on}}|t)$$

$$= C\ \delta(t-t_{\text{on}})\frac{t^{-\alpha}}{\Gamma(1-\alpha)} + \delta(t_{\text{on}})\frac{t^{-\beta}}{\Gamma(1-\beta)}. \tag{4.12}$$

4.1.4.1 *Total fluorescence time distribution*

It is known (e.g., see [Sibatov & Uchaikin (2010)]), that the fraction

$$\frac{s^{\beta-1}}{Cs_{\text{on}}^{\alpha}+s^{\beta}}$$

is the double Laplace transform of a one-sided fractional stable distribution:

$$\mathsf{L}_s^{-1}\mathsf{L}_{s_{\text{on}}}^{-1}\left\{\frac{s^{\beta-1}}{Cs_{\text{on}}^{\alpha}+s^{\beta}}\right\} = (Ct^{\beta})^{-1/\alpha}\ q^{(\alpha,\beta)}\left(t_{\text{on}}(Ct^{\beta})^{-1/\alpha}\right),$$

where the fractional stable distribution $q^{(\alpha,\beta)}(t)$ is expressed in terms of one-sided stable distributions $g_+(t;\alpha)$ and $g_+(t;\beta)$ [Sibatov & Uchaikin (2010)]:

$$q^{(\alpha,\beta)}(t) = \int_0^{\infty} d\tau\ g_+(\tau;\beta)g_+(t\tau^{-\beta/\alpha};\alpha)\tau^{\beta/\alpha}.$$

Taking the inverse Laplace of (4.11)

$$p(t_{\text{on}}|t) = [C(t-t_{\text{on}})^{\beta}]^{-1/\alpha}\ q^{(\alpha,\beta)}\left(t_{\text{on}}[C(t-t_{\text{on}})^{\beta}]^{-1/\alpha}\right)$$

$$+(C^{-1}t_{\text{on}}^{\alpha})^{-1/\beta}\ q^{(\beta,\alpha)}\left((t-t_{\text{on}})(C^{-1}t_{\text{on}}^{\alpha})^{-1/\beta}\right) \tag{4.13}$$

and using the relation between fractional stable distributions,

$$Q^{(\alpha,\beta)}\left(\xi(C\tau^{\beta})^{-1/\alpha}\right) = 1 - Q^{(\beta,\alpha)}\left(\tau(C^{-1}\xi^{\alpha})^{-1/\beta}\right),$$

we write the solution as

$$p_{\text{on}}(t_{\text{on}}|t) = \left[1+\frac{\alpha}{\beta}\frac{(t-t_{\text{on}})}{t_{\text{on}}}\right](C^{-1}t_{\text{on}}^{\alpha})^{-1/\beta}\ q^{(\beta,\alpha)}\left((t-t_{\text{on}})(C^{-1}t_{\text{on}}^{\alpha})^{-1/\beta}\right). \tag{4.14}$$

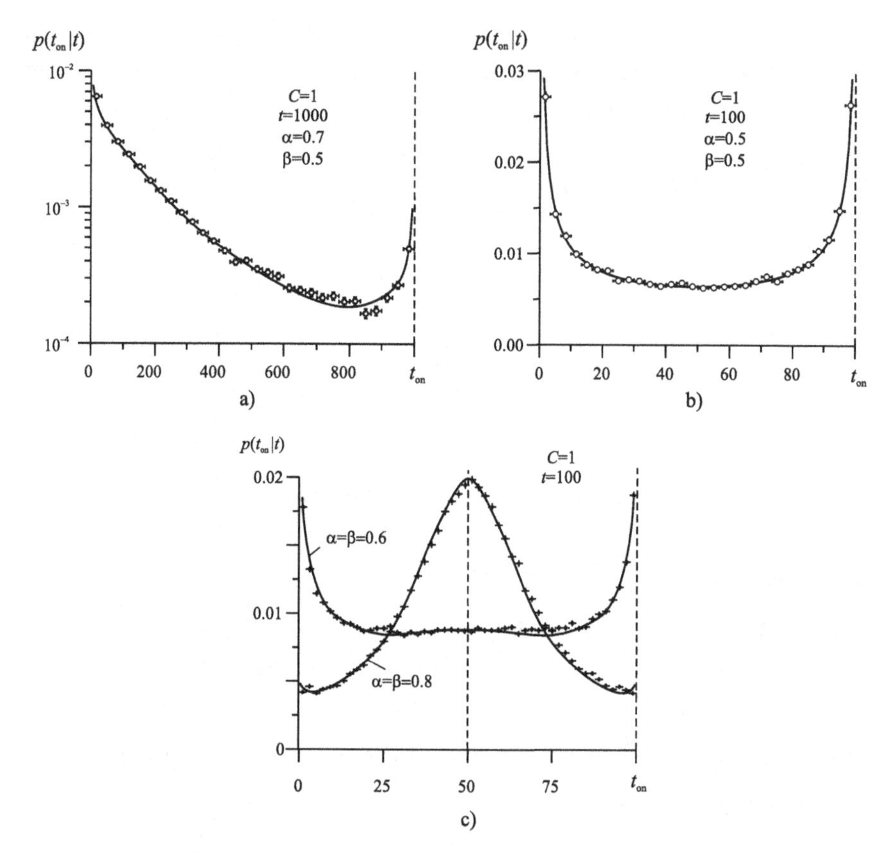

Figure 4.3 Comparison of the analytical on-time distributions with the results of Monte-Carlo simulation.

In the special case of $\alpha = \beta$, the property of fractional stable distributions formulated in Section A.8 of the Appendix is invoked to find Lampertis PDF [Lamperti (1958)], which was used to describe non-ergodic processes in [Bel & Barkai (2005)]

$$p(t_{\text{on}}|t) = \frac{\sin \pi\alpha}{\pi} \cdot \frac{t\; t_{\text{on}}^{\alpha-1}(t - t_{\text{on}})^{\alpha-1}}{C(t - t_{\text{on}})^{2\alpha} + C^{-1}t_{\text{on}}^{2\alpha} + 2t_{\text{on}}^{\alpha}(t - t_{\text{on}})^{\alpha}\; \cos \pi\alpha}.$$

The probability of counting n fluorescence photons over a time interval t is

$$P(n,t) = \int_0^t dt_{\text{on}}\; P(n, t_{\text{on}})p_{\text{on}}(t_{\text{on}}|t) \tag{4.15}$$

In the model discussed in [Jung et al. (2002)]: $P(n, t_{\text{on}}) \propto t_{\text{on}}$.

When the on-time has a finite expectation, distribution (4.14) reduces to the following ($\alpha \to 1,\; g_+(t;\alpha) \to \delta(t-1)$):

$$p_{\text{on}}(t_{\text{on}}|t) = \left(1 + \frac{t - t_{\text{on}}}{\beta t_{\text{on}}}\right)(Ct_{\text{on}})^{-1/\beta}\; g_+\left((t - t_{\text{on}})(Ct_{\text{on}})^{-1/\beta}; \beta\right). \tag{4.16}$$

As $\beta \to 1$, we have

$$p_{\rm on}(t_{\rm on}|t) = \left[1 + \frac{t_{\rm on}}{\alpha(t - t_{\rm on})}\right] \left(\frac{t - t_{\rm on}}{C}\right)^{-1/\alpha} g_+ \left(t_{\rm on}\left(\frac{t - t_{\rm on}}{C}\right)^{-1/\alpha}; \alpha\right). \quad (4.17)$$

These analytical solutions are in a good agreement with the results of Monte Carlo simulations of the process (see Fig. 4.3). In each simulation, starting from an on→off or off→on transition with probability 1/2, on- and off-times are sampled by the algorithm for generating fractional Poisson processes proposed in [Uchaikin et al. (2008b)]. Each sample path has been computed until the total duration of on and off-times exceeds the observation time. The total on-time is calculated for each path. After an ensemble of paths of the two-state renewal process has been simulated, the probability density function of the total on-time is calculated.

Recall that Eq. (4.11) is valid in the long-time asymptotic regime $t \gg c_{\rm on}^{-1}, c_{\rm off}^{-1}$). A random-walk representation of the process (Fig. 4.2), suggests that it can be interpreted as a competition between diffusive and ballistic regimes [Sibatov & Uchaikin (2010)]. When $\alpha = 1$, $\beta < 1$, we have a subdiffusive regime. When $\beta = 1$, $\alpha < 1$ superdiffusive behavior (mostly ballistic regime) is observed. When $\alpha < 1$, $\beta < 1$ the probability density function of the total on-time exhibits singular behavior at the endpoints of the interval.

4.1.5 *Photon counts distribution*

Suppose that the photon emission from a quantum dot is a Poisson process. Then, the probability of counting n photons emitted by the nanocrystal over a time interval t is

$$P(n,t) = \int_0^t \frac{(\mu t_{\rm on})^n}{n!} \exp(-\mu t_{\rm on}) p_{\rm on}(t_{\rm on}|t) dt_{\rm on}.$$

This expression is the Poisson transform of the on-time probability density function. Mandel's parameter is

$$Q = \frac{\langle n^2 \rangle - \langle n \rangle^2}{\langle n \rangle} - 1 = \frac{\sum\limits_{n=1}^{\infty} n^2 P(n,t) - \left(\sum\limits_{n=1}^{\infty} nP(n,t)\right)^2}{\sum\limits_{n=1}^{\infty} nP(n,t)} - 1 = \mu \frac{\langle t_{\rm on}^2 \rangle - \langle t_{\rm on} \rangle^2}{\langle t_{\rm on} \rangle}. \quad (4.18)$$

To determine the first and second moments $m_1(t) = \langle t_{\rm on} \rangle$ and $m_2(t) = \langle t_{\rm on}^2 \rangle$ of the total on-time conditioned on the observation time of the total on-time conditioned on the observation time t, we calculate the first and second derivatives of expression (4.11) for the Laplace transform of $p(t_{\rm on}|t)$,

$$\widehat{m}_1(s) = -\left.\frac{\partial p(s_{\rm on}|s)}{\partial s_{\rm on}}\right|_{s_{\rm on}=0}, \qquad \widehat{m}_2(s) = \left.\frac{\partial^2 p(s_{\rm on}|s)}{\partial s_{\rm on}^2}\right|_{s_{\rm on}=0}.$$

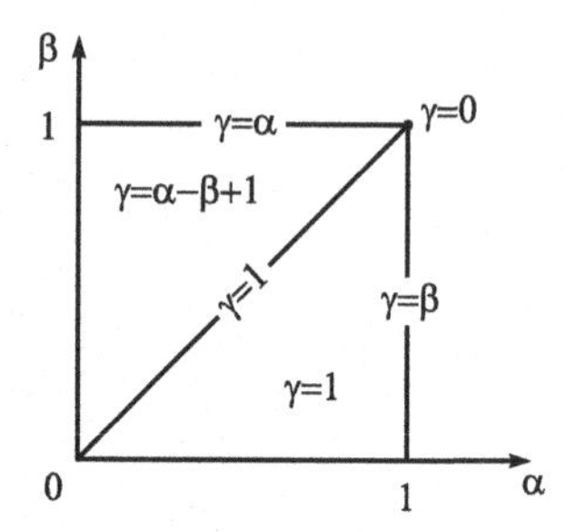

Figure 4.4 The diagram for parameter γ depending on α and β values.

The resulting expressions

$$\widehat{m}_1(s) = \frac{C^2 s^{2\alpha-1} + C s^{\alpha+\beta-1}}{s(Cs^\alpha + s^\beta)^2},$$

$$\widehat{m}_2(s) = \frac{2C(1-\alpha)s^{\alpha+2\beta-1} + 2C^3 s^{3\alpha-1} + 2(2-\alpha)C^2 s^{2\alpha+\beta-1}}{s^2(Cs^\alpha + s^\beta)^3},$$

are used to analyze the long-time asymptotic behavior of Mandel's Q parameter. By the Tauberian theorem, the Laplace-domain behavior at small s determines the long-time behavior of the original time-domain function (see [Feller (1967)] for details).

In case $\alpha = \beta$, we have

$$\widehat{m}_1(s) = \frac{C}{C+1} s^{-2}, \quad \widehat{m}_2(s) = \frac{2(1-\alpha+C)C}{(C+1)^2} s^{-3},$$

$$m_1(t) = \langle t_{\text{on}} \rangle = \frac{C}{C+1} t, \quad m_2(t) = \langle t_{\text{on}}^2 \rangle = \frac{C(1-\alpha)+C^2}{(C+1)^2} t^2,$$

$$M(t) = \frac{1-\alpha}{1+C}\,\mu t. \tag{4.19}$$

When $\beta < \alpha \neq 1$,

$$M(t) \sim \frac{2(1-\alpha)}{2+\beta-\alpha}\mu t, \quad t \gg c_{\text{on}}^{-1},\ c_{\text{off}}^{-1}, \tag{4.20}$$

$\beta < \alpha = 1$

$$M(t) \sim C\left[\frac{\Gamma(\beta)}{\Gamma(2\beta)} - \frac{1}{\Gamma(1+\beta)}\right]\mu t^\beta, \quad t \gg c_{\text{on}}^{-1},\ c_{\text{off}}^{-1}. \tag{4.21}$$

When $\beta > \alpha$

$$M(t) \sim \frac{2C^{-1}(2-\alpha)}{\Gamma(\alpha-\beta+3)}\,\mu t^{\alpha-\beta+1}, \quad t \gg c_{\text{on}}^{-1},\ c_{\text{off}}^{-1}. \tag{4.22}$$

The analysis presented in [Jung et al. (2002)] shows that Mandel's M parameter increases linearly with time when $\alpha = \beta < 1$. According to (4.19)–(4.22), the asymptotic time dependence of Mandel's Q parameter follows a power law:

$$M(t) \propto t^\gamma, \qquad t \gg c_{\text{on}}^{-1},\ c_{\text{off}}^{-1}, \tag{4.23}$$

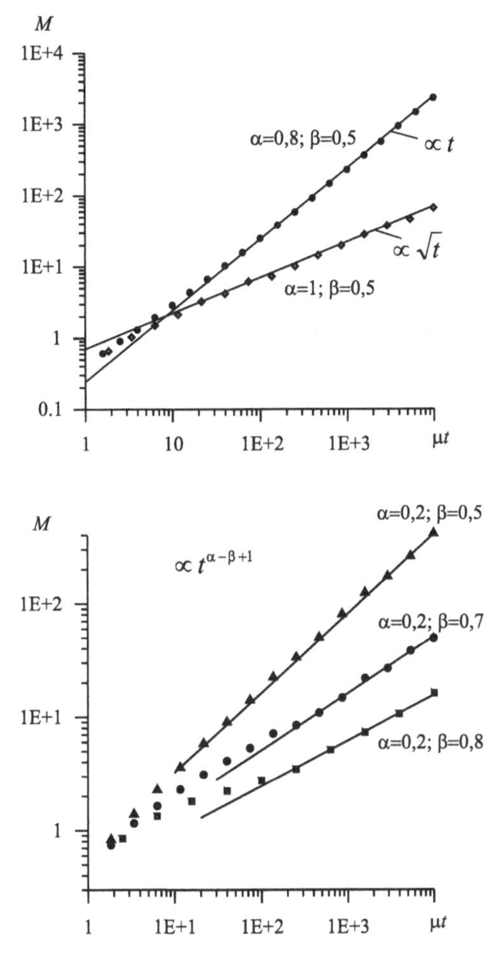

Figure 4.5 Time dependencies of the Mandel parameter in case $\alpha > \beta$ (upper panel) and $\alpha < \beta$ (lower panel).

where γ is a function of α and β. It is clear from Fig. 4.4 that Mandel's M parameter increases with time unless $\alpha = \beta = 1$ its time dependence is sublinear. When $\beta > \alpha$ and $\alpha = 1$ its time dependence is sublinear. When $\beta \leq \alpha \neq 1$, Mandel's Q parameter increases linearly with time in the long-time limit. Power-law dependence (4.23) is supported by Monte Carlo simulations (Fig. 4.5).

Fluctuations of the total on-time are quantified by the relative variance

$$\delta_{\text{on}}^2 = \frac{\langle T_{\text{on}}^2 \rangle - \langle T_{\text{on}} \rangle^2}{\langle T_{\text{on}} \rangle^2} \propto \begin{cases} t^{\alpha-\beta}, \ \alpha < 1, \ \beta \leq 1, \\ t^0, \quad \beta < \alpha = 1, \\ t^{-1}, \quad \alpha = \beta = 1. \end{cases} \quad t \gg c_{\text{on}}^{-1}.$$

One can see that δ_{on}^2 increases with time when $\beta < \alpha < 1$, approaches a constant value when $\alpha = \beta < 1$ or $\beta < \alpha = 1$, and decays inversely as time when $\alpha = \beta = 1$.

A remarkable result is observed for $\beta > \alpha$: on-time fluctuations decay, but at a rate slower than in the case of $\alpha = \beta = 1$.

4.2 Charge kinetics in colloidal quantum dot arrays

4.2.1 *Fractional currents in colloidal quantum dot array*

Researchers appealing to a discrete electron spectrum of quantum dots (QDs) call them an "artificial atoms". An array of identical semiconductor QDs can be considered as artificial solid. Fundamental conceptions of solid state physics can be studied on the base of such systems. Understanding of charge and spin transport processes in QD arrays could lead to applications in spintronics and quantum computation. Despite the sufficient progress in synthesis, description of charge transport in QD arrays is not quite satisfactory [Novikov (2003); Novikov et al. (2005); Morgan et al. (2002)].

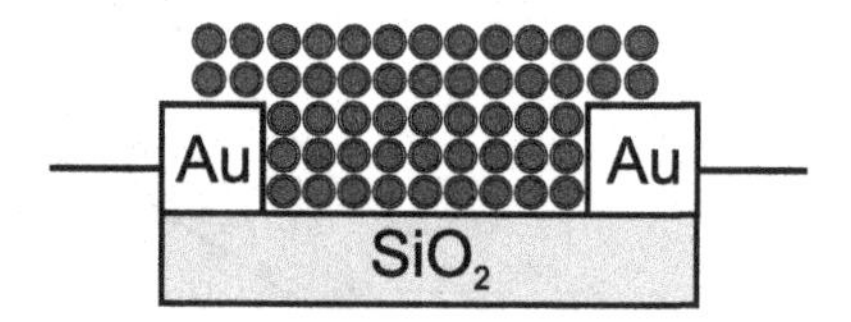

Figure 4.6 A schematic representation of a nanocrystal array located on a silicon oxide substrate between two gold electrodes.

In many samples of colloidal *QD arrays* (in the lateral geometry, Fig. 4.6), power law current decay

$$I(t) \propto t^{-\alpha}, \quad 0 < \alpha < 1, \tag{4.24}$$

is observed after applying of a large constant voltage $V(t) = V_0\mathbf{1}(t)$ [Morgan et al. (2002); Novikov (2003)]. The exponent α is less than 1 and in the general case its value depends on nanocrystal size and temperature. Novikov et al. [Novikov et al. (2005)] assert that (4.24) is not a bias current, it is a true current from source to drain due to the integral of Eq. (4.24) is charge and it tends to infinity

$$Q = \int_0^\infty I(t)dt \to \infty.$$

The observed non-exponential relaxation of current can be explained by time dependence of the state of the system. Ginger & Greenham (2000) suggested that the charge flow decreases due to suppression of injection from the contact. This suppression arises because electrons trapped in a nanocrystal prevent transport of other electrons through this QD, flow is jammed. Morgan et al. [Morgan et al. (2002)] explain power law decay of current $I(t)$ assuming that non-equilibrium electrons are distributed over QD array as the Coulomb glass. Novikov et al. [Novikov et al.

(2005)] proposed the model based on a stationary random process as authors assert. An array consists of $N \gg 1$ identical independent channels operating in the parallel regime. Each channel opens in random time moments and conducts a current pulse. These channels are completely characterized by the distribution of waiting times T between successful pulses. Authors [Novikov et al. (2005)] postulated that this distribution has a heavy tail of the power law kind

$$\Psi(t) = \text{Prob}(T > t) \propto t^{-\nu}, \qquad 0 < \nu < 1, \qquad t \to \infty. \tag{4.25}$$

The mean value of such random variable diverges and this fact provides specific statistical properties of the process. In particular, memory effects arise.

The model [Novikov et al. (2005)] satisfactorily explains power law current transients and power law noise spectrum but it does not reveal the physical mechanisms of the process, they postulate the distribution (4.25). As the authors assert, the base of their model is the stationary stochastic process and this stationarity contradicts to the time dependence of the state of the system, in particular, to the hypothesis about injection blocking from the contact that must occur due to the charge balance conditions. The main idea proposed in Refs. [Novikov (2003); Novikov et al. (2005)] for explanation of power law current transients concludes in the assumption that time intervals between successful current pulses in conduction channels are independent random variables with distribution having heavy power law tails. Nevertheless, some questions remain unanswered. What is the nature of these channels, why does the distribution between successful pulses has power law asymptotics, why are current pulses discrete and identical in values? Furthermore, if memory of the process is explained in frameworks of the hidden variable conception [Uchaikin (2008)], how can this approach be combined with stationarity of the process, or may be is it non-stationary? In [Uchaikin & Sibatov (2009); Sibatov (2011b)], we tried to find the answers to some of them.

Further, a new statistical model that describes power law relaxation of current and memory phenomena is proposed. We call it the nonlinear modification of the Scher-Montroll model (the NSM-model). We shall not obtain transport equation for this process, we shall investigate it numerically. It is shown that the basic random process is non-stationary. The NSM-model leads on the one hand to the idea of charge injection blocking, and it conforms to Novikov's model on the other.

4.2.2 *Modification of the Scher-Montroll model*

The conduction between isolated nanoparticles can be treated as a single-electron process in the framework of the perturbation theory [Suvakov & Tadić (2010)]. The Hamiltonian of the system, $H = H_1 + H_2 + H_{12}$, the spectrum of eigenvalues and eigenfunctions vector operators H_1 and H_2 are known: $H_1\Psi_1 = E_1\Psi_1$, $H_2\Psi_2 = E_2\Psi_2$. In second quantization terms, with the involving creation and annihilation operators $H = \sum_{\mathbf{k}_1} E_{\mathbf{k}_1} c^\dagger_{\mathbf{k}_1} c_{\mathbf{k}_1} + \sum_{\mathbf{k}_2} E_{\mathbf{k}_2} c^\dagger_{\mathbf{k}_2} c_{\mathbf{k}_2} + \sum_{\mathbf{k}_1,\mathbf{k}_2} T_{\mathbf{k}_1,\mathbf{k}_2} c^\dagger_{\mathbf{k}_2} c_{\mathbf{k}_1}$ $+ \sum_{\mathbf{k}_1,\mathbf{k}_2} T_{\mathbf{k}_2,\mathbf{k}_1} c^\dagger_{\mathbf{k}_1} c_{\mathbf{k}_2}$, at nonzero temperatures mean values of occupation numbers

$N_{\mathbf{k}_1} = c^\dagger_{\mathbf{k}_1} c_{\mathbf{k}_1}$, $N_{\mathbf{k}_2} = c^\dagger_{\mathbf{k}_2} c_{\mathbf{k}_2}$ are given by Fermi-Dirac distribution. Particularly, for nanoparticle 1: $f(E_{\mathbf{k}_1}) = [1 + \exp(E_{\mathbf{k}_1} - E_F^1)/kT]^{-1}$. The tunneling rate $\Gamma^+(V) = 2\pi\hbar^{-1}\sum_{\mathbf{k}_1,\mathbf{k}_2} |T_{\mathbf{k}_1,\mathbf{k}_2}|^2 f(E_1)(1 - f(E_2))\delta(E_1 - E_2)$, where $T_{\mathbf{k}_1,\mathbf{k}_2} = \langle \mathbf{k}_2|H_{12}|\mathbf{k}_1\rangle$ is a matrix element of tunneling Hamiltonian.

The Coulomb interaction leads to collective effects in the charge distribution over the array. Long-range interaction should manifest itself in the case of relatively small values of the external voltage U. To study the transport in arrays without long-range correlations, voltage U must be sufficiently large. In some experiments [Morgan et al. (2002); Drndic et al. (2002)], the voltage of about 100 V was applied to samples, which corresponds to the energy transfer between neighboring QDs of the order of tenths of eV, which is the value of the same order as the separating potential barrier between neighboring nanocrystals and Coulomb interaction energy of the electrons localized in neighboring QDs [Novikov (2003)].

In second quantization terms, the system state is characterized by the number of electrons in each QD $n \equiv \{n_1, n_2, \ldots, n_N\}$. The change in energy of the entire system after a single act of tunneling from i-th QD into the j-th one has the following form

$$\Delta E_{i\to j} = E\{n_1, ..., n_i - 1, ..., n_j + 1, ..., n_N\} - E\{n_1, ..., n_i, ..., n_j, ..., n_N\}.$$

For the tunneling rate, [Suvakov & Tadić (2010)] has obtained the following

$$\Gamma_{i\to j}(V) = \frac{\Delta E_{i\to j}}{e^2 R_{i\to j}[1 - \exp(-\Delta E_{i\to j}/kT)]}, \tag{4.26}$$

where $R_{i\to j}$ is the local tunneling resistance. Matrix $R_{i\to j}$ depends on material, size and shape of nanoparticles, the temperature of the external magnetic field, the distance between the nanoparticles and the energy of separating barrier. Eq. (4.26) shows that for sufficiently high temperatures, the tunneling rate is inversely proportional to $R_{i\to j}$, and the Coulomb blockade has a little effect on transfer characteristics. In our case, it must be taken into account. The distribution of the heights of the tunneling barrier is assumed to be Gaussian:

$$\rho(\mathcal{E}) = (2\pi\sigma_{\mathcal{E}})^{-1/2}\exp[-(\mathcal{E} - \bar{\mathcal{E}})^2/2\sigma_{\mathcal{E}}^2].$$

Monte Carlo simulation has shown that there are conductive channels associated with the percolation paths, and the time intervals between successive pulses in one channel τ is asymptotically distributed according to a power law: $\mathrm{Prob}(\tau > t) \propto t^{-\nu-1}$,

$$\nu \approx \frac{\left\langle\sqrt{\mathcal{E}}\right\rangle}{\gamma l\langle \mathcal{E} - \langle\sqrt{\mathcal{E}}\rangle^2\rangle}.$$

To answer the questions listed above we use a modification of the Scher-Montroll model. The classical version of this model explains successfully mean features of dispersive transport in disordered semiconductors [Scher & Montroll (1975)]. In Ref. [Novikov et al. (2005)], the authors provide arguments that the standard Scher-Montroll model does not describe power law decay of current in QD arrays. This model predicts unlimited accumulation of charge in a sample if an injection rate from the contact is constant. Injection blocking takes place in the modified model taking into account the Coulomb blockade effect.

Coulomb interaction is long-range and this leads to collective phenomena of charge distribution over a sample [Novikov (2003)]. These effects become apparent in the case of small values of voltage. To study transport in arrays without taking into account the long-range character of Coulomb interaction one has to apply large voltages. In the experiments described in Ref. [Morgan et al. (2002); Drndic et al. (2002)], values of voltage between source and drain were large (of the order 100 V) and they correspond to several hundred meV between neighboring QDs that is of the order of the interdot Coulomb energy and the nanocrystal charging energy.

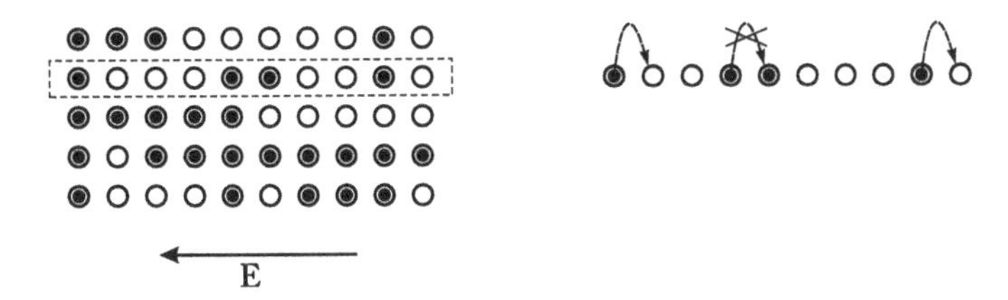

Figure 4.7 The simplest presentation of a quantum dot array as a set of nanocrystal rows. Electrons move in one direction.

In framework of the NSM-model a QD array is represented as two or three dimensional lattice, QDs are situated in nodes of this lattice. In strong electric fields, the latter can be considered as a set of parallel one-dimensional nanocrystal rows (conduction channels). Electrons perform one-sided random walk in the direction opposite to the applied field. The proposed model is qualitative and reflects main statistical properties of the process without long-range correlations. Nevertheless it allows us to interpret the power law decay of current, the presence of memory in nanocrystal arrays, to substantiate charge injection blocking, and it is agree with Novikov's phenomenological model.

In the NSM-model, channels are associated with one-dimensional nanocrystal rows in ordered array. Let us show that if sojourn times in QDs are distributed according to the asymptotic power law with the exponent $0 < \nu < 1$, then time intervals between successful electron jumps from array to drain in one row have the same power law asymptotics in distribution.

Tunnelling from one nanocrystal to another, electrons follow each other. Coulomb repulsion between electrons allows no multiple occupancy of nanocrystals. Let at the moment t_j, j-th electron of some channel has jumped from array to drain. Let us find a distribution of the time interval $\theta = t_{j+1} - t_j$ between exits of this electron (j) and the next one ($j+1$) in the channel.

The next carrier ($j+1$) can be trapped in any nanocrystal of the channel except the last QD. Let p_n are probabilities to occupy the n-th QD at the moment

t_j, where n is a number of nanocrystal in the channel. For the times $t > t_j$, in front of the $(j+1)$-th carrier there are no non-equilibrium electrons trapped in the channel. In other words the carrier will not be influenced by Coulomb repulsion from the side of electrons going ahead. Random walk of the carrier will not be blocked. The exit time of the $(j+1)$-th electron counted since the moment t_j is summed up of sojourn times in nanocrystals which the carrier has to visit before leaving the array,

$$T = \tau_n' + \sum_{k=n+1}^{N} \tau_k. \tag{4.27}$$

Here τ_k is a sojourn time in the k-th QD. The stroke of the time τ_n' signifies that the $(j+1)$-th electron has spent part of its waiting time in the n-th QD till the moment of exit of the j-th carrier.

It is known, that if two random variables having distributions with identical power law asymptotics are summed up, then the distribution of a resultant variable has asymptotics of the same order. Indeed, the asymptotic form $(\lambda \to 0)$ of the Laplace transformation of a PDF with heavy power law tail is as follows,

$$\widehat{\psi}(\lambda) = \int_0^\infty e^{-\lambda t}\psi(t)\,dt \sim 1 - (\lambda/c)^\mu, \quad 0 < \alpha < 1, \quad \lambda \to 0,$$

where c is a scale constant. The distribution of the sum of two random variables is expressed through the convolution of their distributions. The Laplace transformation of the convolution of two functions is a product of their Laplace images. For PDFs with identical power law asymptotics, we have

$$\left(1 - \frac{\lambda^\mu}{c_1^\mu}\right)\left(1 - \frac{\lambda^\mu}{c_2^\mu}\right) = 1 - \left(\frac{\lambda}{b}\right)^\mu + \frac{\lambda^{2\mu}}{(c_1 c_2)^\mu} \sim 1 - (\lambda/b)^\mu,$$

$$\lambda \to 0, \quad b = \left(\frac{1}{c_1^\mu} + \frac{1}{c_2^\mu}\right)^{-1/\mu}.$$

If distributions of sojourn times τ_k and τ_n' are asymptotical power laws with the exponent $0 < \nu < 1$, the random variable (4.27) has a distribution with asymptotics of the same order. The time $\tau_n' = \tau_n - \theta$, where θ is the exit time of the j-th electron counted since the moment of trapping of the $(j+1)$-th electron into n-th nanocrystal. The random time θ has some PDF $p_\theta(t)$. Then

$$\mathcal{P}\nabla\wr\lfloor(\tau_n' > t) = \int_0^\infty \mathcal{P}\nabla\wr\lfloor(\tau_n > t + t')\,p_\theta(t')\,dt' \sim$$

$$\sim \frac{c^{-\nu}}{\Gamma(1-\nu)} \int_0^\infty (t+t')^{-\nu} p_T(t')\,dt' \sim \frac{(ct)^{-\nu}}{\Gamma(1-\nu)}, \quad t \to \infty.$$

Thus, the hypothesis about sojourn times distributed according to asymptotical power law conforms to Novikov's model assuming power law distributions of intervals between successful current pulses in conduction channels.

4.2.3 *Current decay in the modified model*

In the framework of the renewal model, the distribution of pulse number in some channel is given by

Distribution of number of pulses in some channel is as follows,

$$p_n = \mathrm{Prob}(N(t) = n) = \mathrm{Prob}(T_{n+1} > t) - \mathrm{Prob}(T_n > t),$$

where T_n is the time of the n-th pulse. According to the Generalized Limit Theorem (for more details, see [Uchaikin & Zolotarev (1999)]),

$$\mathrm{Prob}(T_n < t) \sim G_+(cn^{-1/\nu}t;\nu), \qquad t \to \infty.$$

Here $G_+(t;\nu)$ is a distribution function of stable random variables. Thus, we have

$$p_n \sim G_+(cn^{-1/\nu}t;\nu) - G_+(c(n+1)^{-1/\nu}t;\nu) \sim \nu^{-1}n^{-1-1/\nu}ct\; g_+(cn^{-1/\nu}t;\nu),$$

where $g_+(t;\nu)$ is the stable density. The current is determined by the expression

$$i(t) = \frac{d\langle Q\rangle}{dt} = eZ\frac{d}{dt}\sum np_n \sim eZ\frac{d}{dt}\left[\nu^{-1}(ct)^{\nu}\int\limits_0^\infty \xi^{-1/\nu}g_+(\xi^{-1/\nu};\nu)d\xi\right] =$$

$$= eZ\nu c(ct)^{\nu-1}\int\limits_0^\infty s^{-\nu}g_+(s;\nu)ds = \frac{eZc^{\nu}}{\Gamma(\nu)}\,t^{\nu-1}, \qquad t \gg c^{-1}, \quad 0<\nu<1,$$

where Z is the number of channels.

Thus, the exponent α of a power-law decay of current is linked to the NSM-model parameter ν by the relation $\alpha = 1-\nu$. The results of Monte Carlo simulation confirming analytical calculations are presented in Sec. 5.

4.2.4 *Interdot disorder*

Different physical mechanisms may lead to the asymptotic power-law distribution of waiting times (see [Sibatov & Uchaikin (2009)] and references therein). The most popular of them are related to disorder of the medium. Due to the disordered structure of interdot space, the energetic disorder always exists in colloidal QD arrays even in the case of ideal arrangement of nanocrystals in the coordinate space. Tunneling probabilities from one nanocrystal to another are determined by the height and weight of the dividing energy barrier.

The random waiting time T in QD is characterized by the probability

$$\mathrm{Prob}\{T > t\} = \exp(-t/\theta), \tag{4.28}$$

where the parameter θ represents the mean waiting time in this nanocrystal if the next one is empty. According to the Bethe-Sommerfeld formula quasi-classical formula for tunneling [Tunaley (1972a)]:

$$\theta = \beta[\exp(\gamma d\sqrt{W}) - 1], \tag{4.29}$$

where d is the distance between neighboring lattice points, W is the work function of electron transfer from one QD to another, and the parameter θ is inversely proportional to the electric field intensity. As we see, the parameter θ depends exponentially on the width d and height W of the dividing barrier, which have dispersion due to disorder. Fluctuations in $d\sqrt{W}$ produce a wide dispersion of θ. Following [Sibatov & Uchaikin (2009)], we choose the gamma density to model the distribution of the quantity $Y = d\sqrt{W}$. After averaging over Y, the pdf of θ has the form of the asymptotic power-law dependence multiplied by a slowly varying function

$$p_\theta(t) \propto \left(\ln \frac{t}{\beta}\right)^{-1+\frac{\langle d\sqrt{W}\rangle^2}{\mathrm{D}[d\sqrt{W}]}} \left(\frac{t}{\beta}\right)^{-1-\frac{\langle d\sqrt{W}\rangle}{\gamma \mathrm{D}[d\sqrt{W}]}}.$$

Thus the power-law asymptotics is characterized by the parameter

$$\nu = \frac{\left\langle d\sqrt{W}\right\rangle}{\gamma\sigma^2[d\sqrt{W}]},$$

where $\sigma^2[d\sqrt{W}]$ is the variance of the quantity $d\sqrt{W}$. As shown in [Sibatov & Uchaikin (2009)], the waiting time distribution has the same power law asymptotics. The mean waiting time diverges in the case $\nu < 1$, in other words when spread of the quantity y is large enough, $\sigma^2[d\sqrt{W}] > \gamma^{-1}\left\langle d\sqrt{W}\right\rangle$.

4.2.5 *Monte Carlo simulation*

Recall that we consider only one-dimensional motion of carriers performing one-sided jumps between neighboring nods of a lattice at random time intervals. Let $i = 1, 2, ...N$ be the lattice point numbers. According to the reasonings of the previous section, the simulation scheme can be realized in the following way. The set of random jump rates μ_j is generated for all electrons trapped in QDs of the array. The probability of jump during a small time dt is determined by the product $\mu_j dt$. In addition, the set of random variables U_j uniformly distributed in the interval (0,1) is generated. If the relation $U_j < \mu_j dt$ is satisfied and the next QD is empty, the j-th electron jumps from the i-th QD to the $(i+1)$-th one and then new jump rate is generated for this electron. If the relation is not satisfied or the next QD contains trapped electron, then the electron stays put.

The jump rates μ_j must be distributed with the following PDF

$$\rho(\mu_j) = \frac{\nu}{\mu_j}\left(\frac{\mu_j}{\mu_{\max}}\right)^{\nu}, \qquad 0 < \mu_j < \mu_{\max}.$$

Indeed, the sojourn time in a chosen QD before jump is distributed according to the exponential law

$$\mathrm{Prob}(T_j > t) = \exp(-\mu_j t),$$

and after averaging over the QD ensemble, we obtain the distribution of waiting times with power law tails

$$\mathrm{Prob}(\tau > t) = \langle \exp(-\mu_j t) \rangle$$

$$= \int_0^{\mu_{\max}} \rho(\mu) \exp(-\mu t) d\mu \sim \Gamma(\nu + 1)(\mu_{\max} t)^{-\nu}, \quad t \to \infty.$$

When electron performs a jump into the drain, the current pulse is registered. Observed current is averaged over all channels of the nanocrystal array. The current calculated in such way is presented in Fig. 4.8, a in reduced coordinates. It decays according to the power law with the exponent $\alpha = 1 - \nu$. Numeric calculations of the distribution of waiting times T between successful current pulses (Fig. 4.8, b) confirm the analytical results.

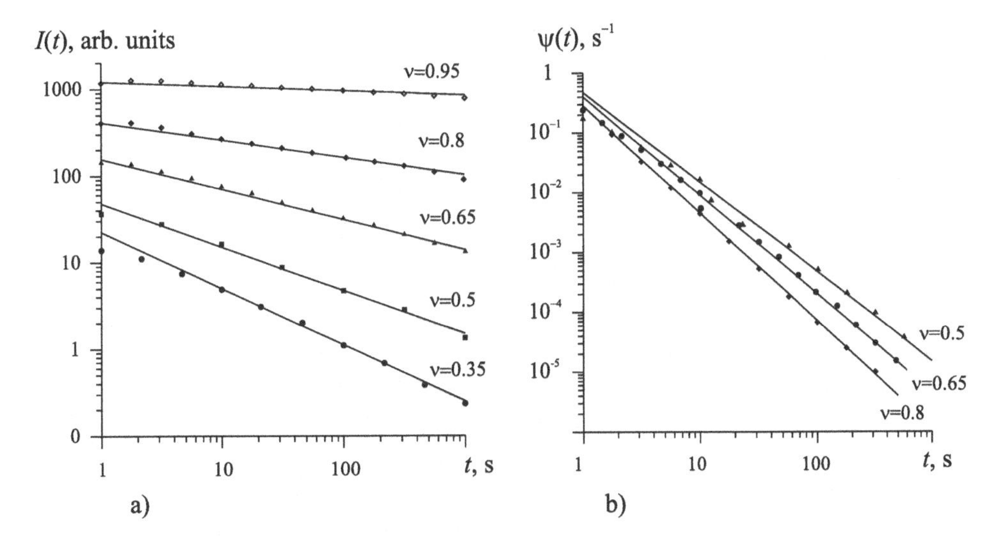

Figure 4.8 (a) The simulated decay of current (points), the lines are power law dependencies with exponents $-\alpha = -(1 - \nu)$. (b) The pdfs of waiting times between successful current pulses in a channel. Slopes correspond to exponents $-1 - \nu$. The points are the result of Monte Carlo simulation.

The NSM-model considered here neglects long-range correlations caused by the Coulomb interaction, and takes into account only the correlations due to the Coulomb blockade effect forbidding trapping of more than one non-equilibrium electron by a QD. This model is in essence a nonlinear modification of the Scher-Montroll model. In the new model, electron trajectories are correlated. Power-law asymptotics in the waiting time distribution is a consequence of the fluctuations in interdot energy barriers related to energetic disorder of interdot space. The dispersion of jump rates increases if the arrangement of nanocrystals in the matrix is not ordered.

The fractional current-voltage relation (the fractional Ohm law) can be obtained phenomenologically (see the next chapter). Due to nonlocality of the fractional derivative operator, the relation describes a process with power-law memory. Due to their small sizes, QD arrays are attractive as memory elements.

Although the NSM-model does not take into account some particular qualities of the process; for example, long-range correlation, it is in agreement with experimental data obtained at suitable voltages ($\sim$ 100 V) and with phenomenological models proposed earlier. Thereby, it allows us to solve the contradictions indicated in [Novikov (2003); Novikov et al. (2005)]. The model reflects two important aspects of the process: energetic disorder of interdot space; and forbiddance of multiple occupancy of QDs due to the Coulomb blockade. The agreement with experiments and other models indicates the adequacy of these two positions at large enough voltage values. At the same time, including effect of long-range Coulomb interactions into the model can be made, at least, via numerical simulation.

4.3 Conductance through fractal quantum conductors

Conductance distribution is usually studied in three different transport regimes: metallic ($\xi \gg L$), insulating ($\xi \ll L$) and crossover ($\xi \sim L$). Here, ξ is localization length and L typical size of the system. Most detailed theoretical results have been obtained for one-dimensional and quasi one-dimensional disordered systems in frames of two approaches to description of phase-coherent conduction and localization in disordered wires: the field theory [Efetov & Larkin (1983); Efetov (1983)], and the transfer matrix approach (see for review [Beenakker (1997)]). Frahm (1995) and Brouwer (1996) have demonstrated the equivalence of these two theories in case of large number of transverse modes $N \gg 1$ in all symmetry classes.

In the transfer matrix theory, the electronic transport is formulated as a scattering problem. The zero-temperature conductance g is related to the quantum-mechanical transmission matrix t through the Landauer formula. [Dorokhov (1982)] and [Mello et al. (1988)] have derived an equation known as the DMPK-equation for the joint probability distribution of the transmission eigenvalues in the weak localization regime. [Muttalib & Klauder (1999)] have proposed the generalized DMPK-equation to describe the electron transport in strongly disordered systems. These equations of the Fokker-Planck type in both cases (weak and strong localization) have been derived on the assumption of regular space distribution of scatterers. However, experimental works [Kohno & Yoshida (2004); Kohno (2008); Barthelemy et al. (2008); Hegger et al. (1996)] and numerical simulations [Leadbeater et al. (1998)] have shown that disorder in mesoscopic systems can be of fractal (self-similar) type. It is obvious that the standard DMPK equation approach and its multidimentional generalizations are not applicable in this case.

[Kohno & Yoshida (2004)] have demonstrated the scaling behavior of diameter modulations in semiconductor nanowires SiC grown via a self-organized process.

To reveal the property of digitized diameter fluctuation, the probability distribution function of the fluctuation displacements has been plotted in [Kohno (2008)] and it has been shown that the tail of this function coincides with the Lévy stable distribution function. Another example of fractal disorder in nanostructures is presented in [Barthelemy et al. (2008)]. The authors have demonstrated that it is possible to engineer an optical material in which light waves perform a Lévy flight characterized by path length distributions of power law type. This system has been called "Lévy glass".

4.3.1 *Weak localization (scattering)*

In this subsection, we study the conductance through quasi-one-dimensional quantum wires with fractal disorder characterized by a heavy-tailed distribution of distances between scatterers. Such type of disorder in quantum wires has already been considered in [Beenakker et al. (2009); Falceto & Gopar (2010); Burioni et al. (2010)]. We shall derive a new family of universal conductance distributions that are related to Lévy stable statistics. These distributions obey fractional equations and we propose a fractional generalization of the DMPK-equation. The presence of the fractional operator in the equation reflects a self-similar structure of the scatterers arrangement. The order of fractional operator coincides with the fractal dimension.

The standard *DMPK-equation* for the evolution of the eigenvalue distribution function $P(\{\lambda\}, L)$ in the ensemble of long ($L \gg l$) wires with increasing length L has the following form:

$$l\frac{\partial P}{\partial L} = \frac{2}{\beta N + 2 - \beta}\sum_{j=1}^{N}\frac{\partial}{\partial \lambda_j}\left[\lambda_j(1+\lambda_j)J(\lambda)\frac{\partial}{\partial \lambda_j}\frac{P}{J(\lambda)}\right].$$

Here, λ_j are eigenvalues of quantum-mechanical *transfer matrix* t, N number of parallel channels, $P(\{\lambda\}, L)$ the N-dimensional probability distribution function of random vector $(\lambda_1, \lambda_2, \dots, \lambda_N)$. The integer parameter β equals 1 in a zero magnetic field and 2 in a time-reversal-symmetry breaking magnetic field. In the case of zero field and strong spin-orbit scattering $\beta = 4$. The Jacobian $J = \prod_{i<j} |\lambda_j - \lambda_i|^{\beta}$ corresponds to the transformation from the transfer matrix space to the eigenvalue space.

For a one-dimensional chain ($N = 1$, $J \equiv 1$), the DMPK-equation takes the simple form

$$l\frac{\partial P(\lambda, L)}{\partial L} = \frac{\partial}{\partial \lambda}\left[\lambda(1+\lambda)\frac{\partial}{\partial \lambda}P(\lambda, L)\right].$$

It is derived by computing the change in the scattering matrix S when the length of the wire is increased from L to $L + \delta L$, under the condition that the mean free path l exceeds the wavelength λ. Another essential assumption is the isotropy assumption about the statistical equivalence of scattering from one channel to another. It is correct when the length of the wire is much greater than the width, and time for transverse diffusion can be neglected.

The standard DMPK-equation can be generalized for random fractal distribution of scatterers in different ways. The direct way is based on the algebraic dependence of number of scatterers in the part of wire of macroscopic length δl. The second way is the maximum entropy principle with fractal condition.

Mello and Shapiro (1988) obtained the following equation

$$w_{n+m}(\lambda) - w_n(\lambda)$$

$$= \frac{m}{N}\langle \mathrm{tr}\lambda' \rangle_1 \, \widehat{\mathcal{D}} \, w_n(\lambda) + O\left(m(m-1)\langle \lambda_c' \rangle_1^2\right) + O\left(m\langle \lambda_a' \lambda_b' \rangle_1\right),$$

where operator $\widehat{\mathcal{D}}$ has the form

$$\widehat{\mathcal{D}} = \frac{2}{\beta N + 2 - \beta} \sum_a \frac{\partial}{\partial \lambda_a} \lambda_a (1 + \lambda_a) J(\lambda) \frac{\partial}{\partial \lambda_a} \frac{1}{J(\lambda)}.$$

For fractal case under consideration, according to the generalized limit theorem, the length of the wire corresponding to the large number $m \gg 1$ of scatterers is distributed according to the Lévy stable law. Hence, in the limit of weak scattering ($m^{1/\alpha} l \to \infty$), we have

$$\lim_{m^{1/\alpha} l \to \infty} \int_0^L \left[P(\{\lambda\}, L) - P(\{\lambda\}, \xi) \right] \frac{1}{l m^{1/\alpha}} g_+ \left(\frac{L - \xi}{l m^{1/\alpha}}; \alpha \right) d\xi$$

$$= \frac{m}{N} \langle \mathrm{tr}\lambda' \rangle_1 \, \widehat{\mathcal{D}} \, P(\{\lambda\}, L).$$

The characteristic exponent $\alpha < 1$ of the one-sided Lévy stable pdf coincides with a fractal dimension of the structure. From the asymptotical relation for the one-sided stable density (see [Uchaikin (2003b)])

$$g_+(s;\alpha) \sim \frac{\alpha}{\Gamma(1-\alpha)} s^{-1-\alpha}, \quad s \to \infty,$$

it follows

$$\frac{\alpha}{\Gamma(1-\alpha)} \int_0^L \frac{P(\{\lambda\}, L) - P(\{\lambda\}, \xi)}{(L-\xi)^{\alpha+1}} d\xi = \frac{\langle \mathrm{tr}\lambda' \rangle_1}{N l^\alpha} \widehat{\mathcal{D}} P(\{\lambda\}, L). \tag{4.30}$$

This integral operator with the singular kernel in this equation is related to the fractional Marshaud derivative which can be replaced for good functions by the Riemann-Liouville derivative

$${}_a\mathsf{M}_x^\alpha f(x) = \frac{1}{\Gamma(1-\alpha)} \left[\frac{f(a)}{(x-a)^\alpha} + \alpha \int_a^x \frac{f(x) - f(\xi)}{(x-\xi)^{\alpha+1}} d\xi \right]$$

$$= \frac{1}{\Gamma(1-\alpha)} \frac{d}{dx} \int_a^x \frac{f(\xi)}{(x-\xi)^\alpha} d\xi = {}_a\mathsf{D}_x^\alpha f(x).$$

Taking into account the initial condition $P(\{\lambda\}, L \to 0) = \delta(\lambda)$ and introducing notation $K = \langle \mathrm{tr}\lambda' \rangle_1 / N$, we arrive at the generalized DMPK-equation for quantum wire with fractal disorder,

$$l_0^\alpha D ... \mathsf{D}_L^\alpha P(\{\lambda\}, L) = K \frac{2}{\beta N + 2 - \beta}$$

$$\times \sum_{j=1}^{N} \frac{\partial}{\partial \lambda_j} \left[\lambda_j (1 + \lambda_j) J(\lambda) \frac{\partial}{\partial \lambda_j} \frac{P(\{\lambda\}, L)}{J(\lambda)} \right] + \frac{(L/l)^{-\alpha}}{\Gamma(1 - \alpha)} \delta(\lambda).$$

Using the link between solutions of fractional and integer-order equations [Uchaikin (1999); Uchaikin & Zolotarev (1999)], we obtain

$$P_\alpha(\{\lambda\}, L) = \frac{L}{l\alpha} \int_0^\infty P_1(\{\lambda\}, \tau) \; g_+ \left(\frac{L}{l} \tau^{-1/\alpha}; \alpha \right) \tau^{-1-1/\alpha} d\tau$$

Here, $P_1(\{\lambda\}, \tau)$ is the solution for regular distribution of scatterers. They can be found in [Cassele (1995); Beenakker & Rejaei (1993)].

Analogous formula can be derived for conductance distribution,

$$p_\alpha(g, L) = \frac{L}{l\alpha} \int_0^\infty p_1(g, \tau) \; g_+ \left(\frac{L}{l} \tau^{-1/\alpha}; \alpha \right) \tau^{-1-1/\alpha} d\tau. \tag{4.31}$$

This formula coincides with expression obtained by [Falceto & Gopar (2010)]. They use the standard DMPK-equation modifying the solution by changing the number of scatterers according to fractal distribution.

Change of integration variable leads to

$$p_\alpha(g, L) = \int_0^\infty p_1 \left(g, (L/ly)^\alpha \right) \; g_+ (y; \alpha) \, dy = \left\langle p_1 \left(g, \frac{L^\alpha}{l^\alpha S_+(\alpha)^\alpha} \right) \right\rangle, \tag{4.32}$$

where

$$S_+(\alpha) = \frac{\sin(\alpha \pi U_1) [\sin((1 - \alpha) \pi U_1)]^{1/\alpha - 1}}{[\sin(\pi U_1)]^{1/\alpha} [\ln U_2]^{1/\alpha - 1}} \tag{4.33}$$

is the stable random variable with parameter $\alpha \in (0, 1)$. Independent random variables U_1 and U_2 are uniformly distributed in (0,1]. So, formulas (4.32) and (4.33) represent the probabilistic scheme for calculating the conductance distribution in fractal quantum wire.

For a regular distribution of scatterers, the relative fluctuations of their number in the wire vanish when $L \to \infty$. Parameter $s = L/l$ represents the number of scatterers in the long wire and it is the unique parameter of the universal conductance distribution. In fractal case, the relative fluctuations tend to a non-zero value, and the behavior depends on α. In that case, parameter $s = L/l$ is not applicable, because l for $\alpha < 1$ does not have the sense of the average distance between scatterers, this average is infinite. [Falceto & Gopar (2010)] use $\xi = \langle -\ln G \rangle$ as the second

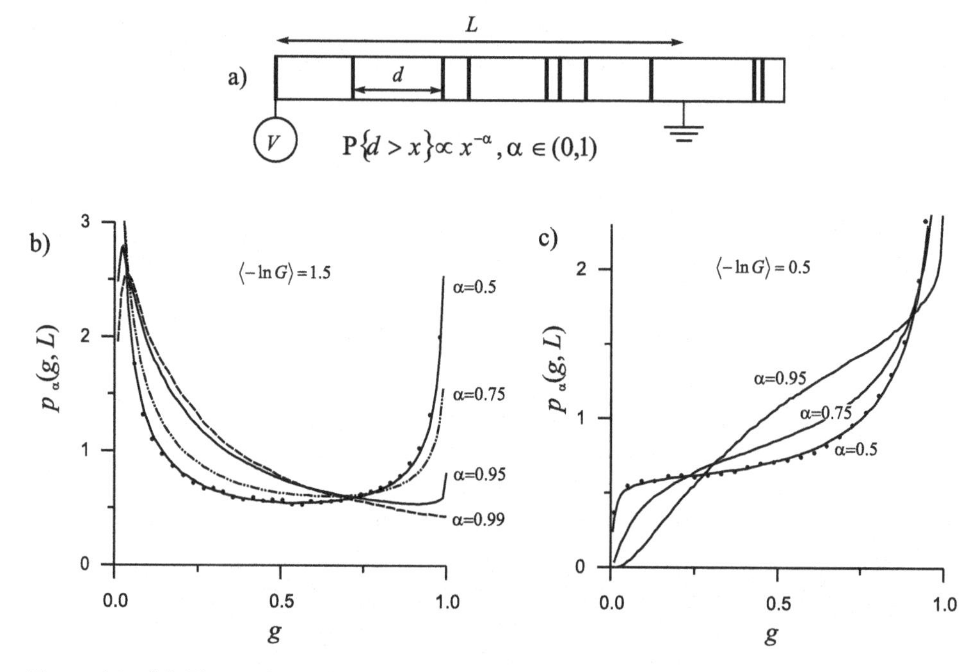

Figure 4.9 (a) The model of fractal distribution of barriers. (b) Conductance distribution for different values of α and fixed value of $\langle -\ln G\rangle = 1.5$. Points are the result of Monte Carlo simulation for $\alpha = 0.5$. (c) The same for $\langle -\ln G\rangle = 0.5$.

(scale) parameter. Indeed, it is related to l through $\langle -\ln G\rangle = (L/l)^\alpha/\Gamma(1+\alpha)$. So, $(L/l)^\alpha$ in formula (4.32) can be replaced by $\langle -\ln G\rangle\Gamma(1+\alpha)$. The conductance distributions have meaningful values in all region $(0,1)$ and can have two peaks. Such behavior is related to large fluctuations of the number of scatterers.

Fig. 4.9 demonstrates conductance distributions for different values of α. Points are the results of Monte Carlo simulation of the quantum wire ensemble with given distributions of scatterers. Conductance distributions have effective non-zero values in all interval $(0,1)$ and can have two peaks. This behavior is related to large fluctuations of number of scatterers.

4.3.2 *Sequential incoherent tunneling*

Let us consider, how these results are related to these obtained by Beenakker et al. [Beenakker et al. (2009)]. They analyse a linear chain of tunnel barriers, distances between which are independent and distributed according to asymptotical power law. The barriers have identical transmission probabilities $\Gamma \ll 1$ and resistance $\rho = h/e^2N\Gamma$, where N is the number of transverse modes. The authors consider the case $N\Gamma \ll 1$, that is $\rho \ll h/e^2$, so that the Coulomb blockade of single-electron tunneling can be neglected. They deals with the regime of incoherent sequential tunneling (no resonant tunneling), that is why the resistance of the chain is equal to the sum of resistance of each barrier $R = \rho K$, where $K(L)$ is the random number of barriers in the wire of length L. This regime is realized in cases of large number of transverse modes N or of a short phase coherence length. Beenakker et al. (2009) have obtained asymptotic formulas for moments of conductance distribution and relative fluctuations, average spectrum of shot noise and Fano factor. They also investigated the asymptotical ($L \to \infty$) behavior of negative moments of the resistance $\langle R(L)^{-k}\rangle$, where $k = 1,\ 2,\ 3, ...$, which correspond to positive moments of the conductance $G = 1/R$. Here, we'll find the asymptotical conductance distribution, and show that these distributions conform to the moments behavior obtained by [Beenakker et al. (2009)].

The Lévy-Lorentz gas considered in Refs. [Barkai et al. (2000a); Uchaikin & Sibatov (2004)] may serve as the mathematical model for a system of tunnel barriers. Barkai et al. (2000a) investigate the scattering of particles on a fixed point scatterers located on line. The intervals ξ between them are independent random variables identically distributed according to the law with power asymptotics. Let, the distribution function $Q(x)$ satisfies the following relation

$$Q(x) = \mathrm{Prob}(d > x) \sim \frac{(x/l)^{-\alpha}}{\Gamma(1-\alpha)}, \quad l > 0, \quad x \to \infty.$$

In case of finite average value of ξ (i. e. if $\alpha > 1$), the average number of scatterers on the segment of large length L is proportional to L, and relative fluctuations tends to zero [Uchaikin (2003b)]. It means, that any smooth function $f(N(L), L)$ of $N(L)$ is self-averaged, i. e. $\langle f(N(L), L)\rangle \to f(\langle N(L)\rangle, L)$, when $L \to \infty$. In case $\alpha < 1$ it is necessary to fix the first scatterer in the origin, then the mean number of scatterers $\langle N(L)\rangle \propto L^\alpha$ for $L \to \infty$, and relative fluctuations $\Delta N/\langle N\rangle$ do not decrease with L, but tends to constant value, and the medium do not possess the property of self-averaging. In Refs. [Uchaikin (2004); Uchaikin & Sibatov (2004)], random walks on a fractal Lorentz gas was considered, the role of path length correlations related to quenched disorder was demonstrated. Also it was shown that walks on random fractal differ from Lévy flights by limit distributions [Uchaikin (2004)] and transport equations [Uchaikin & Sibatov (2004)].

Beenakker et al. (2009) present resistance R of a quantum wire with length L in the form

$$R(L) = \rho + \rho \sum_{j=1}^{\infty} \theta(x_j)\ \theta(L - x_j), \tag{4.34}$$

where $\theta(x)$ is the Heavyside step function, i. e. $R = \rho N(L)$, and pdf (probability density function) of $w_R(r; L)$ is directly related to the distribution $w_N(n; L)$ of number of barriers $N(L)$. Using the relation:

$$w_N(n; L) = Q^{*n}(L) - Q^{*(n+1)}(L),$$

and passing to the Laplace image $\widetilde{w}_N(n; s) = \int\limits_0^\infty e^{-sL} w_N(n; L)dL$, we obtain

$$\widetilde{w}_R(r; s, \alpha) = \rho^{-1}\ w_N\left(\frac{r}{\rho}, L\right) = \frac{[\widetilde{q}(s)]^{r/\rho} - [\widetilde{q}(s)]^{1+r/\rho}}{s\rho}$$

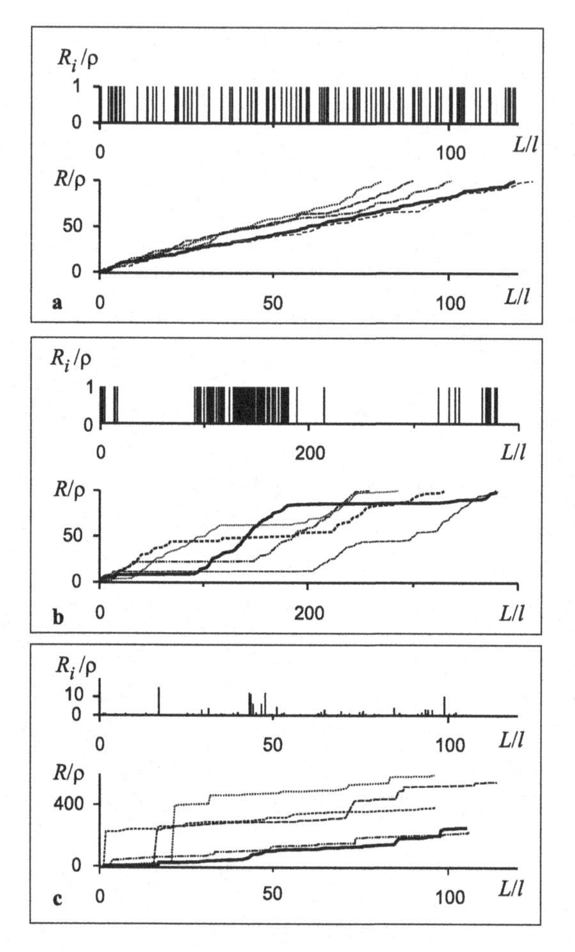

Figure 4.10 Random realizations of distributions of tunnel barriers for three cases and corresponding random trajectories $R(L) = \sum_{x_i<L} R_i$. Solid lines correspond to the trajectories for the resistance distribution presented in figure. Other lines are for other realizations. In case (a) intervals between barriers are distributed according to the exponential law and resistances of each barrier is equal to ρ, the asymptotical ($L \to \infty$) distribution of the resistance has the Gaussian form. Case (b) presents 1D stochastic fractal, where distances between barriers are distributed according to the Lévy stable law with parameter $\alpha = 3/4$. In case (c), fractal space disorder ($\alpha = 3/4$) is accompanied by wide (power law) distribution of individual barrier resistance ($\gamma = 3/4$). In cases (b) and (c) asymptotical distribution of R tends to the one-dimensional fractional stable law characterized by parameters γ and α.

where $\widetilde{q}(s)$ is the Laplace transform of pdf of ξ.

When $L \to \infty$ and $r/\rho \gg 1$, difference in the numerator of the fraction may be considered as a differential,

$$\widetilde{w}_R(r; s, \alpha) = \frac{1}{s\rho}\frac{\partial}{\partial x}\left[\widetilde{q}(s)\right]^x\bigg|_{x=r/\rho} = \frac{\ln \widetilde{q}(s)}{s\rho}\left[\,\widetilde{q}(s)\,\right]^{r/\rho}.$$

Taking the one-sided stable density $q(x) = l^{-1}g_+(x/l;\alpha)$ having the Laplace image $\widetilde{q}(s) = e^{-(ls)^\alpha}$ [Uchaikin (2003b)], we obtain

$$\widetilde{w}_R(r;s,\alpha) = \frac{l}{\rho}(ls)^{\alpha-1}\exp\left(-\frac{r}{\rho}\,l^\alpha s^\alpha\right).$$

The inverse Laplace transform of this expression gives

$$w_R(r;L,\alpha) = \frac{L}{\alpha\rho\, l}\left(\frac{r}{\rho}\right)^{-1-1/\alpha} g_+\left(\frac{L}{l}\left(\frac{r}{\rho}\right)^{-1/\alpha};\alpha\right). \tag{4.35}$$

This is the fractional stable density [Kolokoltsov et al. (2001)], which satisfies the fractional equation [Uchaikin (2003b)]:

$${}_0\mathsf{D}_L^\alpha\, w_R(r;L,\alpha) + \rho l^{-\alpha}\,\frac{\partial w_R(r;L,\alpha)}{\partial r} = \delta(r-\rho)\frac{L^{-\alpha}}{\Gamma(1-\alpha)}.$$

Here

$$\mathsf{D}_x^\alpha f(x) = \frac{1}{\Gamma(1-\alpha)}\frac{d}{dx}\int_a^x \frac{f(\xi)}{(x-\xi)^\alpha}d\xi.$$

is the RiemannLiouville fractional derivative.

The density $w_G(y;L,\alpha)$ of the conductance $G(L) = 1/R(L)$ is given by the expression

$$w_G(y;L,\alpha) = y^{-2}\, w_R(r;L,\alpha)\Big|_{r=1/y} \tag{4.36}$$

$$= \frac{L}{\alpha\rho l y^2}(y\rho)^{1+1/\alpha}\, g_+\left(Ll^{-1}(y\rho)^{1/\alpha};\alpha\right),\quad 0<y<\rho^{-1}. \tag{4.37}$$

This density satisfies the fractional equation

$${}_0\mathsf{D}_L^\alpha\, w_G(y;L,\alpha) - \rho l^{-\alpha}\,\frac{\partial}{\partial y}\left[y^2 w_G(y;L,\alpha)\right] = \frac{L^{-\alpha}}{y\,\Gamma(1-\alpha)}\delta(y-\rho^{-1}). \tag{4.38}$$

In special case of $\alpha = 1/2$,

$$g_+(x;1/2) = \frac{1}{2\sqrt{\pi}}\,x^{-3/2}e^{-1/4x},$$

$$w_G(y;L,\alpha) = \frac{1}{\rho y^2\sqrt{\pi L/l}}\exp\left(-\frac{l}{4L\rho^2 y^2}\right),\quad y\in(0,\rho^{-1}). \tag{4.39}$$

The comparison of the obtained distributions with the Monte Carlo simulation results is given in Fig. 4.11,(a).

In [Beenakker et al. (2009)] the scaling behavior in the asymptotic form ($L\to\infty$) was found for the negative moments of the resistance $\langle R(L)^{-k}\rangle$, where $k = 1,\ 2,\ 3,...$, which corresponds to the scaling of the positive moments of the conductance $G = 1/R$. By using the asymptotic expression for the one-sided stable density

$$g_+(s;\alpha) \sim \frac{\alpha}{\Gamma(1-\alpha)}s^{-1-\alpha},\quad s\to\infty$$

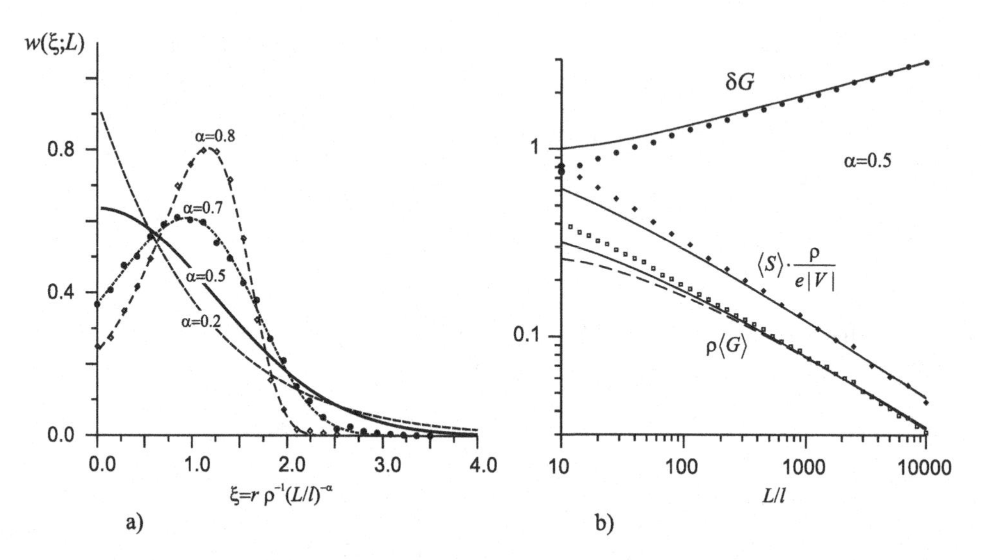

Figure 4.11 (a) Distribution density $w(\xi; L)$ of the reduced resistance $R\,l/\rho L$ for a quantum wire with the length L and different α values. The lines are calculated by Eq. (4.35), and the points are the Monte Carlo simulation results. (b) The average value $\langle G \rangle$, the relative fluctuations δG of the conductance and the average power of the shot noise $\langle S \rangle$ as functions of the chain length for $\alpha = 0.5$. The solid lines are calculated by Eq. (4.39), the dashed line is the plot of the function $\propto L^{-1/2} \ln L$, and the points are the Monte Carlo simulation results.

and formula (4.35), we find

$$\langle G(L) \rangle \propto \rho^{-1}(L/l)^{-\alpha} \ln L, \quad \langle G^k(L) \rangle \propto \rho^{-k}(L/l)^{-\alpha}$$

for $L \gg l$. The relative fluctuations of the conductance increase with the wire length:

$$\delta G(L) = \sqrt{\operatorname{Var} G(L)}/\langle G(L) \rangle \propto (L/l)^{\alpha/2}/\ln L.$$

If the moments are known, it is possible to calculate the average power of the shot noise

$$\langle S \rangle = \frac{2}{3} e|V|\rho^{-1}\,[\rho\langle G \rangle + 2\rho^3\langle G^3 \rangle].$$

The analytical results (solid lines) for the average value and relative fluctuations of the conductance, as well as the average power of the shot noise, is shown in Fig. 4.11(b) in comparison with the Monte Carlo simulation results.

In the regime of incoherent sequential tunneling,

$$p_1(g, \tau) = \rho^{-1} g^{-2} \delta\left(\tau - (g\rho)^{-1}\right).$$

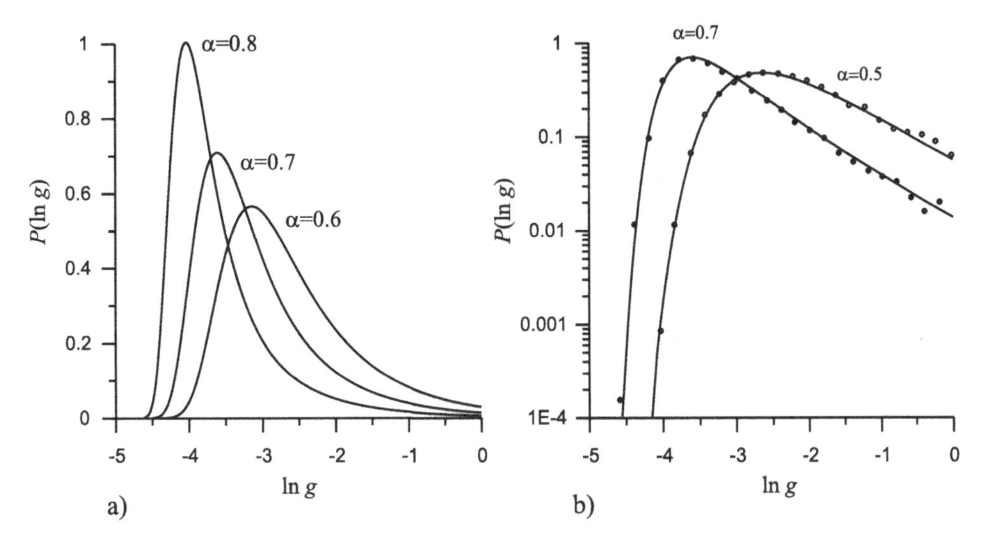

Figure 4.12 (a) Distribution density of $\ln G$ for different α values according to Eq. (4.36). (b) The analytically calculated densities $P(\ln g)$ in comparison with (points) the Monte Carlo simulation results.

Substituting this function in Eq. (4.32), we obtain the conductance distribution in the regime of incoherent sequential tunneling

$$p_\alpha(g,L) = \frac{cL}{\alpha\rho g^2}\,(g\rho)^{1+1/\alpha}\; g_+\left(cL\,(g\rho)^{1/\alpha};\alpha\right),\;\; g\in(0,\rho^{-1}).$$

For the distribution density of resistance $R = 1/G$, we have

$$p_R(r,L) = \frac{cL}{\alpha\rho}\left(\frac{r}{\rho}\right)^{-1-1/\alpha}\; g_+\left(cL\left(\frac{r}{\rho};\alpha\right)^{-1/\alpha};\alpha\right). \tag{4.40}$$

This is a fractional stable density. Relative fluctuations, average spectrum of shot noise and Fano factor calculated with the help of these densities agree with results of [Beenakker et al. (2009)] (see details in [Sibatov (2011a)]).

It is known, that the standard DMPK-equation is applicable only for quasi-one-dimensional systems in the case of weak scattering. Muttalib & Klauder (1999) and Muttalib & Gopar (2002) have proposed generalizations, which should be applicable to strongly localized systems and beyond the quasi-1D regime. All these equations were derived under the assumption that the small changes in scale parameters of the system (for example, length) can be expected to lead to a small change in the transmission parameters λ_i. This assumption is a controversial point for strongly disordered systems. It is well known for quantum tunneling that small fluctuations of a potential barrier can generate very large fluctuations of the quantum transmission. Bardou (1997) writes that "the large fluctuations modify the scaling properties of the tunneling transmission with the sample size". Evidently, the small-parameter expansion used by [Muttalib & Klauder (1999)] and [Muttalib

& Gopar (2002)] for the joint probability distribution of the transmission eigenvalues is not applicable in this case. The effect is also manifested at the macroscopic scale, for example, in the phenomenon of the dispersive transport in disordered semiconductors. If it is assumed that fluctuations of the transport parameters are self-similar, then it is possible to avoid the expansion over a small parameter and to generalize the DMPK method to the case of the strong localization and higher dimensions within the fractional differential approach.

Chapter 5

Fractional relaxation in dielectrics

In this chapter, we consider the fractional differential approach describing anomalous dielectric relaxation, discuss interrelation between different time and frequency domain responses, some physical mechanisms and corresponding fractional relaxation equations and describe some memory phenomena in disordered dielectrics.

5.1 The relaxation problem

5.1.1 *The relaxation functions*

In the framework of the simplest phenomenological model of polarization processes in dielectrics, the polarization (the dipole moment of a unit volume or mass) P consists of two parts P_1 and P_2 which are connected with two different processes, for example, with the electron displacement and atom orientation. The first follows an external field proportionally its strength $E(t)$,

$$P_1(t) = \chi_1 E(t),$$

whereas the second retards in such a way, that if $\chi_2 E$ is its maximal possible value for fixed E, then P_2 tends to reach it with the speed proportional to remaining difference $\chi_2 E$ and P_2:

$$\frac{dP_2}{dt} = \frac{1}{\tau}(\chi_2 E - P_2). \tag{5.1}$$

If at $t = 0$ the external field is suddenly switched off,

$$E(t) = \begin{cases} E_-(t), & t < 0; \\ 0, & t \geq 0. \end{cases} \tag{5.2}$$

the equation yields

$$P(t) = P_2(t) = P_2(0)e^{-t/\tau}, \quad t > 0.$$

On the contrary, if

$$E(t) = \begin{cases} 0, & t < 0; \\ E_+ = \text{const}, & t \geq 0, \end{cases}$$

then

$$P(t) = P_1(t) + P_2(t) = [\chi_1 + (1 - e^{-t/\tau})\chi_2]E_+.$$

In both cases, the polarization reaches its limit value according to exponential law with the relaxation constant (*relaxation time*) τ. This law of relaxation is known as the *Debye relaxation*. It will be convenient for what follows to use two normalized relaxation functions,

$$\Phi(t) = \frac{P_2(t)}{P_2(0)}$$

and

$$\phi(t) = -\frac{d\Phi(t)}{dt},$$

describing the relaxation process after the field being constant at $t < 0$ is switched off at $t = 0$. The latter function $\phi(t)$ relates to the case of the delta-pulse of the field and is usually called the *response function*. In the framework of the Debye model,

$$\Phi(t) = e^{-t/\tau}1(t),$$

and the response function

$$\phi(t) = \frac{1}{\tau}e^{-t/\tau}$$

satisfies the equation

$$\tau\frac{d\phi(t)}{dt} + \phi(t) = \delta(t).$$

In the frequency domain, the equation is reduced to the algebraic one,

$$[i\omega\tau + 1]\tilde{\phi}(\omega) = 1,$$

the solution of which has the form

$$\tilde{\phi}(\omega) = \frac{1}{1 + i\omega\tau}. \tag{5.3}$$

The Debye relaxation is predominantly observed in gaseous or ordered (say, crystalline) dielectrics. Debye has derived Eq. (5.3) considering the rotational Brownian motion of noninteracting dipoles in the presence of an external time-dependent field, and solving the corresponding Fokker-Planck equation [Debye (1954)]. The elementary entities of dipole moment participating in the thermal motion behave similar to what we observe in a boiled boullon. There exist other physical mechanisms leading to the Debye process. Jonscher lists five of them [Jonscher (1981a)]:

(1) the movement of independent (noninteracting) identical dipoles in a viscous fluid with heat fluctuations [Debye (1954)];
(2) identical noninteracting dipoles jumping between the preferred orientations with respect to positions of the nearest neighbors;
(3) charge jumps between identical double-well potential [Fröhlich (1949); Argall & Jonscher (1968); Mott & Davis (1971)];
(4) sequential combination of the resistance R and capacitances C;

(5) the generation-recombination of electron-hole pairs with a simple mono-energetic levels in a forbidden zone.

5.1.2 *Non-Debye empirical laws*

Most disordered dielectrics (polymers, biomolecules, colloids, porous materials, doped ferroelectric crystals, etc.) manifest nonexponential (*non-Debye*) relaxation laws, which can not be expressed in terms of an exponential function with a single relaxation time τ. This fact has been recognized over a hundred years ago [Curie (1889); von Schweidler (1907)] and continues to attract attention throughout the century [Mandelstam & Leonyovich (1937); Cole & Cole (1941); Fröhlich (1949); Debye (1954); Hippel (1954); Davidson & Cole (1951); Smyth (1955); Brown (1956); Williams & Watts (1970); Rudenko (1977); Blumen et al. (1986); Ngai (1986); Kliem & Arlt (1987); Scaife (1998)].

A set of experimental data presented, for example, in books [Jonscher (1983, 1986); Ramakrishnan & Lakshmi (1987)], convincingly testify that the Debye theory is not in a position to describe the relaxation processes in most solids. For this reason, empirical approximations different from the Debye law were proposed: the Cole-Cole response function [Cole & Cole (1941)]

$$\tilde{\phi}_{\mathrm{CC}}(\omega) = \frac{1}{1+(i\omega\tau)^{\alpha}}, \quad 0 < \alpha < 1,$$

the Cole-Davidson function [Davidson & Cole (1951)]

$$\tilde{\phi}_{\mathrm{CD}}(\omega) = \frac{1}{[1+(i\omega\tau)]^{\beta}}, \qquad 0 < \beta < 1,$$

and the Havriliak-Negami formula ([Havriliak & Negami (1966)]) united both of them:

$$\tilde{\phi}_{\mathrm{HN}}(\omega) = \frac{1}{[1+(i\omega\tau)^{\alpha}]^{\beta}}, \qquad 0 < \alpha, \beta < 1.$$

Jonsher [Jonscher (1986)]) indicates the following mechanisms, which may be responsible for the non-Debye polarization in solid dielectrics.

(1) The energy losses are associated with overcoming the binding forces and the internal viscosity.
(2) Generation and recombination of electron-hole pairs involving the levels in the forbidden band of the semiconductor.
(3) Hopping of electrons or ions in disordered solid under the action of an electric field.
(4) Macroscopic impurities and inclusions are oriented under the influence of electrical forces.

(5) The spontaneous polarization which causes the presence of a macroscopic electric moment in a sample, even in the absence of an external electric field. In this case, the non-linearity and hysteresis properties are observed. For this type of polarization, high values of dielectric permittivity are typical.
(6) The polarization caused by the presence of layers with different conductivities in the material. This leads to the generation of volume charges and high voltage gradients. The high losses and long relaxation times are typical for this type of mechanisms.

Jonscher has also noticed a common property of the above mentioned approximations and many other experimental data, namely algebraic asymptotically behaviour at large and small arguments [Jonscher (1977, 1986)]:

$$\tilde{\phi}(\omega) \propto \begin{cases} (i\omega\tau)^{\nu-1}, & \omega \gg 1/\tau; \\ (i\omega\tau)^{\mu}, & \omega \ll 1/\tau, \end{cases}$$

or in the time domain

$$\phi(t) \propto \begin{cases} (t/\tau)^{-\nu}, & t \ll \tau; \\ (t/\tau)^{-\mu-1}, & t \gg \tau, \end{cases}$$

where $0 < \mu, \nu < 1$. This property was assumed as a basis for definition of the *universal relaxation law* [Jonscher (1986)].

5.1.3 *Superposition model*

Many works (see [Jonscher (1977); ?); Weron, (1991); Weron & Jurlewicz (1993); Jonscher (1986); Novikov & Privalko (2002); Raju (2003); Novikov et al. (2005)] and reference therein) are devoted to the problem of interpretation of the physical causes of the observed universality. Early attempts to reconcile the observed non-exponential relaxation of the classical Debye process was based on the assumption that there exists not a single relaxation time τ but a whole set of them. Denoting by $w(\tau)$ the relaxation times density, normalized to 1.

$$\int_0^\infty w(\tau)d\tau = 1,$$

one can write the disordered solid response as

$$\tilde{\phi}(\omega) = \int_0^\infty \frac{w(\tau)d\tau}{1 + i\omega\tau}.$$

or, in the time domain,

$$\phi(t) = \int_0^\infty \exp(-t/\tau)w(\tau)d\tau/\tau,$$

and correspondingly,

$$\Phi(t) = \int_t^\infty \phi(t)dt = \int_0^\infty e^{-t/\tau} w(\tau) d\tau.$$

In particular, for the Debye response function with $\tau = \tau_{\rm D}$,

$$\phi_{\rm D}(t) = \frac{1}{\tau_{\rm D}} e^{-t/\tau_{\rm D}}, \quad g_{\rm D}(\tau) = \delta(\tau - \tau_{\rm D}), \quad \tilde{\phi}_{\rm D}(\omega) = \frac{1}{1 + i\omega\tau_{\rm D}},$$

for the CC response function with $\tau = \tau_{\rm CC}$,

$$g_{\rm CC}(\tau) = \frac{1}{\pi} \frac{(\tau/\tau_{\rm CC})^{2\alpha} \sin(\alpha\pi)}{1 + (\tau/\tau_{\rm CC})^{2\alpha} + 2(\tau/\tau_{\rm CC})^{\alpha} \cos(\alpha\pi)},$$

and for the HN function with $\tau = \tau_{\rm HN}$,

$$g_{\rm HN}(\tau) = \frac{1}{\pi} \frac{(\tau/\tau_{\rm HN})^{\alpha\beta} \sin(\beta\Theta)}{[1 + (\tau/\tau_{\rm HN})^{2\alpha} + 2(\tau/\tau_{\rm HN})^{\alpha} \cos(\alpha\pi)]^{\beta/2}},$$

$$\Theta = \arctan\left[\frac{\sin(\alpha\pi)}{(\tau/\tau_{\rm HN})^{\alpha} + \cos(\alpha\pi)}\right].$$

Although, there is not found a simple conventional physical explanation for such forms of the distribution, the fact that a similar feature is observed in polar and non-polar materials, in electronic semiconductors with hopping conductivity, in ionic conductors, etc., continues to encourage research in this direction.

5.1.4 *Stochastic interpretations of the universal relaxation law*

The Jonscher analysis of numerous experimental data led him to the conclusion that the explanation for deviation from the pure Debye law should be sought not in terms of a distribution of relaxation times, but rather in terms of manifestation of a universal law.

The situation with the relaxation universality looks similar to the diffusion process, when the random motion of small particles of pollen, neutrons in nuclear reactors, electrons in semiconductors being different by nature are approximately modeled by the same random process of Brownian motion. This analogy encourages the search for a suitable stochastic model for the universal relaxation law.

One of the first versions of such stochastic models proposed in [Weron, (1991)] was based on the Förster mechanism of excitation transport giving an example of a relaxation process with parallel channels [Forster (1949)] (see also [Shlesinger & Klafter (1986)]). Applying this model to dielectric relaxation, Weron et al. [Weron & Jurlewicz (1993); Weron et al. (1994)] define the relaxation function $\Phi_N(t)$ for a system consisting of a large number N of dipoles as the probability that the system as a whole has not changed its initial state, imposed by the external field, until the time t. As a result, the following assumptions were underlain this version:

(1) the system under consideration consists of N noninteracting independent entities (dipoles, for instance);
(2) random variables β_j and Θ_j related to j-th dipole are independent and identically distributed;
(3) the random waiting time Θ_j for the i-th dipole in its aligned position after the removal of the polarizing field is exponentially distributed:
$$\mathrm{Prob}(\Theta_j \geq t|\beta_j = b) = \exp(-bt) \equiv \Phi_{\mathrm{D}}(t; b), (11.35)$$
where $\beta_j = 1/\tau_j$ is the relaxation rate of j-th dipole;
(4) the distribution law of random rate β_j belongs to the domain of attraction of a one-sided α-stable law:
$$\mathrm{Prob}(\beta_j > bt) \sim t^{-\alpha}\mathrm{Prob}(\beta_j > b), \quad b \to \infty, \quad 0 < \alpha < 1.$$
(5) the relaxation function $\Phi(t)$ of the system is determined as the limit probability
$$\Phi(t) = \lim_{N\to\infty} \mathrm{Prob}(A_N \mathrm{min}\{\Theta_1, ..., \Theta_N\} > t).$$

The scaling expresses self-similar behavior of the set of relaxation rates in the limit of large values taken of them. When all β_j are equal, the stable distribution becomes δ-distribution ($\alpha = 1$), and we observe the purely Debye relaxation. The case $\alpha \neq 1$ leads to the relaxation function which contains only one parameter α, although the more general approximation of experimental data requires a two-parameter form. The second parameter appears by adopting the concept of clusters developed by Dissado (1987) and Dissado and Hill (1983, 1989). The cluster is an abstract concept representing an aligned dipole for which the other dipoles in the local environment constitute a viscose medium with a damping rate β_j.

Recall, that in solids, dipole may take up discrete orientations determined by by nearest-neighbor dispositions [Fröhlich (1949)]. A cluster can also be considered as a group of relaxing entities having common motion resulting from mutual interactions and individual clusters are separated by energy barriers preventing their merging together. The presence of clusters prevents the free reorientation of individual dipoles. They may remain in some intermediate positions before reaching full equilibration.

Let us return to the corresponding relaxation function which can be represented in the framework of the cluster concept as
$$\Phi(t) = \lim_{N\to\infty} \left\langle \Phi_{\mathrm{D}} \left(t; \sum_{j=1}^{\infty} \beta_j / A_N \right) \right\rangle .$$
After the averaging one gets [Shlesinger & Klafter (1986)]:
$$\Phi(t) = \exp\left[-k^{-\alpha/\gamma} S(t)\right],$$
where
$$S(t) = \int_0^{(k^{1/\gamma} A t)^\alpha} \left[1 - \exp(-s^{-\gamma/\alpha})\right] ds, \quad \alpha > 0, \ \gamma > \alpha.$$

The new parameter $k > 0$ represents the effect of clusters: it determines how fast the structural reorganization of clusters spread out; $k = 0$ corresponds to absence of clusters effect. The asymptotical analysis of this form yields the expression

$$\phi(t) \propto \begin{cases} (At)^{-(1-\alpha)}, & At \ll 1; \\ (At)^{-(\gamma-\alpha)-1}, & At \gg 1; \end{cases}$$

compatible with the universal relaxation law.

Describing this approach, Jonscher remarks, that "the macroscopic response is based on a very limited knowledge of the stochastic microscopic properties of a relaxing system. It is not necessary to know the detailed stochastic nature." [Jonscher (1986)]. Studies of this type were continued in the works [Weron & Kotulski (1996); Jurlewicz & Weron (2000); Coffey et al. (2002); Déjardin (2003); Aydiner (2005)] and others. Weron and Jurlewicz (2005) showed how to modify this scheme in order to obtain Havriliak-Negami process. Moreover, they derived a formula for simulation of random variables with PDF of the HN process including the algorithm of Lévy-stable random variables. Coffey et al. (2002) reformulated the Debye relaxation theory based on a polar molecules ensemble by embedding a nonlinear generalization of fractional Fokker-Planck equation. Déjardin (2003) considered the fractional approach to orientational motion of polar molecules under action of external perturbations. This problem was treated in terms of non-inertial rotational motion has been led the fractional Smoluchowski equation. This model proved to be in good agreement with experimental data for the spectra of the third order nonlinear dielectric relaxation of ferroelectric liquid crystals.

Summing up the discussion of common causes of non-Debye relaxation, one can refer to the paper [Niss (2003)], the author of which believes that there may be three scenarios for the relaxation. In the first ("homogeneous", by author's terminology) case, all the microscopic quantities relax in the same way, i.e., behave like an average. This means that the cause of anomalous behaviour should be sought at the microscopic level. In the second ("non-homogeneous") case, the microscopic characteristics of relaxation independent of each other and behave in very different ways, so although the individual microscopic relaxation can be of exponential type, averaging over the ensemble changes it to non-Debye type. Two examples of such transformation we saw above. The third scenario suggests that the non-Debye relaxation is a consequence of the correlated dynamics of microscopic elements, and therefore can hardly be described in the framework of the standard statistical approach.

We are developing here the approach based on the second scenario, where the crucial role belongs to non-Gaussian stable statistics, directly leading, as we saw in the first chapter, to differential equations of fractional orders.

5.1.5 *Random activation energy model*

An accepted explanation of the long-time asymptotics of the Jonscher universal law is based on the random activation energy model. Semiclassical calculations show that for the particles, which may leave a potential well through an energy barrier of a fixed height ϵ by tunneling:

$$p_T(t|\epsilon) = \frac{1}{\tau}e^{-t/\tau}, \quad \frac{1}{\tau} = \lambda(\epsilon).$$

At the same time the hopping rate $\lambda(\epsilon)$ is determined by the Arrhenius law

$$\lambda(\epsilon) = \nu \exp(-\beta\epsilon)$$

with the exponent β being inversely proportional to the absolute temperature τ. However, a strong interaction with the environment in a dense medium disrupts this relationship [Vlad (1992)]. Assuming that the activation energy ϵ is an exponentially distributed random variable:

$$p_E(\epsilon) = \alpha\beta e^{-\alpha\beta\epsilon},$$

one obtains

$$\phi(t) = \int_0^\infty \phi_D(t;\tau)w(\tau)d\tau = \int_0^\infty p_T(t|\epsilon)p_E(\epsilon)d\epsilon$$

$$= \frac{\alpha\beta}{\nu^\alpha}\int_0^\infty \lambda(\epsilon)e^{-\lambda(\epsilon)t}(\nu e^{-\beta\epsilon})^\alpha d\epsilon = -\frac{\alpha\beta}{\nu^\alpha}\int_0^\nu e^{-\lambda t}\lambda^{\alpha+1}\epsilon'(\lambda)d\lambda$$

$$= \frac{\alpha}{\nu^\alpha}\int_0^\nu e^{-\lambda t}\lambda^\alpha d\lambda = \left[\frac{\alpha}{\nu^\alpha}\int_0^{\nu t} e^{-z}z^\alpha dz\right] t^{-\alpha-1} \sim \frac{\alpha\Gamma(\alpha+1)}{\nu^\alpha}t^{-\alpha-1}, \quad t \to \infty.$$

Overview of reactions in condensed media with rates of power type, associated with thermally driven tunneling, can be found in [Plonka (2000)].

5.2 Fractional approach

5.2.1 *Fractional derivatives for relaxation problem*

Beginning since sixteens, the fractional calculus is more and more widely embedding in description of the non-Debye relaxation processes [Nigmatulin & Vyaselev (1964); Nigmatullin & Belavin (1964); Nigmatullin (1986); Glöckle & Nonnenmacher (1993); Nigmatullin & Ryabov (1987); Mainardi & Gorenflo (2000); Hilfer (2000a); Coffey et al. (2002); Novikov & Privalko (2002); Aydiner (2005); Tarasov (2008a); Novikov et al. (2005); Uchaikin & Uchaikin (2007)].

Glökle and Nonnenmacher (1994) showed that the fractional equation appear in a natural way to describe the time evolution of *dissipative statistical systems*. This was done using the technique of Zwanzig projection operator in the framework of linear response theory, was obtained hereditary equation of the form

$$\frac{\Phi(t)}{dt} = -\int_0^t K(t-\tau)\Phi(\tau)d\tau.$$

Nonnenmacher and Metzler (2000) examined different types of the $K(t)$ memory kernel, and noted that the form $K(t) = K_0\delta(t)$, relating to the Markov process, yields the solution in the exponential form,

$$\Phi(t) = \exp(-Ct).$$

For $K(t) = C$, oscillating solutions are obtained:

$$\Phi(t) = \cos\left(\sqrt{C}\,t\right).$$

When $K(t) \propto t^{\gamma}$, $t \to \infty$, we get the *Kohlrausch-Williams-Watts relaxation* expressed through the stretched exponential function

$$\Phi(t) = \exp\left(-Ct^{\gamma+2}\right)$$

In the opposite case, when $K(t) \sim Ct^{\alpha-2}$, $t \to \infty$, we arrive at a fractional relaxation equation of the form

$$\frac{d\Phi(t)}{dt} = -\tau^{-\alpha}\ {}_0\mathsf{D}_t^{1-\alpha}\Phi(t), \quad \tau = [C\,\Gamma(\alpha-1)]^{-1/\alpha}.$$

In [Hilfer (2000a)], the anomalous relaxation is considered in the framework of a generalized concept of evolution with a fractal time. This concept indicates the possible existence of an equilibrium state with the time distributed by law with power-law asymptotics and shows that the infinitesimal generator of time evolution operator is proportional to the fractional time derivative ${}_{-\infty}\mathsf{D}_t^{\alpha}$.

We consider the fractional approach to description of non-Debye relaxations including the dipole orientation mechanism of the fractional subdiffusion over a spherical surface and two new relaxation mechanisms: the displacement mechanism caused by dispersive transport and the formation of percolation channels with irregular dynamics of the delocalization processes. These models lead to the fractional differential relations between current and voltage and explain the hereditary effects of polarization in organic dielectrics. The dependence of polarization currents on the temperature and amount of moisture for the paper-oil capacitor are investigated.

5.2.2 *Polar dielectrics: model of rotational subdiffusion*

In classical approach, a polar dielectric is considered as a set of polar molecules freely moving in a neutral liquid simulating an influence of the thermal motion. Each molecule has a constant dipole moment, and in the absence of an external electrical field, their angular distribution is isotropic. The applied electrical field add a deterministic component $\mathbf{p}$ proportional to the field strength. This addition is much smaller than the main component of the dipole moment, but namely it is responsible for the polarization of the material. After switching off the electric field, **p** is exposed to a heat reservoir, the direction changes at random dipoles and as a result, the average value of its projection on the $\mathbf{E} = E\mathbf{e}_z$ direction, $\langle p_z(t) \rangle$, decreases with time.

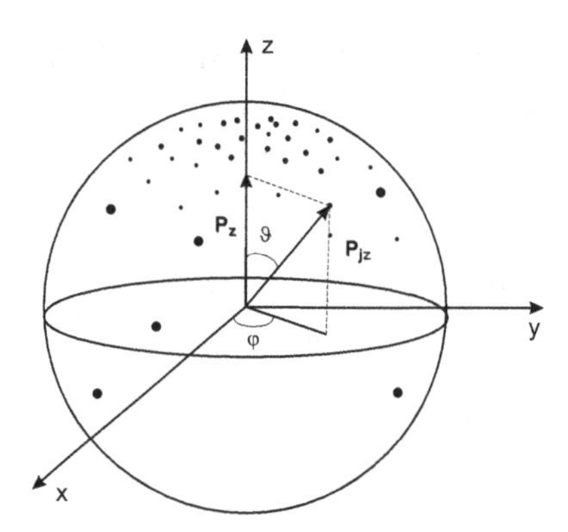

Figure 5.1 The random distribution of dipole orientation.

The basic proposition of the model is the assumption of independent behavior of the molecules. The relaxation process is presented as a random motion of single molecules in the angular space, or in other words, as the Brownian motion of independent points over a sphere of a fixed radius. The diffusion packet $f(\theta, \varphi)$, initially concentrated at the "North pole" ($\theta = 0$), gradually spreads over the sphere and eventually fills it uniformly. The relaxation function is proportional to the average z-coordinate of this distribution,

$$\langle p_z(t) \rangle = p_0 \int_0^{\pi} \int_0^{2\pi} \cos\theta f(\theta, \varphi, t) \sin\theta d\theta d\varphi,$$

which tends to 0 as $t \to \infty$ defining the relaxation law. This process is described by the diffusion equation, in which the spherical coordinates θ and φ are used instead of Cartesian ones:

$$\frac{\partial f(\theta, \varphi, t)}{\partial t} = K \triangle_{\theta,\varphi} f(\theta, \varphi, t) + \frac{\delta(\cos\theta - 1)}{2\pi}\delta(t), \quad K = \text{const} > 0, \tag{5.4}$$

where

$$\triangle_{\theta,\varphi} = \frac{1}{\sin\theta}\frac{\partial}{\partial\theta}\left(\sin\theta\frac{\partial}{\partial\theta}\right) + \frac{1}{\sin^2\theta}\frac{\partial^2}{\partial\varphi^2}$$

stands for the angular part of Laplacian. Multiplying both sides of the equation by $p_0 \cos\theta \sin\theta d\theta d\varphi$ and integrating them over the total solid angle, we arrive at the equation for $p(t) \equiv \langle p_z(t) \rangle$,

$$\frac{dp(t)}{dt} = -Kp(t) + p_0\delta(t), \tag{5.5}$$

Figure 5.2 Trajectories on the sphere, modeling the rotation of diffusing dipole moments.

leading to the Debye relaxation.

Discussing why the relaxation in solids is different from the relaxation in gases, Jonsher rightly notes that the dipole moment carriers can not be regarded as free and independent any more. This can take place, for example, when the molecules are combined into clusters. It is natural to assume that clumsy clusters interfere with each other during the movement and have to wait until there is a room for rotation. The dipole moment remains immobile as if staying in traps (in the orientational space). After some random time interval T, it makes a turn at a small angle and remains in a rest again. Thus, we deal with a CTRF process. If $\langle T\rangle < \infty$, the asymptotical long-time behavior of the process is a normal diffusion process governed by Eq. (1), and the relaxation is of the Debye type. However, if $\langle T\rangle = \infty$ but $\mathsf{P}(T > t) \propto t^{-\alpha}$, we observe the non-Debye relaxation of the delayed type. The relaxation in disordered solids is characterized by a power-type behavior in time, due to energy and/or structural disorder, which retards the motion of particles and causes the delay. As we saw in Chapt. 1, these effects are taken into account by means of fractional time-derivatives in the kinetic equations,

$$ {}_0\mathsf{D}_t^\alpha f(\theta,\varphi,t) = K\triangle_{\theta,\varphi} f(\theta,\varphi,t) + \frac{\delta(\cos\theta - 1)}{2\pi}\frac{t^{-\alpha}}{\Gamma(1-\alpha)}, \tag{5.6} $$

and relaxation equations,

$$ {}_0\mathsf{D}_t^\alpha p(t) = -Kp(t) + p_0\frac{t^{-\alpha}}{\Gamma(1-\alpha)}, \quad t > 0, \quad 0 < \alpha < 1, \tag{5.7} $$

generalizing Eqs. (5.4) and (5.5), respectively.

5.2.3 *A prehistory contribution*

In order to deeper understand the problem with an influence of a process prehistory on the solutions of Eqs. (5.6–5.7), we should keep in mind that these equations describe long-time asymptotic behavior of a CTRF process. The latter is a Markov process if only its waiting time distribution is an exponential function. In this case, the CTRF integral equation is strictly converted into a first-order integro-differential equation which describes the same Markov process. The differential term relates to the exponential waiting time distribution, but the integral term corresponds to random flights: the kernel of the operator is a probability density function for the particle displacements. To find an appropriate solution of this equation, one has to impose on its general solution an only condition, say its initial value at $t = 0$. Observe, that original integral CTRF equations do not require any additional conditions for their solutions: the need for additional conditions arises in the course of the asymptotic transition to long time scales.

When the waiting time distribution density $q(t)$ is not an exponential function but has a finite first moment, the CTRF is not a Markov process but becomes such in the long-time asymptotics. Non-Markovian process is a *process with memory*, but CTRF is a special kind of such processes: it loses the memory of prehistory in each jump and begins a new life independent of a previous one. When $q(t)$ has a power tail $t^{-\alpha-1}$ with $\alpha \in (0,1)$, the CTRF equation becomes fractional one such as Eqs.(5.6–5.7). However, Eqs. (5.6–5.7) assume that $t = 0$ is not an arbitrary moment but a start time for the first waiting duration T. Choosing t_0 instead of 0, we rewrite these equations as

$$ {}_{t_0}\mathsf{D}_t^{\alpha} f(\theta,\varphi,t) = K\triangle_{\theta,\varphi} f(\theta,\varphi,t) + \frac{\delta(\cos\theta - 1)}{2\pi}\frac{(t-t_0)^{-\alpha}}{\Gamma(1-\alpha)}, \tag{5.8} $$

$$ {}_{t_0}\mathsf{D}_t^{\alpha} p(t) = -Kp(t) + p_0\frac{(t-t_0)^{-\alpha}}{\Gamma(1-\alpha)}, \quad t > t_0, \quad 0 < \alpha < 1, \tag{5.9} $$

Thus, if we know that the process under investigation did not exist till say θ and at this moment all possible trajectories (the statistical ensemble) of the process appear and begin their motion with the start time of the first waiting, we have reason to put $t_0 = \theta$. However, if we do not know this exactly, we can resort to the usual trick in such cases: considering t_0 as a random variable to perform averaging over this parameter. Denoting the corresponding pdf by $W(t_0)$, $-\infty < t_0 < t$, transforming the derivative term as

$$ \int_{\infty}^{t} {}_{t_0}\mathsf{D}_t^{\alpha} f(\theta,\varphi,t) W(t_0) dt_0 = \int_{\infty}^{t} {}_{-\infty}\mathsf{D}_t^{\alpha} f(\theta,\varphi,t) 1(t-t_0) W(t_0) dt_0 $$

$$ = {}_{-\infty}\mathsf{D}_t^{\alpha}\left[\int_{\infty}^{t} {}_{-\infty}\mathsf{D}_t^{\alpha} f(\theta,\varphi,t) 1(t-t_0) W(t_0) dt_0\right] = {}_{-\infty}\mathsf{D}_t^{\alpha}\langle f(\theta,\varphi,t)\rangle $$

and leaving the same notation for the averaged solution, we represent these equations as

$$ {}_{-\infty}\mathsf{D}_t^{\alpha} f(\theta,\varphi,t) = K\triangle_{\theta,\varphi} f(\theta,\varphi,t) + F(\theta,\varphi,t), \tag{5.10} $$

and

$$ {}_{-\infty}\mathsf{D}_t^{\alpha} p(t) = -Kp(t) + P(t), \quad t > -\infty, \quad 0 < \alpha < 1, \tag{5.11} $$

with

$$F(\theta,\varphi,t) = \int\limits_{-\infty}^{t} \frac{\delta(\cos\theta - 1)}{2\pi} \frac{(t-t_0)^{-\alpha}}{\Gamma(1-\alpha)} W(t_0)dt_0,$$

and

$$P(t) = \int\limits_{-\infty}^{t} p_0 \frac{(t-t_0)^{-\alpha}}{\Gamma(1-\alpha)} W(t_0)dt_0.$$

When fractional equation

$$_{-\infty}\mathsf{D}_t^\alpha f(\theta,\varphi,t) = K\triangle_{\theta,\varphi} f(\theta,\varphi,t) + F(\theta,\varphi,t), \tag{5.12}$$

or

$$_{-\infty}\mathsf{D}_t^\alpha p(t) = -Kp(t) + P(t), \quad t > -\infty, \quad 0 < \alpha < 1, \tag{5.13}$$

describing a real physical process is involved irrelatively of any prelimit process like CTRF, and the system under consideration existed till t_0, we should know its prehistory. Formally, the initial conditions for the unknown functions $f(\cos\theta,\varphi,t)$ and $p(t)$ are given by fractional integrals, which do not imply a direct physical interpretation. To resolve these contradictions, some authors extend the functions on the interval $t < t_0$ by involving values of the physical quantity undergoing to operator action [Ochmann & Makarov (1993); Zhang & Shimizu (1999); Lorenzo & Hartley (2000); Fukunaga (2002)]. Lorenzo and Hartley [Lorenzo & Hartley (2000)] expressed the influence of a prehistory in terms of so-called initialization functions and applied this method to various types of fractional equations (see, e.g., [Lorenzo & Hartley (2000); Fukunaga (2002)]).

5.2.4 *Green's function*

The solution $p_\alpha(t)$ of the fractional equation

$$_{-\theta}\mathsf{D}_t^\alpha p_\alpha(t) = -Kp_\alpha(t) + P(t), \quad t > -\infty, \quad 0 < \alpha < 1, \tag{5.14}$$

provided $p_\alpha(-\theta) = 0$ can be expressed through the Green's function $G_\alpha(t)$ by means of integral

$$p_\alpha(t) = \tau^\alpha \int\limits_{-\theta}^{t} G_\alpha((t-t')/\tau)P(t')dt', \quad \tau^\alpha = 1/K. \tag{5.15}$$

This function obeys the equation with the dimensionless time-argument,

$$_0\mathsf{D}_t^\alpha G_\alpha(t) + G_\alpha(t) = \delta(t),$$

the solution of which can be represented in a few equivalent forms. We consider here two of them.

5.2.4.1 *The first representation*

The first of them uses the two-parameter Mittag-Leffler function [Podlubny (1999)]

$$E_{\alpha,\beta}(z) = \sum_{j=0}^{\infty} \frac{z^j}{\Gamma(\alpha j + \beta)}$$

and has the form

$$G(t) = t^{\alpha-1} E_{\alpha,\alpha}(-(t/\tau)^{\alpha}). \tag{5.16}$$

Here are the simplest properties of the functions:

$$E_{1,1}(z) = E_1(z) = e^z, \quad E_{1,2}(z) = \frac{e^z - 1}{z}, \quad E_{2,2}(z) = \frac{\sinh(\sqrt{z})}{\sqrt{z}}.$$

(see in detail [Gorenflo & Mainardi (1997); Podlubny (1999)]).

5.2.4.2 *The second representation*

The second form is associated with the one-sided Lévy stable density defined by the Laplace transform:

$$g_+(t;\alpha) = \frac{1}{2\pi i} \int_{\sigma-i\infty}^{\sigma+i\infty} e^{\lambda t - \lambda^{\alpha}} d\lambda.$$

The Laplace transform of the Green function

$$\hat{G}_{\alpha}(\lambda) = \frac{1}{\lambda^{\alpha} + \tau^{-\alpha}} \equiv \int_0^{\infty} e^{-(\lambda^{\alpha}+\tau^{-\alpha})x} dx$$

leads to the following expression for it original:

$$G_{\alpha}(t) = \frac{1}{2\pi i} \int_{\sigma-i\infty}^{\sigma+i\infty} d\lambda \, e^{\lambda t} \int_0^{\infty} e^{-(\lambda^{\alpha}+\tau^{-\alpha})x} dx$$

$$= \int_0^{\infty} \exp\left(-\tau^{-\alpha} x\right) x^{-1/\alpha} \, g_+\left(t x^{-1/\alpha}; \alpha\right) dx \tag{5.17}$$

$$= \alpha t^{\alpha-1} \int_0^{\infty} \exp\left(-\frac{t^{\alpha}}{(\tau z)^{\alpha}}\right) g_+(z;\alpha) z^{-\alpha} dz. \tag{5.18}$$

Recall that $g_+(z;1) = \delta(z-1)$, and (5.18) is followed by $G_1(t) = e^{-t/\tau}$, as it should be.

5.3 The Cole-Cole kinetics

5.3.1 *Fractional generalization of the Ohm's law*

Fractional generalization of Ohm's law is based on the empirical law of Curie-von Shweidler [Curie (1889); von Schweidler (1907)], which states that the displacement current $\Delta I(t-t')$ caused in a capacitor by the voltage jump $\Delta U(t')$, contains a component, decaying by a power law [Westerlund (1991)]

$$\Delta I(t) = C_0 \Delta U(t')\delta(t-t') + \frac{C_\alpha}{\Gamma(1-\alpha)}\frac{\Delta U(t')}{(t-t')^\alpha}, \quad t \geq 0, \quad \alpha \in (0,1).$$

In case of an arbitrary function $U(t)$, representable as a superposition of jumps, $U(t) = \sum_j \Delta U(t_j), \quad \{j : t_j < t\}$, we have

$$I(t) = C_0 \sum_j \Delta U(t_j)\delta(t-t_i) + \frac{C_\alpha}{\Gamma(1-\alpha)}\sum_j \frac{\Delta U(t_j)}{(t-t_j)^\alpha}.$$

Passing to the continuous limit yields

$$I(t) = C_0 \int\limits_{-\infty}^{t} \frac{du}{dt'}\,\delta(t-t')\,dt' + \frac{C_\alpha}{\Gamma(1-\alpha)}\int\limits_{-\infty}^{t} \frac{dU(t')}{dt'}\frac{dt'}{(t-t')^\alpha}.$$

Rewritten in the form

$$I(t) = C_0 \frac{dU}{dt} + C_\alpha \ {}_{-\infty}\mathsf{D}_t^\alpha U(t), \quad 0 < \alpha < 1,$$

this expression can be accepted as a fractional generalization of Ohm's law. The classical Kirchhoff equation

$$I(t)R + U(t) = V(t)$$

takes the fractional form

$$C_0 R \frac{dU}{dt} + C_\alpha R \ {}_{-\infty}\mathsf{D}_t^\alpha U(t) + U(t) = V(t).$$

Applying the Laplace transform with an imaginary parameter,

$$f(t) \mapsto \tilde{f}(\omega) = \int\limits_0^\infty e^{-i\omega t} f(t)dt$$

to the reduced version of this equation

$$C_\alpha R \ {}_{-\infty}\mathsf{D}_t^\alpha U(t) + U(t) = V(t),$$

we arrive at the algebraic equation

$$[1 + (i\omega\tau)^\alpha]\,\tilde{U}(\omega) = \tilde{E}(\omega), \quad \tau^\alpha = C_\alpha R, \quad 0 < \alpha < 1.$$

Its solution is written as

$$\tilde{U}(\omega) = \tilde{\phi}_{\mathrm{CC}}(\omega)\tilde{E}(\omega),$$

where

$$\tilde{\phi}_{CC}(\omega) = \frac{1}{1 + (i\omega\tau)^\alpha}.$$

As we saw above, such equation describes the rotational subdiffusion of molecules (or, more generally, of dipole moment carriers in solid dielectrics). Here, we touch on another version of the CC relaxation.

The well-known deterioration of the insulation in time is usually associated with an increase in direct conductivity (see, e.g. [Saha (2003); Saha & Purkait (2008)]). It was established experimentally that in some cases the *percolation channels* are responsible for this conductivity. Such a mechanism provides directed movement of charge carriers, but as in the case of the rotational relaxation, it is discontinuous with a broad distribution of waiting times in the localized states. The localization processes lead to the the charge accumulations blocking the traffic. The blocking times distributions are of the inverse power type, and consequently, the same distributions describe time-intervals between successive pulses of current in the individual channels. Let p_n be discrete distribution of the number of current pulses due to transitions of carriers into an electrode. Then, the current is given by

$$I_{\text{perc}}(t) = \frac{d\langle Q\rangle}{dt} = eZ\frac{d}{dt}\sum np_n \sim eZ\nu c(ct)^{\nu-1}\int\limits_0^\infty s^{-\nu}g_+(s;\nu)ds = \frac{eZc^\nu}{\Gamma(\nu)}\,t^{\nu-1},$$

$$t \gg c^{-1}, \quad 0 < \nu < 1,$$

where Z is the number of channels.

5.3.2 *Numerical demonstration of the memory effect*

5.3.2.1 *Mittag-Leffler representation*

To demonstrate the influence of a prehistory in the capacitor charging-discharging, we consider the solution of the Eq. (5.14) under condition that $p(t) = 0$ for all $t < -\theta$ and

$$P(t) = \begin{cases} 0, \; t < -\theta; \\ 1, \; -\theta < t < 0; \\ 0, \; t > 0. \end{cases}$$

Inserting equations (5.16) and (5.18) into Eq. (5.15), we obtain

$$p_\alpha(t) = \int\limits_{-\theta}^{0}\sum_{j=0}^{\infty}\frac{(t-t')^{\alpha-1}}{\Gamma(\alpha j+\alpha)}\left[-\left(\frac{t-t'}{\tau}\right)^\alpha\right]^j dt', \quad t \geq 0.$$

Interchanging integration and summation and using the gamma function property yield

$$p_\alpha(t) = \tau^\alpha\sum_{j=0}^{\infty}\frac{(-1)^j}{\Gamma(\alpha j+\alpha+1)}\left[\left(\frac{t+\theta}{\tau}\right)^{\alpha j+\alpha} - \left(\frac{t}{\tau}\right)^{\alpha j+\alpha}\right]. \tag{5.19}$$

Making the change of summation index $k = j + 1$ and using the relation

$$-\sum_{k=1}^{\infty} \frac{(-1)^k (t/\tau)^{\alpha k}}{\Gamma(\alpha k + 1)} = 1 - E_\alpha\left(-(t/\tau)^\alpha\right),$$

we have got for $t \geq 0$:

$$\frac{p_\alpha(t)}{p_\alpha(0)} = \frac{E_\alpha\left(-t^\alpha/\tau^\alpha\right) - E_\alpha\left[-(t+\theta)^\alpha/(\tau)^\alpha\right]}{1 - E_\alpha\left[-(\theta/\tau)^\alpha\right]}. \tag{5.20}$$

Each value of the parameter $\alpha \in (0,1)$ generates a family of curves relating to different times of charging θ, but $\alpha = 1$ produces a single curve, which is independent of θ. Indeed, applying the formula for the Mittag-Leffler function with this integer parameter, we obtain the equation (5.20):

$$p_1(t) = \frac{E_1(-t/\tau) - E_1[-(t+\theta)/\tau]}{1 - E_1(-\theta/\tau)}$$

$$= \frac{1 - \exp(-\theta/\tau)}{1 - \exp(-\theta/\tau)} \exp(-t/\tau) = \exp(-t/\tau), \quad t \geq 0.$$

Let us discuss the asymptotic behavior of solutions (5.19). At long times $t \gg \theta$, the arguments θ and $t + \theta$ in (5.19) can be considered as dt and $t + dt$ respectively. Then, Eq. (5.19) is followed by

$$p_\alpha(t) \sim \theta t^{\alpha-1} \sum_{j=0}^{\theta} \frac{(-1)^j}{\Gamma(\alpha j + \alpha)} \left(\frac{t}{\tau}\right)^{\alpha j} = \theta t^{\alpha-1} E_{\alpha,\alpha}[-(t/\tau)^\alpha], \quad t \gg \theta.$$

The latter function is characterized by a long tail, $p_\alpha(t) \propto t^{-\alpha-1}$, $t \to \infty$. In case $t \ll \theta$, the second term in Eq. (5.20) is negligible and for $t \gg \tau$

$$p_\alpha(t) \propto t^{-\alpha}, \quad \tau \ll t \ll \theta.$$

5.3.2.2 *Monte Carlo calculations*

According to our calculations [Uchaikin et al. (2009)], some problems occur when computing the Mittag-Leffler functions at large values of their argument (the phenomenon of numerical noise). To avoid these difficulties, we have developed another way of calculating, based on the second presentation of the Green function. Rewriting Eq. (5.15) in the form

$$p_\alpha(t) = \int_0^t G_\alpha(\xi) E(t - \xi) d\xi$$

and setting here Eq. (5.18), we get

$$p_\alpha(t) = \int_0^\infty g_+(z;\alpha)\ h(t;z,\alpha), \tag{5.21}$$

where

$$h(t;z,\alpha) = z^{-\alpha} \int_0^\infty dy\ \exp(-y/(\tau z)^\alpha) E(t - y^{1/\alpha}). \tag{5.22}$$

The function $g_+(z;\alpha)$ is the probability density for the stable random variable $S(\alpha)$; so the $p_\alpha(t)$ in Eq. (5.21) can be interpreted as an expectation of the function of the random variable $h(t; S(\alpha), \alpha)$. According to the law of large numbers, it can be approximated by sample mean:

$$p_\alpha(t) \approx \frac{1}{N}\sum_{j=1}^{N} h(t; S_j(\alpha), \alpha).$$

Independent realizations of the random variable $S_j = S_j(\alpha)$ are simulated according to the Kantor algorithm

$$S(\alpha) = \frac{\sin(\alpha\pi U_1)\left[\sin((1-\alpha)\pi U_1)\right]^{-1+1/\alpha}}{\left[\sin(\pi U_1)\right]^{1/\alpha} |\ln U_2|^{-1+1/\alpha}},$$

where U_1 and U_2 are random numbers uniformly distributed in $(0, 1)$ (see for detail [Uchaikin & Zolotarev (1999)]).

In case of (5.18), we have

$$p_\alpha(t) \approx \frac{\tau^\alpha}{N}\sum_{j=1}^{N}\left\{\exp\left[-\left(\frac{t}{\tau S_j}\right)^\alpha\right] - \exp\left[-\left(\frac{t+\theta}{\tau S_j}\right)^\alpha\right]\right\}. \qquad (5.23)$$

This algorithm produces stable results for all values of time t.

5.3.3 *Polarization-depolarization currents*

Diagnosis of insulation plays an important role in the maintenance of transformers and other expensive equipment of high voltage. The diagnostic method based on the measurement of polarization and depolarization currents (PDC) allows us to distinguish the effect of material properties and geometric structure, due to the high sensitivity to changes in temporal response of the system. PDC measurements were used to test engines, generators, capacitors, cables and transformers [Saha (2003); Saha & Purkait (2008)].

[Gafvert et al. (2000)] argue that the analysis of PDC allows to estimate separately the properties of oil and paper. In particular, the authors of [Kuchler & Bedel (2001)] show that this method is suitable for estimation of moisture content in the barriers and determination of the transformer oil conductivity.

In [Uchaikin et al. (2009)], we investigated the temperature dependence of the PDC for paper-oil capacitors and the effect of the background (time polarization) on the current depolarization. The results are described in terms of the fractional differential approach, which can form a mathematical basis of the PDC technique.

It is known that the transformer oil is characterized by a nonlinear conductivity, which depends on the geometry and the voltage prehistory while the oil does not accumulate the charge. Its properties are determined by the movement, release and generation of ions.

Using the algorithms, described in previous section, we calculated the charge-discharge curves for $\theta = 10, 7.5, 5, 2.5$ s.

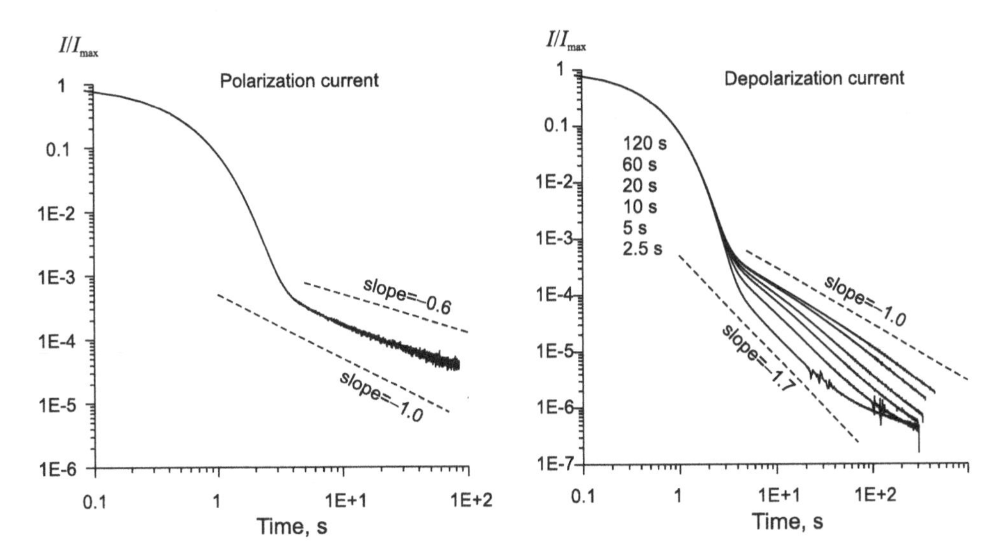

Figure 5.3 Experimental currents of polarization-depolarization at room temperature for six polarization times $\theta = 2,5;\ 5;\ 10;\ 20;\ 60;\ 120$ s [S. A. Ambrozevich, 2006].

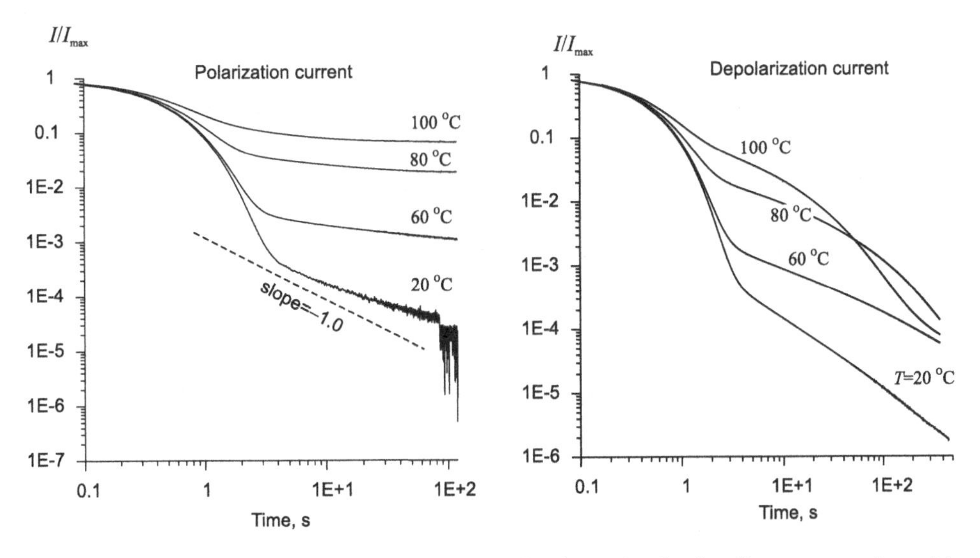

Figure 5.4 Experimental currents of polarization-depolarization in the oil-paper capacitor with 2 μF capacity and 250 V maximal vaoltage (R=200 kOhm) at different temperatures [S. A. Ambrozevich, 2006]. Asymptotical behavior conforms to theoretical one.

Then, we repeated such calculations for $\alpha = 0.998$ and observed that the discharge voltage obviously changed its behaviour: now it drops according to the exponential (Debye) law, but some time later it returned to power type relaxation. Because the first type of relaxation is associated with absence of memory, and the second with its presence, it can seem that we observe a process with memory

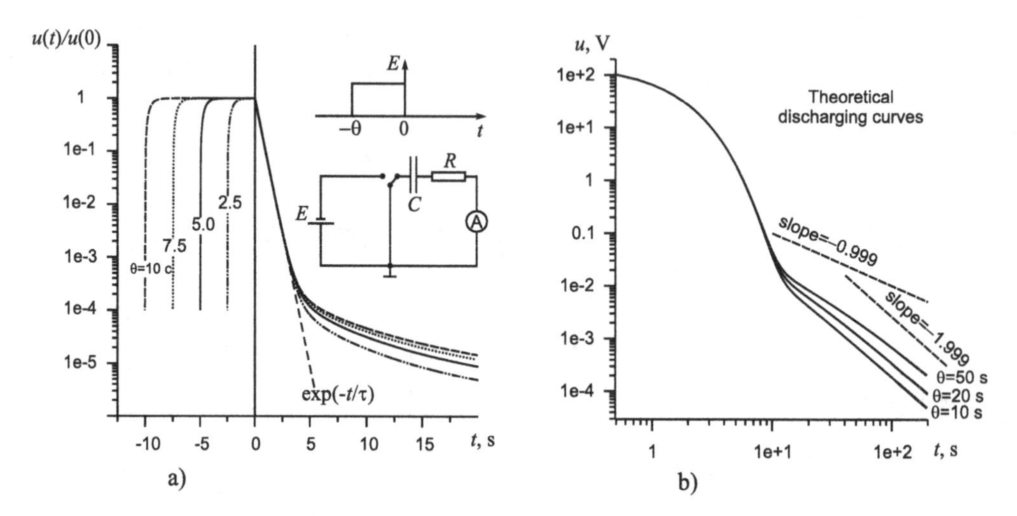

Figure 5.5 (a) Polarization-depolarization curves for different values of θ, calculated via solution of the fractional relaxation equation by the finite difference method. (b) Theoretical discharging curves in log-log scale.

recovering with time [Uchaikin & Uchaikin (2005)]. We shall return to discussion of this conclusion later.

Experiments were carried out as follows. The paper-oil capacitor with a fixed capacity of $2 \cdot 10^{-6}$ Farad was bypassed for a long time with a resistor and an amperemeter . Then a power supply was applied to the capacitor during time θ . The charge-discharge current (CDC) was measured during the experiment. The resistor was of $2 \cdot 10^5\ \Omega$, the power supply voltage 200 V. The experimental CDC curves at the room temperature for six charging times $t = 2,5;\ 5,\ 10,\ 20,\ 60,\ 120$ s are shown in Fig. 5.3. Observe that the discharge tails of the curves do not depend on θ at the initial stage, but some time later begin to depend on the charging history reflected by the parameter θ (Fig. 5.5), as the fractional model predicts. Figure. 5.6, b shows a comparison of the calculated relaxation curves (dots) with the experimental data (solid lines). We used only one adjustable parameter, α. The theoretical solution and experimental result are in good agreement. A slight deviation is observed at large times. For the process of charge-discharge transition from one power-law decay ($t^{-\alpha}$) to another ($t^{-\alpha-1}$) is predicted for sufficiently large charging times θ. This transition is difficult to observe experimentally, since it occurs at long times, when the current signal is very small. However, this change in the power-law decay was observed experimentally for electrolytic capacitor (Fig. 5.7).

Let us say a few words how one can interpret these results. As a matter a fact, the term *memory regeneration* does not reflect the sense of the phenomenon. Imagine a solution consisting of two components

$$\frac{U(t)}{U(0)} = \exp(-t/\tau) + \eta_\alpha(t),$$

first of which does not depend on the prehistory, and the second ($\eta_\alpha(t)$) does.

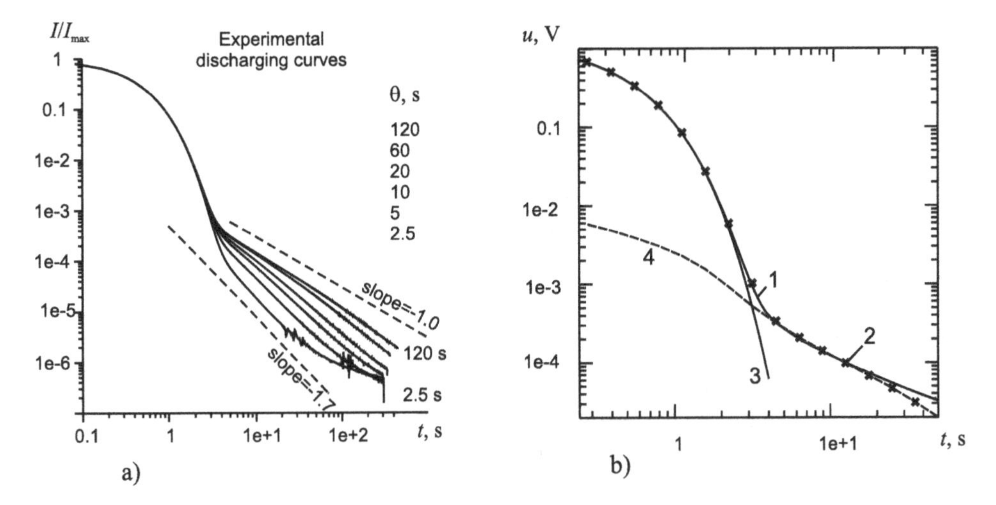

Figure 5.6 (a) Experimental curves of polarization-depolarization in the oil capacitor for different values of charging time θ. (b) Comparison of the numerical result (points, 2) with the experimental curve (solid line, 1) ($\theta = 300$ s, $\tau \approx RC = 0,4$ s^{-1}). Parameter $\alpha = 0,998$, curve 3 is a graph of the exponential function, curve 4 corresponds to the difference between the theoretical solution and the exponential function.

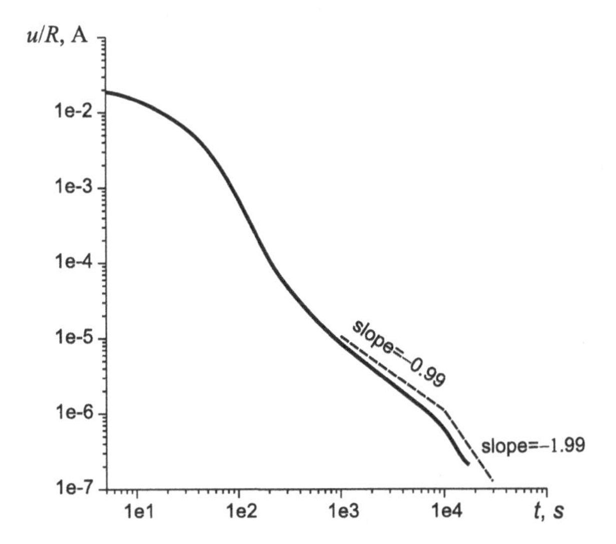

Figure 5.7 Experimental curves of electrolytic capacitor ($\theta = 3600$ s) [S. A. Ambrozevich, 2006].

Both term decay, the first of them initially dominates ($U(t)/U(0) \approx \exp(-t/\tau)$ at $t < t^*$) but decays faster then the second one and at long time becomes negligible $U(t)/U(0) \approx \eta_\alpha(t), \quad t > t^*$. This is a reason, for which the relaxation functions relating to difference prehistories look coinciding on not long time scales.

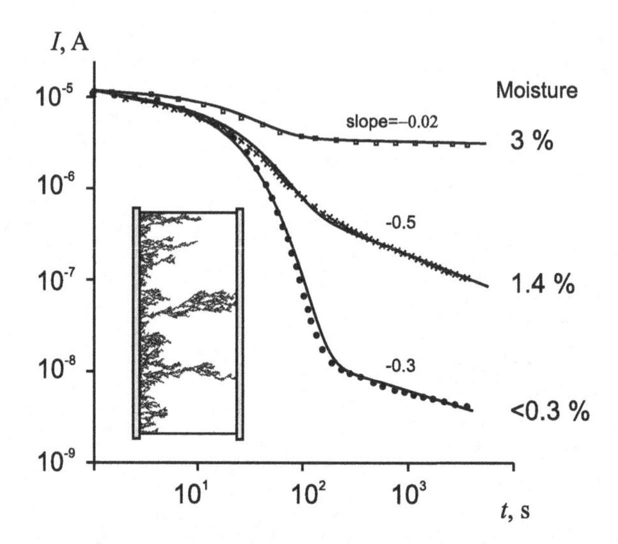

Figure 5.8 Influence of moisture content on depolarization current in oil-paper capacitor: points are experimental data digitized from [G. Frimpong, U. Gafvert, J. Fuhr, 1997], lines are the results of calculations in frames of the model of percolation channels with irregular dynamics of delocalization processes.

5.3.4 *Radiation-induced dielectric effect in polymers*

Irradiation of polymers produce isolated ion pairs, blobs and tracks, uncharged radiolysis products (free radicals, excited molecules and the stable chemical compounds), as well as free charges [Vannikov et al. (1982); Tyutnev et al. (2005)]. All of these agents may contribute to the radiation-induced dielectric effect (RDE). All of them, except free charges, are easily polarized in an external electric field. The free charges cause the redistribution of the electric field and the creation of a volume charge field. As a result, the inhomogeneous polarization of the Maxwell-Wagner type appears. Further, Tyutnev et al. (2005) tried to take into account the contribution of uncharged radiolysis products by using the simplified Debye approach. We use the fractional model of relaxation.

Assuming that the dynamics of the dipole moment $p(t)$ of an individual micro-relaxator is described by the following equation

$$\tau^{\alpha}\ {}_0\mathsf{D}_t^{\alpha}p(t) + p(t) = \beta E(t),$$

where $E(t) = E_0$ is a homogeneous stationary electric field in the sample, we write its solution as

$$p(t) = \beta E_0\{1 - E_{\alpha}[(-t/\tau)^{\alpha}]\}.$$

Following [Tyutnev et al. (2005)], we replace $p(t)$ by $P(t)$ for a macroscopic system consisting of n microrelaxators:

$$P(t) = n\frac{\beta}{\tau}\int_0^t E(t-t')(t'/\tau)^{\alpha-1}E_{\alpha,\alpha}[-(t'/\tau)^{\alpha}]\,dt'.$$

In case of a uniform volume generation of microrelaxators with the rate g_0, the polarization response to a sinusoidal field in the quasi-stationary regime ($t \gg \tau$) is given by

$$P(t) = \frac{g_0 \beta F_0}{1 + (i\omega\tau)^\alpha} \exp(i\omega t)$$

In case $\alpha = 1$, this coincides with the expression obtained in [Tyutnev et al. (2005)].

At a constant rate $g(t) = g_0$, the microrelaxators component of the current has the form

$$j_r(t) = g_0 \beta E_0 \{1 - E_\alpha[(-t/\tau)^\alpha]\}.$$

Tyutnev et al. (2005) consider the contribution to the RDE of the isolated ion pairs, following [Zubov et al. (1972)], while retaining only the linear approximation. In the framework of fractional differential kinetics, it is formulated as

$$\mathsf{D}_t^\alpha P(t) + \frac{p(t)}{\tau_i^\alpha} = \beta_1 E(t)\delta(t),$$

where $t = 0$ relates to the ion pair generation. Another approach belonging to the same authors assumes that the geminate pair polarization is instantaneous, and $p(t) = e\bar{\mu}_0\bar{\tau}_0 E(t)$, meanwhile the decay after the generation is described by a relaxation function $f_{\text{rel}}(t - t')$, where t' relates to the birth of the pair. Then

$$P(t) = \gamma_p F_0 \int_0^t \exp(i\omega t) f_{rel}(t - t')dt'.$$

Zubov et al. [Zubov et al. (1972)] use in their model the Debye relaxation function, Tyutnev et al. [Tyutnev et al. (2005)] use the function $f_{rel} = \exp(-t^2/\zeta^2)$. The choice of a fractional exponent leads to results very close to the reported earlier in this section.

As Tyutnev et al. (2005) report, the detailed consideration of the nonstationary radiation electric alternating current is connected with taking into account the kinetics of geminate pair polarization and relaxation in a dispersive transport regime. A part of this gap is covered by the research presented in [Nikitenko (2006)].

5.3.5 *Hysteresis in ferroelectric ceramics*

In the works [Guyomar et al. (2007, 2008)], the fractional model is used to describe some characteristics of the processes observed in ferromagnetic materials. Thus, the polarization response to a time varying electric field forms the basis for potentially important computer engineering applications, such as ferroelectric random access memory. A common method of observing these property employs hysteresis loops of polarization field versus electric field, $P = P(E)$. The fractional approach developed in [Guyomar et al. (2007)] associates static hysteresis and a time-dependent losses term due to the resistive term introduced as the product of a resistor ρ and a fractional polarization derivative term $d^\alpha P/dt^\alpha$.

Below the Curie temperature, the major hysteresis $P(E)$ is usually approximated by a translation of the anhysteretic curve as

$$E - E_c \mathrm{sign}\left(\frac{dP}{dt}\right) + \rho\frac{dP}{dt} = f^{-1}(P),$$

where E_c is the coercive field and $f^{-1}(P)$ represents the behaviour of a perfect dielectric. However the comparison of the calculations results with experimental data manifests discrepancies. As supposed in (Guyomar et al., 2007) this is related to the fact that the resistive term $\rho dP/dt$ leads to an overestimation of the high frequency component of the polarization signal. To overcome this problem, the authors of the cited work have made the replacement

$$\rho\frac{dP}{dt} \quad \rightarrow \quad \rho\ {}_0\mathsf{D}_t^\alpha P.$$

To identify each parameter (α, ρ) of the dynamical term one only needs two major hysteresis loops (static and 10 Hz, for instance). For a typical soft piezoelectric transducer composition studied in the cited work, $\rho = 0.01$ and $\alpha = 0.3$.

5.4 The Havriliak-Negami kinetics

5.4.1 *The Cole-Davidson response*

We start this section with consideration of Cole-Davidson response function. Its frequency representation

$$[1 + i\omega\tau]^\beta\, \tilde{\phi}(\omega) = 1$$

may suggest that the time counterpart of this function is written as

$${}_0\left[1 + \tau\mathsf{D}\right]_t^\beta\, \phi(t) = \delta(t).$$

In order to explain this type of relaxation, Nigmatullin and Ryabov [Nigmatullin & Ryabov (1987)] represent the ordinary differential equation for the Debye relaxation as form

$$\exp(-t/\tau)\mathsf{D}_t \exp(t/\tau)\phi(t) = 0. \tag{5.24}$$

The authors interpret a micro-level relaxation by means of the Frölich relaxator that is a system with two equilibrium states ([Fröhlich (1949)]). In this model, the relaxation process begins when the external field induces energy difference between states in each domains of the double potential well. The Fröhlich model is known to be used for a wide class of dielectrics. [Nigmatullin & Ryabov (1987)] modify this model and consider the function $G(t) = \exp(t/\tau)\phi(t)$. If $\phi(t)$ is a solution of Eq. (5.24), then $G(t) = \phi(0)$ is constant and

$$\mathsf{D}_t G(t) = 0. \tag{5.25}$$

In some periods, the system occurs in an equilibrium so that the function $G(t)$ equals zero but is not equal to $\phi(0)$. Such a behavior can be cause by the thermal fluctuations of local fields, shielding the external field. In addition, the authors suggest that times when $G(t) = \phi(0)$ form a fractal set. In other words, at times, coinciding with the points of a self-similar sets $G(t) = \phi(0)$, whereas at other times $G(t) = 0$. The authors report that integrating $G(t)$ over this fractal set and averaging over some ensemble of its realizations yields [Nigmatullin & Ryabov (1987)]:

$$ {}_0\mathsf{D}_t^{-1} G(t) = A \; {}_0\mathsf{D}_t^{-\beta}[\phi(0)], \tag{5.26} $$

where A is a constant depending on the fractal set structure and $\beta \in (0, 1]$ is a fractal dimensionality. Using the properties of the operators of fractional integration and differentiation, one can rewrite the equation (5.26) as

$$ {}_0\mathsf{D}_t^{\beta}[G(t)] = 0. \tag{5.27} $$

In other words, if we consider a system that interacts with an external field during the relaxation process, and this interaction has an intermittent self-similar pattern, then Eq. (5.25) for the function $G(t)$ is replaced by Eq. (5.27). The equation for the response function $\phi(t)$ can be written in the form

$$ \exp(-t/\tau) \; {}_0\mathsf{D}_t^{\beta}[\exp(t/\tau)\phi(t)] = 0. $$

By using the operator interrelation [Nigmatullin & Ryabov (1987)]

$$ \exp(-\Omega u \; {}_0\mathsf{D}_t^{1-\varepsilon}) \; {}_0\mathsf{D}_t^{\alpha} \exp(\Omega u \; {}_0\mathsf{D}_t^{1-\varepsilon}) = ({}_0\mathsf{D}_t^{\varepsilon} + \Omega\varepsilon)^{\alpha/\varepsilon}, $$

$$ 0 < \varepsilon \leq 1, \quad \alpha \leq \varepsilon, $$

Nigmatullin and Ryabov represent this equation as

$$ (\mathsf{D}_t + \tau^{-1})^{\beta}\phi(t) = 0. $$

Calderwood [Calderwood (2004)] considering the CD relaxation in a liquid also believes that the physical reason of such a behavior is in the spatio-temporal fluctuations which continuously occur in the liquid. Assuming existence of two types of dipoles, one of each are able to vibrate more easily than others and so make larger swings when an a.c. field is applied, Calderwood has showed that this assumption predicts a skewed arc of the CD type. But the CD equation itself is considered by Calderwood as a consequence of the existence of a wide distribution of relaxation times rather than that of the single relaxation time.

5.4.2 *Fractional kinetics and Havriliak-Negami response*

As noted in section above, the most popular and quite common approximation for the frequency response function is given by a two-parameter formula proposed in [Havriliak & Negami (1966)]. The solution of the corresponding fractional equation

$$ [1 + (\tau \; {}_0\mathsf{D}_t)^{\alpha}]^{\beta} \, \phi(t) = \delta(t), \tag{5.28} $$

is based on the expansion of fractional power of the operator sum in power series

$$[1 + (\tau\ {}_0\mathsf{D}_t)^\alpha]^\beta = \sum_{n=0}^{\infty} \binom{\beta}{n} (\tau\ {}_0\mathsf{D}_t)^{\alpha(\beta-n)} .$$

The result has the form:

$$\phi(t) = -\frac{1}{\Gamma(\beta)} \sum_{n=0}^{\infty} \frac{(-1)^n \Gamma(n+\beta)}{n! \Gamma(\alpha(n+\beta))} \left(\frac{t}{\tau}\right)^{\alpha(n+\beta)} .$$

Note, that the HN operator

$$\mathsf{W}^{\alpha,\beta}_{\tau^{-1}} \equiv [1 + \tau^\alpha ({}_0\mathsf{D}_t)^\alpha]^\beta$$

can be written as

$$\mathsf{W}^{\alpha,\beta}_{\omega_p} = \omega_p^{-\alpha\beta} \exp\left(-\frac{\omega_p^\alpha t}{\alpha}\ {}_0\mathsf{D}_t^{1-\alpha}\right)\ {}_0\mathsf{D}_t^{\alpha\beta} \exp\left(\frac{\omega_p^\alpha t}{\alpha}\ {}_0\mathsf{D}_t^{1-\alpha}\right) .$$

The Havriliak-Negami function is an explicit version of the universal relaxation law [Jonscher (1986)]. This analogy encourages the search for a suitable stochastic model for the universal relaxation law. Studies of this type were carried out in many papers (see, e.g., [Weron & Kotulski (1996); Weron, (1991); Nigmatullin (1984c); Nigmatullin & Ryabov (1987); Glöckle & Nonnenmacher (1993); Jurlewicz & Weron (2000); Coffey et al. (2002); Déjardin (2003); Aydiner (2005)]). Weron et al(2005) have demonstrated how the random walk scheme underlying the Debye response should be modified to get the empirical function Havriliak-Negami function. In addition, they obtained a formula for generating random variables with the corresponding probability density. To get these results, Weron et al consider the process in terms of the subordinated processes theory. Starting with the one-sided stable process, that is, with a Markov process $S(t)$ obeying the condition

$$S(t) \stackrel{d}{=} t^\alpha S_+(\alpha),$$

they take it as an *operational time*, and $T(s)$ as its inverse, that is

$$T(S(t)) = t,$$

while $S(T(s)) > s$. The random variable $T(s)$ corresponds to the first passage time of the strictly increasing process $S(t)$ above t, and the subordinator $S(T(s))$ results in stretching the real time s. With reference to the theory of subordination, the CC process shows that the dipoles tend to equilibrium via motion alternating with stops so that the temporal intervals between them are random. Formally [Stanislavsky (2003)], it has been revealed in passage from the Debye relaxation equation

$$[\mathsf{D}_t + \omega_{\mathrm{D}}]\, \phi_D(t) = \delta(t)$$

to the Cole-Cole one:

$$[\ {}_0\mathsf{D}_t^\alpha + \omega_{\mathrm{CC}}^\alpha]\, \phi_D(t) = \delta_\alpha(t).$$

By analogy with quantum-mechanical quantization, this operation can be named the *primary subordination*, whereas the next similar operation leading to the HN-process equation,

$$[\,{}_0\mathsf{D}_t^{\alpha} + \omega_{\mathrm{CC}}^{\alpha}]^{\beta}\,\phi_D(t) = \delta_{\alpha\beta}(t),$$

may be referred to as a secondary subordination. An alternative (in some sense) stochastic interpretation of HN-process was proposed in [Weron, 2005] on the base of a space-time coupled model of a special kind. Namely, considering a random walk around a regular spatio-temporal lattice, they start with random displacements and trapping times as

$$R_j = M_j \Delta R, \quad T_j = M_j \Delta R,$$

where $\Delta R > 0$ and $\Delta T > 0$ are constant and M_j are the common multipliers forming a sequence of positive integer-valued identically distributed independent random variables. Replacing deterministic steps ΔR and ΔT by their random counterparts δR and δT and assuming that the conditions

$$x^{\alpha}\mathrm{Prob}(\Delta R > x) \propto 1$$

and

$$x^{\alpha}\mathrm{Prob}(\Delta T > t) \propto 1$$

as $x \to \infty$, and moreover that the random number M_i has also a heavy tile with exponent $\beta \in (0, 1)$, i.e.

$$x^{\beta}\mathrm{Prob}(M > x) \propto 1,$$

the authors show that the relaxation function $\phi(t)$ corresponding to the diffusion front of this process is related to the Havriliak-Negami function. At the same time, they have found that the classical relaxation equation

$$\frac{d\phi(t)}{dt} = -r(t)\phi(t)$$

with time-dependent transition rate

$$r(t) = \frac{\sum_{k=0}^{\infty}(-1)^k(\gamma + k)\alpha a_k \omega_{\mathrm{HN}}(\omega_{\mathrm{HN}}t)^{\alpha(\gamma+k)-1}}{1 - \sum_{k=0}^{\infty}(-1)^k a_k \omega_{\mathrm{HN}}(\omega_{\mathrm{HN}}t)^{\alpha(\gamma+k)-1}}.$$

In [Coffey et al. (2002)], the Debye theory of dielectric relaxation in polar dielectrics was reformulated in terms of fractional Fokker-Planck equation for rotational diffusion. Déjardin (2003) considered the fractional differential approach to the orientational motion of polar molecules, which occurs in accordance with the external perturbation. The problem was treated in terms of rotational diffusion of inertialess in the configuration space, which led to the fractional Smoluchowski equation. This model showed a good agreement with experimental data for the spectrum of the nonlinear dielectric relaxation of third order in the ferromagnetic liquid crystal.

5.4.3 *Stochastic inversion of the Havriliak-Negami operator*

Now, we describe a Monte Carlo algorithm for inverting the HN operator which can be taken as a basis for numerical method to solve fractional equations related to non-Debye relaxation problems with various initial pre-histories.

Recalling that

$$\hat{g}_+(\lambda;\alpha) = \int_0^\infty e^{-\lambda t} g_+(t;\alpha)dt = e^{-\lambda^\alpha}, \quad \alpha \in (0,1),$$

we rewrite this expression as

$$e^{-\lambda^\alpha t} = \int_0^\infty e^{-\lambda t^{1/\alpha} s} g_+(s;\alpha)ds = \int_0^\infty e^{-\lambda t'} g_+(t^{-1/\alpha}t';\alpha)t^{-1/\alpha}dt'.$$

Replacing the number variable $-\lambda$ by the operator one A leads us to the relation

$$e^{-(-\mathsf{A})^\alpha t} = \int_0^\infty e^{\mathsf{A}t'} g_+(t^{-1/\alpha}t';\alpha)t^{-1/\alpha}dt'.$$

In terms of functional analysis, operators $\mathsf{T}_\alpha(t) \equiv e^{-(-\mathsf{A})^\alpha t}, \quad \alpha \in (0,1]$, form semigroups generated by infinitesimal operators $-(-\mathsf{A})^\alpha$. For different α values, they are linked by interrelation

$$\mathsf{T}_\alpha(t) = \int_0^\infty \mathsf{T}_1(t')g_+(t^{-1/\alpha}t';\alpha)t^{-1/\alpha}dt',$$

which is known as the *Bochner-Phillips theorem* [Yosida (1980)]. To make it more clear for applications, we apply both sides to some function φ belonging to the domain of the operator A. Introducing the notation

$$f_1(t) = \mathsf{T}_1(t)\varphi, \quad f_\alpha(t) = \mathsf{T}_\alpha(t)\varphi,$$

we obtain

$$f_\alpha(t) = \int_0^\infty f_1(t')g_+(t^{-1/\alpha}t';\alpha)t^{-1/\alpha}dt',$$

where $f_\alpha(t)$ is the solution of the Cauchy problem

$$\frac{df_\alpha(t)}{dt} = -(-\mathsf{A})^\alpha f_\alpha(t) \quad f_\alpha(0) = \varphi.$$

Returning to variable s,

$$f_\alpha(t) = \int_0^\infty f_1(t^{1/\alpha}s)g_+(s;\alpha)ds,$$

we represent it as a mean value,

$$f_\alpha(t) = \left\langle f_1(t^{1/\alpha}S(\alpha))\right\rangle$$

where $S(\alpha)$ is the random variable distributed according to one-sided stable density $g_+(t;\alpha), \quad \alpha \in (0,1)$.

The infinitesimal operator $-[1 +_{-\infty} \mathsf{D}_t^\alpha]$ generates the semigroup

$$T(t) = e^{-t} e^{-t_{-\infty}\mathsf{D}_z^\alpha},$$

According to the Bochner-Phillips relation, the semigroup generated by this infinitesimal operator $[1 +_{-\infty} \mathsf{D}_t^\alpha]^\beta$, where $\beta < 1$, has the form

$$\widehat{T}_t\, f = \int_0^\infty t^{-1/\beta} g_+(t^{-1/\beta}\tau;\alpha)\, \widetilde{T}_\tau f\, d\tau =$$

$$\int_0^\infty d\tau\, e^{-\tau}\, t^{-1/\beta} g_+(t^{-1/\beta}\tau;\beta) \int_0^\infty \tau^{-1/\alpha} g_+(\tau^{-1/\alpha}u;\alpha)\, T_u f\, du$$

Considering this integral as an average over ensemble of stable variables, we obtain the following equation

$$\widehat{T}_t\, f = \left\langle \exp\left(-t^{1/\beta} S_\beta\right) f\left(z - \left[t^{1/\beta} S_\beta\right]^{1/\alpha} S_\alpha\right)\right\rangle$$

Knowing this semigroup, we can find the corresponding infinitesimal operator $[1 +_{-\infty} \mathsf{D}_t^\alpha]^\beta$.

To find the inverse operator

$$[1 +_{-\infty} \mathsf{D}_t^\alpha]^{-\beta}, \qquad 0 < \alpha, \beta < 1,$$

we use the relation for a potential operator

$$A^{-1} f = \int_0^\infty T_s f ds.$$

Hence,

$$[1 +_{-\infty} \mathsf{D}_t^\alpha]^{-\beta}\phi = \left\langle \int_0^\infty \exp\left(-t^{1/\beta} S_\beta\right)\, f\left(z - \left(t^{1/\beta} S_\beta\right)^{1/\alpha} S_\alpha\right) dt\right\rangle.$$

Here S_α and S_β are the one-sided random stable variables with characteristic exponents $0 < \alpha \leq 1 \;\; 0 < \beta \leq 1$. Introducing exponentially distributed random variable U, we arrive at

$$[1 +_{-\infty} \mathsf{D}_t^\alpha]^{-\beta} f = \left\langle \int_0^\infty e^{-\xi}\, f\left(z - S_\alpha \xi^{1/\alpha}\right) \frac{\beta\xi^{\beta-1}}{S_\beta^\beta} d\xi\right\rangle$$

$$= \beta \left\langle S_\beta^{-\beta} U^{\beta-1} f\left(z - S_\alpha U^{1/\alpha}\right)\right\rangle \tag{5.29}$$

We use this formula to solve the fractional relaxation equation for arbitrary prehistory of the charging process.

Substituting frequency dependence of the Havriliak-Negami permittivity

$$\varepsilon^*(\omega) = \varepsilon_\infty + \frac{\varepsilon_s - \varepsilon_\infty}{[1+(i\omega/\omega_p)^\alpha]^\beta}$$

into the Fourier transform of the relation between electric induction and field strength $\widetilde{\mathbf{D}}(\omega) = \varepsilon^*(\omega)\, \widetilde{\mathbf{E}}(\omega)$ and invert it, we obtain

$$\mathbf{D}(t) = \varepsilon_\infty \mathbf{E}(t) + (\varepsilon_s - \varepsilon_\infty)\, \omega_p^{\alpha\beta} \times$$

$$\exp\left(-\frac{\omega_p^\alpha}{\alpha}\, t\ {}_{-\infty}\mathsf{D}_t^{1-\alpha}\right)\ {}_{-\infty}\mathsf{I}_t^{\alpha\beta} \exp\left(\frac{\omega_p^\alpha}{\alpha}\, t\ {}_{-\infty}\mathsf{D}_t^{1-\alpha}\right) \mathbf{E}(t).$$

Here, we arrive at fractional operators of special form

$$\mathsf{W}_{\omega_p}^{\alpha,\beta} f(t) = [1 + \omega_p^{-\alpha}\ {}_{-\infty}\mathsf{D}_t^\alpha]^\beta f(t)$$

$$\omega_p^{-\alpha\beta} \exp\left(-\frac{\omega_p^\alpha t}{\alpha}\ {}_{-\infty}\mathsf{D}_t^{1-\alpha}\right)\ {}_{-\infty}\mathsf{D}_t^{\alpha\beta} \exp\left(\frac{\omega_p^\alpha t}{\alpha}\ {}_{-\infty}\mathsf{D}_t^{1-\alpha}\right) f(t).$$

The inverse operator has form

$$\mathsf{W}_{\omega_p}^{\alpha,-\beta} f(t) = [1 + \omega_p^{-\alpha}\ {}_{-\infty}\mathsf{D}_t^\alpha]^{-\beta} f(t)$$

$$= \omega_p^{\alpha\beta} \exp\left(-\frac{\omega_p^\alpha}{\alpha}\, t\ {}_{-\infty}\mathsf{D}_t^{1-\alpha}\right)\ {}_{-\infty}\mathsf{I}_t^{\alpha\beta} \exp\left(\frac{\omega_p^\alpha}{\alpha}\, t\ {}_{-\infty}\mathsf{D}_t^{1-\alpha}\right) f(t).$$

Following asymptotical relations result from this expression:

$$\mathsf{W}_{\omega_p}^{\alpha,\beta} f(t) \sim \begin{cases} [1 + \beta\omega_p^{-\alpha}\ {}_{-\infty}\mathsf{D}_t^\alpha] f(t),\ t \gg 1/\omega_p, \\ \\ \omega_p^{-\alpha\beta}\ {}_{-\infty}\mathsf{D}_t^{\alpha\beta} f(t), \quad t \ll 1/\omega_p, \end{cases} \quad \alpha < 1. \tag{5.30}$$

5.4.4 *Three-power term approximation of the HN-relaxation*

There exists a simple but approximate way to avoid the trouble with the complicated generalization of fractional operator, at least for the less typical region L ($\beta > 1$). Indeed, the HN-function is an empirical one and as a result it is an approximate expression for real dependence. Thus, it is not necessary to work with namely this expression: one can find another approximate formula fitting the same experimental data, for example

$$[c_0 + c(i\omega)^\gamma + d(i\omega)^\delta + e(i\omega)^\varepsilon]\tilde{\phi}_{\gamma\delta\varepsilon}(i\omega) = 1. \tag{5.31}$$

Numerical calculations show that the HN-function[1].

$$\tilde{\phi}_\alpha^\beta(z) = \frac{1}{[1+z^\alpha]^{\beta/\alpha}} \tag{5.32}$$

can be replaced by the function

[1]Observe, we use further β/α instead of β

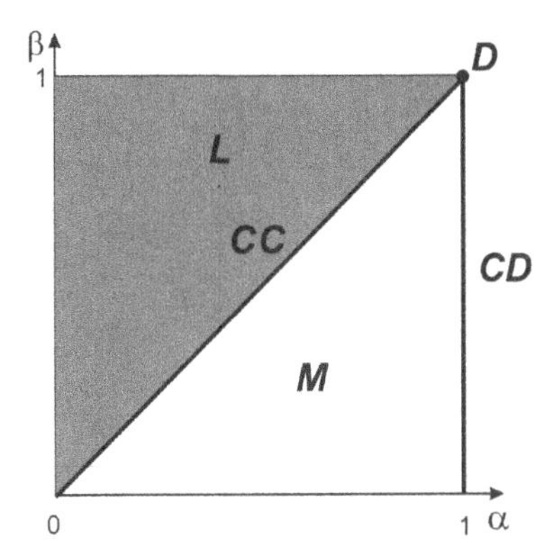

Figure 5.9 Various types of dielectric response. Abbreviation: D – Debye law, CC – Cole-Cole law, CD – Cole-Davidson law, M – region of more typical behavior, L – region of less typical behavior.

$$\tilde{\phi}_{\gamma\delta\varepsilon}(z) = \frac{1}{c_0 + cz^\gamma + dz^\delta + ez^\varepsilon}, \quad \gamma < \delta < \varepsilon \tag{5.33}$$

in the case $\alpha < \beta$. The constants $c_0, c, d, e, \gamma, \delta$ and ε can be determined from the conditions

$$\lim_{z\to\infty} \left[\tilde{\phi}_{\gamma\delta\varepsilon}(z) \Big/ \tilde{\phi}_\alpha^\beta(z)\right] = 1$$

and

$$\lim_{z\to 0} \left[1 - \tilde{\phi}_{\gamma\delta\varepsilon}(z)\right] \Big/ \left[1 - \tilde{\phi}_\alpha^\beta(z)\right] = 1\,,$$

$$\tilde{\phi}_{\gamma\delta\varepsilon}(1) = \tilde{\phi}_\alpha^\beta(1),$$

and

$$\tilde{\phi}'_{\gamma\delta\varepsilon}(1) = \tilde{\phi}'^\beta_\alpha(1)$$

The two first conditions yield $c_0 = e = 1,\ c = \beta/\alpha$, and $\gamma = \alpha$. From the third we have

$$d = 2^{\beta/\alpha} - 2 - \beta/\alpha,$$

and from the fourth

$$\delta = \beta\left[2^{(\beta-\alpha)} - 2\,/d\right].$$

The case $\beta = 2\alpha$ is obtained by the limit transition leads to the exact result

$$\tilde{\phi}_{\gamma\delta\varepsilon}(z) = \frac{1}{(1+z^\alpha)^2}.$$

Results of comparative calculations for real and imaginary components of function $\tilde{f}(i\omega)$ plotted in Fig. 5.10 show that Eq. (5.33) fit HN-function with an acceptable accuracy [Uchaikin (2003a)].

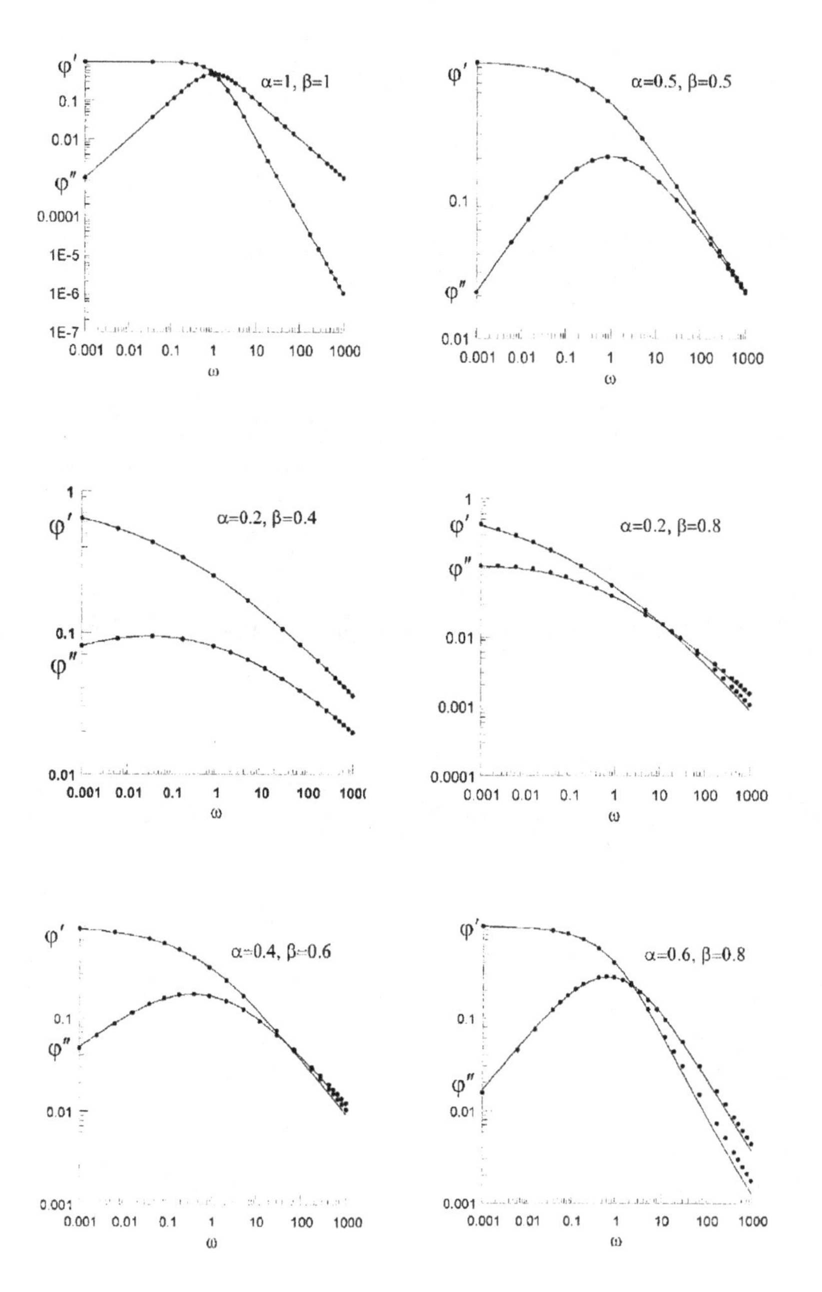

Figure 5.10 Comparison of approximation (5.33) for the Havriliak-Negami law (circles) with exact expression (5.32) (lines), $\varphi' = \mathrm{Re}\phi$, $\varphi'' = \mathrm{Im}\phi$.

5.4.5 *Pass-through conductivity and Raicu's response*

Raicu [Raicu (1999)] proposed an empirical function for the frequency dependence of the dielectric permeability, which summarizes the approximations considered above, satisfies the Jonscher universal response and in addition takes into account the contribution of constant current conductivity:

$$\varepsilon^*(\omega) = \varepsilon_\infty + \frac{\sigma}{i\omega} + \frac{\varepsilon_s - \varepsilon_\infty}{[(i\omega\tau)^{1-\beta} + (i\omega\tau)^\alpha]^\gamma},$$

$$\widetilde{\mathbf{D}}(\omega) = \varepsilon^*(\omega)\, \widetilde{\mathbf{E}}(\omega), \tag{5.34}$$

Interrelation between the electric displacement and electric field strength is written as

$$\mathbf{D}(t) = \varepsilon_\infty \mathbf{E}(t) + \sigma \int\limits_{-\infty}^{t} \mathbf{E}(t')dt' + (\varepsilon_s - \varepsilon_\infty)[\tau^{1-\beta}\ {}_{-\infty}\mathsf{D}_t^{1-\beta} + \tau^\alpha\ {}_{-\infty}\mathsf{D}_t^\alpha]^{-\gamma}\mathbf{E}(t)$$

For a plane capacitor

$$\frac{q}{S} = \frac{\varepsilon_\infty}{d}u(t) + \frac{\sigma}{d}\int\limits_{-\infty}^{t} u(t')dt' + \frac{\varepsilon_s - \varepsilon_\infty}{d}[\tau^{1-\beta}\ {}_{-\infty}\mathsf{D}_t^{1-\beta} + \tau^\alpha\ {}_{-\infty}\mathsf{D}_t^\alpha]^{-\gamma}u(t)$$

Differentiating this equation leads to fractional current-voltage relation

$$i(t) = C_\infty \frac{du(t)}{dt} + \frac{u(t)}{R} + (C_s - C_\infty)\frac{d}{dt}\,[\tau^{1-\beta}\mathsf{D}_t^{1-\beta} + \tau^\alpha \mathsf{D}_t^\alpha]^{-\gamma}u(t),$$

which may be used in the method of polarization-depolarization currents for testing insulating materials. The algorithm for Monte Carlo computing of fractional powers of operators presented in this formula, as described above.

5.4.6 *Fractional waves in the HN dielectrics*

The Maxwell equations

$$\text{rot}\ \mathbf{H} = \frac{4\pi}{c}\mathbf{j} + \frac{1}{c}\frac{\partial \mathbf{D}}{\partial t}, \qquad \text{rot}\ \mathbf{E} = -\frac{1}{c}\frac{\partial \mathbf{B}}{\partial t}$$

together with constitutive relations

$$\mathbf{D} = \varepsilon_\infty \mathbf{E} + (\varepsilon_s - \varepsilon_\infty)[1 + \omega_p^{-1}\ {}_{-\infty}\mathsf{D}_t^\alpha]^{-\beta}\ \mathbf{E}, \quad \mathbf{B} = \mu \mathbf{H}$$

lead to the following fractional wave equation

$$\frac{\mu\varepsilon_\infty}{c^2}\frac{\partial^2 \mathbf{E}}{\partial t^2} + \frac{\mu(\varepsilon_s - \varepsilon_\infty)}{c^2}[1 + \omega_p^{-1}\ {}_{-\infty}\mathsf{D}_t^\alpha]^{-\beta}\ \frac{\partial^2 \mathbf{E}}{\partial t^2} +$$

$$+\nabla(\text{div}\ \mathbf{E}) - \nabla^2 \mathbf{E} = \frac{4\pi\mu}{c^2}\frac{\partial \mathbf{j}}{\partial t}. \tag{5.35}$$

At short times ($t \ll 1/\omega_p$)

$$\frac{\mu\varepsilon_\infty}{c^2}\frac{\partial^2 \mathbf{E}}{\partial t^2} + \frac{\mu(\varepsilon_s - \varepsilon_\infty)}{c^2}\omega_p^{\alpha\beta}\ {}_{-\infty}\mathsf{D}_t^{2-\alpha\beta}\ \mathbf{E} + \nabla(\text{div}\ \mathbf{E}) - \nabla^2 \mathbf{E} = \frac{4\pi\mu}{c^2}\frac{\partial \mathbf{j}}{\partial t},$$

but the long time asymptotic behavior reads as

$$\frac{\mu\varepsilon_s}{c^2}\frac{\partial^2 \mathbf{E}}{\partial t^2} - \frac{\mu(\varepsilon_s - \varepsilon_\infty)}{c^2}\beta\omega_p^{-\alpha}\ {}_{-\infty}\mathsf{D}_t^{2+\alpha}\ \mathbf{E} + \nabla(\text{div}\ \mathbf{E}) - \nabla^2 \mathbf{E} = \frac{4\pi\mu}{c^2}\frac{\partial \mathbf{j}}{\partial t}.$$

These equations are in agreement with those asymptotical ones derived by [Tarasov (2008b)] from the Jonscher universal law. He obtained for short times $t \ll \omega_p^{-1}$,

$$\triangle \mathbf{E} - \text{grad div } \mathbf{E} - \frac{1}{c^2}\frac{\partial^2 \mathbf{E}}{\partial t^2} - \frac{\chi_\mu \omega_p^\mu}{c^2} {}_{-\infty}\mathsf{D}_t^{2-\mu}\mathbf{E} = \mu_0 \frac{\partial \mathbf{j}}{\partial t},$$

and for long times $t \gg \omega_p^{-1}$,

$$\triangle \mathbf{E} - \text{grad div } \mathbf{E} - \frac{1+\chi(0)}{c^2}\frac{\partial^2 \mathbf{E}}{\partial t^2} - \frac{\chi_\nu \omega_p^{-\nu}}{c^2} {}_{-\infty}\mathsf{D}_t^{2+\nu}\mathbf{E} = \mu_0 \frac{\partial \mathbf{j}}{\partial t}.$$

The current density is the sum of the conduction current and the displacement current associated with polarization in the approximation of a homogeneous in a sample electric field $\mathbf{E}(t)$ has the form:

$$\mathbf{j}(t) = \sigma \mathbf{E}(t) + \frac{d\mathbf{D}(t)}{dt}, \quad \mathbf{D}(t) = \varepsilon_\infty \varepsilon_0 \mathbf{E}(t) + \Delta \mathbf{P}(t).$$

Here, σ is the constant current conductivity, $\mathbf{D}(t)$ the electrical displacement. The polarization vector is linked the strength by the hereditary relation:

$$\Delta \mathbf{P}(t) = \varepsilon_0 \int_0^\infty \phi(t')\mathbf{E}(t-t')dt'.$$

The current-voltage in this case is

$$I(t) = K\frac{d}{dt}\int_0^\infty \phi(t')\, U(t-t')dt'.$$

Passing to the Fourier domain yields

$$\frac{\tilde{I}(\omega)}{\tilde{\phi}(\omega)} = K \cdot i\omega \cdot \tilde{U}(\omega).$$

For the HN dielectrics

$$[1 + (j\omega\tau)^\alpha]^\beta \, \tilde{I}(\omega) = K \cdot i\omega \cdot \tilde{U}(\omega)$$

and correspondingly

$$[1 + \tau^\alpha \, {}_{-\infty}\mathsf{D}_t^\alpha]^\beta \, I(t) = g(t), \quad g(t) = K\dot{V}(t)$$

In case of a step-function $V(t) = V_0 1(t)$ we have equation

$$[1 + \tau^\alpha \, {}_{-\infty}\mathsf{D}_t^\alpha]^\beta \, i(t) = KV_0\delta(t),$$

solution of which is expressed through three-parameter Mittag-Leffler function [Prabhakar (1971)],

$$E_{\alpha,\gamma}^\beta(z) = \sum_{n=0}^\infty \frac{\Gamma(\beta+n)}{\Gamma(\beta)\Gamma(\alpha n+\gamma)n!}z^n.$$

Using its property

$$\int_0^\infty e^{-st}\, t^{\gamma-1}E_{\alpha,\gamma}^\beta(at^\alpha)dt = \frac{s^{\alpha\beta-\gamma}}{(s^\alpha - a)^\beta},$$

we arrive at the result

$$f(t) = i(t)/KV_0 = \tau^{-\alpha\beta}t^{\alpha\beta-1}E_{\alpha,\alpha\beta}^\beta\left(-(t/\tau)^\alpha\right).$$

5.5 The Kohlrausch-Williams-Watts kinetics

5.5.1 *The KWW relaxation function*

One can often encounter the non-Debye relaxation law approximated by the stretched exponential function,

$$\phi(t) = \exp[-(t/\tau)^{\beta}], \qquad 0 < \beta \leq 1, \qquad \tau = \text{const}, \tag{5.36}$$

[Jonscher (1983); Raju (2003); Phillips (1996)]. The value of parameter β is usually dependent on the absolute temperature and chemical composition of the material. This law was proposed by Kohlrausch (1854), who considered dynamic processes in complex materials with slow relaxation. For dielectric systems, this relaxation law was discussed by Williams and Watts [Williams & Watts (1970)]. Physical models leading to the KWW dielectric relaxation have been discussed in many works (see [Phillips (1996); Jonscher (1986)] reference therein). Other applications of KWW function can be found in works devoted to relaxation in amorphous and glassy materials and polymers near the glass transition temperature, long-term fading in the process of capturing, nonradiative recombination of excitons, etc.

In the case of electron relaxation, the stretched exponential law is often explained in terms of dispersive transport, discussed above. The index β is regarded as an adjustable parameter, but many researchers belive that there is no microscopic justification for $0 < \beta(T) < 1$, even near the glass transition temperature $T \approx T_g$ [Phillips (1996)]. The Phillips mathematical model treats relaxation kinetics using the Lifshitz-diffusion model of Kac-Luttinger liquids in the configuration space of an effective dimension.

Klafter & Shlesinger (1986) discussed three models: the model of direct Forster transitions, the hierarchical model of dynamics and the limited diffusion of defects. Each model is based on its own physical mechanism, but they have a similar rationale for the stretched exponential law: scale-invariant relaxation rates.

The Forster model of direct transitions appears when considering the excitation of the donor defects in solids [Forster (1949)]. Let $\mathbf{R}_d$ indicates the position of the donor, which can make a direct energy transfer into the defect with the coordinate $\mathbf{R}_i$. There are several defects around the donor. In order to find the relaxation function $\Phi(t)$, it is necessary to calculate the probability that the donor is excited up to time t. For a given configuration of the donor distribution, this probability is of the form $\prod_i \exp\left[-t\ w(\mathbf{R}_d \to \mathbf{R}_i)\right]$, where $w(R_d \to \mathbf{R}_i)$ is the rate of transition from a donor $\mathbf{R}_d$ in a static defect of $\mathbf{R}_i$. Let p be the probability that the node (defect) is occupied. In the case of an empty node $\mathbf{R}_i$, the transition rate $w(R_d \to \mathbf{R}_i)$ is 0. Then, in the case of weak defects in the population of nodes $p \ll 1$, we obtain [Forster (1949); Blumen (1981)]:

$$\Phi(t) = \prod_i{}' \{1 - p + p\ \exp[-t\ w(\mathbf{R}_d \to \mathbf{R}_i)]\}$$

$$\approx \exp\left[-p\sum_i{}' \{1 - \exp[-t\ w(\mathbf{R}_d \to \mathbf{R}_i)]\}\right]$$

$$\approx \exp\left[-p\int d\mathbf{R}\ \rho(\mathbf{R})\left\{1 - \exp[-t\ w(\mathbf{R})]\ \right\}\right].$$

Klafter & Shlesinger (1986) consider two kinds of transition rate: $W(\mathbf{R}) = a|\mathbf{R}|^{-s}$ and $W(\mathbf{R}) = B\exp(-\gamma|\mathbf{R}|)$. In the case of a uniform distribution of defects in d-dimensional space, $\rho(\mathbf{R}) = \rho_0$, calculations lead to the relaxation function, respectively, $\Phi(t) = \exp[(-t/\tau)^{d/s}]$ (stretched exponential) and $\Phi(t) = (Bt)^{-A\ln^{d-1}(Bt)}$ (enhanced power-law relaxation). Palmer et al. (1984) proposed a hierarchical model of dynamics for a limited relaxation in strongly interacting glassy materials. Klafter and Shlesinger (1986) discussed the model of consecutive relaxations. The way to equilibrium consists of a set of correlated successive steps of activation. Relaxation occurs in stages. The model assumes that the time scale of relaxation is subject to the same level of relaxation at the lower levels. Choosing weight functions in a special form, the authors come either to power type, or to the stretched exponential relaxation.

Glarum [Glarum (1960)] considered the model of defect diffusion to explain the dielectric behavior of isoamyl bromide. The model implies that the relaxation of the molecule is more likely after a neighboring molecule to relax. The stretched exponential law with exponent $\beta = 1/2$ arises in the case of one-dimensional space. In three-dimensional space, an exponential (Debye) relaxation is observed under these conditions. In order to obtain the KWW behavior of the relaxation with $0 < \beta < 1$, Shlesinger and Montroll [Schlesinger & Montroll (1984)] proposed the idea of a hierarchy of waiting times for defect jumps. Generalization of this model for the case of fractal media was examined in [Klafter & Blumen (1985); Zumofen et al. (1985)].

Klafter and Shlesinger (1986) have formulated a model of the one-dimensional diffusion of defects in mathematical form, using approximate arguments of Redner and Kang (1984). The target is located at the origin, and the defects are placed randomly around it. Let $f(R_1)$ be the probability of no defects in the circle of radius R_1 centered at the origin. For the case of uniformly distributed defects, $f(R_1) = \exp(-pR_1)$. The probability that the defect located at the R_i at time $t = 0$ does not get the origin to time t, is given by $\exp(-t/4R_i^2)$, where the diffusion coefficient of defects is taken to be unity. The relaxation function has the form

$$\phi(t) = \langle \phi(R_1, t)\rangle_{R_1}$$

$$\approx \left\langle \exp(-pR_1)\prod_{i=1}\exp\left(\frac{t}{4R_i^2}\right)\right\rangle \approx \left\langle \exp(-pR_1)\exp\left(-tp\int_{R_1}^{\infty}\frac{dR}{4R^2}\right)\right\rangle$$

$$= \left\langle \exp\left(-pR_1 + p\frac{t}{4R_1}\right)\right\rangle = \int_0^{\infty}\exp(-pR_1)\exp\left(-\frac{pt}{4R_1}\right)dR_1 = \sqrt{t}\ K_1(p\sqrt{t}),$$

where $K_1(x)$ is the Bessel function of second kind. For $p\sqrt{t} \gg 1$ the function $\phi(t)$ behaves like a stretched exponential with parameter $\beta = 1/2$. A generalization of this result to the other dimension gives

$$\phi(t) = \exp[-p\, S(t)],$$

where $S(t)$ is the average number of nodes visited by a defect during time t. For regular lattices [Klafter & Shlesinger (1986)]:

$$S(t) \propto \begin{cases} t^{1/2}, & d = 1, \\ t/\ln t, & d = 2, \\ t, & d = 3. \end{cases}$$

In case of a fractal structure, with the spectral dimension $\tilde{d}$

$$S(t) \propto \begin{cases} t^{\tilde{d}/2}, & \tilde{d} < 2, \\ t, & \tilde{d} > 2. \end{cases}$$

For waiting time distributions with heavy power tails, $q(t) \propto t^{-\alpha-1}$, $t \gg 1$ with $0 < \alpha < 1$, Shlesinger and Montroll obtained $S(t) \propto t^{\alpha/2}$ for a one-dimensional structure, and $S(t) \propto t^{\alpha}$ for a three-dimensional one. This leads to the KWW relaxation law with $\beta = \alpha/2$ and $\beta = \alpha$, respectively.

As mentioned above, in the framework of the diffusion model the reorientation of a molecule occurs preferably when one of the neighboring molecules have relaxed. Glarum said that is not necessary to postulate an exact way for the cooperative effect. The physical mechanisms may be different in different systems. The simplest interpretation of relaxation via defect is a vacancy. If such a defect reaches the molecule, the increased volume becomes available for the molecule reorientation. The long power-law tail in the distribution of waiting times $q(t)$ can have the same nature as in the case of dispersive transport in disordered semiconductors, and the physical mechanisms for this behavior were discussed in Chapter 2.

5.5.2 *Lévy-stable statistics and KWW relaxation*

5.5.2.1 *Relaxation in glassy materials*

Let us come back to the statistical approach, which interprets the non-exponential relaxation behavior (5.36) of the material in terms of a superposition of exponentially relaxing processes:

$$\Phi(t) = \int_0^\infty e^{-t/\tau} w(\tau)\, d\tau, \tag{5.37}$$

where $w(\tau)$ is the density of a distribution of relaxation times τ across different atoms, clusters, or degrees of freedom. If $\mu = \tau_D/\tau$, then μ is called the *relaxation rate* and is interpretable as dimensionless time. Substituting $s = t/\tau_0$ into (5.37), we obtain

$$\Phi(\tau_0 s) = \int_0^\infty e^{-s\mu} \tau_0 \mu^{-2} w(\tau_0/\mu)\, d\mu = \int_0^\infty e^{-s\mu} v(\mu)\, d\mu, \tag{5.38}$$

where

$$v(\mu) = \tau_0 \mu^{-2} w(\tau_0/\mu) \tag{5.39}$$

is the density of a distribution of dimensionless rates. Since this approach is microscopically arbitrary, one may consider the random variables $\mu_i = \tau_0^i$, $i = 1, \ldots, n$, as the possible relaxation rates of elements in a given complex material. Here n indicates the total number of the elements in the system, and μ_i are independent and identically distributed by (5.39) random variables.

Under these hypotheses,

$$\mu = \sum_{i=1}^{n} \mu_i,$$

and to use the limit theorem in the case of a large number of terms n, we need to introduce the normalization

$$\mu' = \frac{1}{b_n}\left(\sum_{i=1}^{n} \mu_i - a_n\right), \qquad b_n > 0. \tag{5.40}$$

As seen from Eqs. 5.36 and (5.38)

$$\int_0^\infty \mu v(\mu)\, d\mu = \infty.$$

Thus, if the variable μ' has a limit distribution, it should be a stable distribution with $\alpha < 1$ and $\beta = 1$. Then

$$a_n = 0, \qquad b_n = b_1 n^{1/\alpha}$$

and

$$v(\mu) = b_n g_+(b_n \mu; \alpha), \qquad 0 < \alpha < 1. \tag{5.41}$$

Substituting (5.41) into (5.38) and recalling the generalized limit theorem (Section 1.1), we arrive at formula (5.36).

As observed in [Weron & Kotulski (1996)], the statistical approach explains the universal character of formula (5.36) as the consequence of the use of universal limit law in macroscopic behavior of the relaxing system.

5.5.2.2 *Quantum decay theory*

The problem considered above is closely related to the general quantum decay theory.

In quantum mechanics, the state of an unstable physical system is described by the so-called state vector $|\psi(t)\rangle$, which is a solution of the time-dependent Cauchy problem for the Schrödinger equation

$$i\frac{\partial}{\partial t}|\psi(t)\rangle = H|\psi(t)\rangle, \tag{5.42}$$

where H is the Hamiltonian operator (Hermitian operator) corresponding to the system, and $|\psi(0)\rangle$ is a given initial state vector. The units are chosen so that the Planck constant $\hbar = 1$. Let $\{|\varphi_E\rangle, |\varphi_k\rangle\}$ be the complete system of eigenvectors of the operator H ($|\varphi_E\rangle$ corresponds to the absolutely continuous component of its spectrum, and $|\varphi_k\rangle$ corresponds to the discrete component), i.e.,

$$H|\varphi_E\rangle = E|\varphi_E\rangle, \qquad \langle\varphi_{E'}|\varphi_E\rangle = \delta(E' - E),$$
$$H|\varphi_k\rangle = E_k|\varphi_k\rangle, \qquad \langle\varphi_k|\varphi_l\rangle = \delta_{kl},$$

where $\delta(E' - E)$ is the Dirac delta function, and δ_{kl} is the Kronecker symbol.

We are interested in the probability $P(t)$ that at a time t the system is in the initial state $|\psi_0\rangle$. According to the laws of quantum mechanics,

$$P(t) = |\langle\psi(0)|\psi(t)\rangle|^2.$$

Solving the Cauchy problem (5.42) for the Schrödinger equation, we assume that $\langle\psi(0)|\psi(0)\rangle = 1$. In this case, the Fock–Krylov theorem [Krylov & Fock (1947)] yields

$$f(t) = \langle\psi(0)|\psi(t)\rangle = \sum_k |c_k|^2 \exp(-iE_k t) + \int_0^\infty |c(E)|^2 \exp(-iEt)\, dt, \tag{5.43}$$

where c_k and $c(E)$ are the Fourier coefficients in the expansion of the vector $|\psi(0)\rangle$ in the complete system $\{|\varphi_E\rangle, |\varphi_k\rangle\}$ of eigenvectors

$$|\psi(0)\rangle = \sum_k c_k|\varphi_k\rangle + \int_0^\infty c(E)|\varphi_E\rangle dE.$$

Thus, $f(t)$ can be interpreted as the characteristic function of some distribution having discrete components (probabilities of isolated values) $|c_k|^2$ and absolutely continuous component (i.e., density) $|c(E)|^2$. Instability of the system means that the probability $P(t) = |f(t)|^2$ of the system returning to the original state at time t tends to zero as $t \to \infty$.

Since $f(t)$ is a characteristic function, $|f(t)| \to 0$ only if the discrete components of the spectrum of H are missing, i.e., $c_k = 0$. In this case

$$f(t) = \int_0^\infty \rho(E) \exp(-iEt)\, dE, \tag{5.44}$$

where $\rho(E) = |c(E)|^2$ denotes the density of the energy distribution of the decaying physical system described by equation (5.42).

It turns out that for a very broad class of unstable physical systems the densities $\rho(E)$ are meromorphic functions. For a number of reasons, the case of a function $\rho(E)$ having only two simple poles (they are complex conjugated in view of the condition $\rho(E) \geq 0$) is of great interest. In this case it is obvious that

$$\rho(E) = A[(E - E_0)^2 + \Gamma^2]^{-1}, \qquad E \geq 0,$$

where A is a normalizing constant, and E_0 and Γ are the most probable value and the measure of dispersion (with respect to E_0) of the system's energy. For actual unstable systems[2] the ratio Γ/E_0 is very small, as a rule (10^{-15}, or even smaller). Therefore, to compute $P(t)$ we can, without adverse effects, replace the lower limit 0 in integral (5.44) by $-\infty$, after which the density function $\rho(E)$ and the probability $P(t)$ connected with it take the approximate expressions

$$\rho(E) \approx \frac{\Gamma}{\pi}[(E - E_0)^2 + \Gamma^2]^{-1},$$

$$\Phi(t) = |f(t)|^2 \approx \exp(-2\Gamma t).$$

It is clear from the first relation (the Lorentz distribution of the energy of the unstable system) that we are dealing with the Cauchy distribution, and it is clear from the second relation that the lifetime for unstable systems of the type under consideration behaves according to the exponential law.

Thus, the Cauchy law appears here only as a more or less good approximation of the real energy distribution for unstable systems. And there are situations where the replacement of 0 by $-\infty$ in (5.44) is unacceptable, because the corresponding law $P(t)$ of decay of the system differs essentially from the exponential law.

We touch here a result of [Hack (1982)]. Imposing the constraint normally applied in quantum theory that the self-adjoint Hamiltonian H is lower semi-bounded, i.e., that the energy spectrum is bounded below, Hack established that $\Phi(t)$ cannot decay exponentially fast as $t \to \infty$, i.e.,

$$\Phi(t) > Ce^{-at}$$

for $t > T$, where C, a and T are positive constants.

The following theorem is proved in [Weron & Weron (1985)].

Theorem 5.1. *The non-decay probability function for many-body weakly interacting quantum system is of the form*

$$\Phi(t) = \exp\{-at^{\alpha}\}, \quad a > 0, \quad 0 < \alpha < 1.$$

Representing the amplitude $f(t)$ as

$$f(t) = \langle\psi|\exp(-\mathcal{D}t)|\psi\rangle = \int_0^{\infty} \exp(-Et)p(E)\,dE$$

where $\mathcal{D}$ is the development operator governing the dynamic evolution of the quantum system under investigation and $p(E)$ is the probability density of the state $|\psi\rangle$ associated with the continuous spectrum of the development operator $\mathcal{D}$, the authors conclude that we observe an arbitrariness in the specification of ψ and $p(E)$. In general, one considers ψ to represent a decaying state for a many-body system, and therefore the number of components in the system should not influence the decay. In other words, the same decaying law should be obtained for one portion or several portions of the system. Consequently, in a weakly interacting quantum system, microscopic energies can be considered as independent identically distributed

[2] An example of such a system is a neutron with an average lifetime of 18.6 min decaying at the end of its lifetime into a photon, an electron, and a neutrino ($n \to p + e + \nu$).

energy random variables. The microscopic energy distribution $p(E)\,dE$ associated with the decaying system is identified to be the limit distribution of normalized sums of the microscopic energy random variables. By the limit theorem [Gnedenko & Kolmogorov (1954)], it is well known that the limit $p(E)\,dE$ has α-stable distribution $0 < \alpha \leq 2$. Since $p(E)$ is associated from the above construction with the development operator D, it has to have positive support. This holds only when $p(E)dE$ has a completely asymmetric ($\beta = 1$, $0 < \alpha < 1$) stable distribution [Weron & Weron (1985)].

5.5.3 *Fractional equation for KWW relaxation*

Nakhushev (2003) has derived a fractional equation for the stretched exponent

$$\Phi(t) = \exp\left[-\left(\frac{t}{\tau}\right)^{\alpha}\right], \qquad \tau = \text{const}, \quad 0 < \alpha < 1.$$

Passing to the dimensionless variables

$$x = \frac{t}{\tau}, \qquad y = \ln[\Phi(\tau x)]$$

reduces to a simple algebraic form

$$y = -x^{\alpha}.$$

This function satisfies the fractional equation

$${}_0\mathsf{D}_x^{\alpha+1} y(x) = 0. \tag{5.45}$$

Nakhushev (2003) shows that any solution of Eq. (5.45) is representable in the form

$$y(x) = (ax + b)x^{\alpha-1}, \tag{5.46}$$

where a and b are arbitrary constants. Indeed, let $y(x)$ be a solution of Eq. (5.45) from the class $L[0,l]$, then the equation

$$\frac{d^2}{dx^2}\, {}_0\mathsf{D}_x^{\alpha-1} y(x) = 0$$

implies

$${}_0\mathsf{D}_x^{\alpha-1} y(x) = Ax + B,$$

where A and B are constant values. For any function $f(x) \in L(0,l)$ and almost all $x \in [0,l]$, the following equality takes place:

$${}_0\mathsf{D}_x^{\nu}\, {}_0\mathsf{D}_x^{-\nu} f(x) = f(x), \quad \forall \nu > 0.$$

Inserting here $y(x)$ and $\nu = 1 - \alpha$, we get

$$y(x) = {}_0\mathsf{D}_x^{1-\alpha}(Ax + B).$$

For any ν and $\mu > -1$,

$$_0\mathsf{D}_x^{\nu} x^{\mu} = \frac{\Gamma(1+\mu)}{\Gamma(1+\mu-\nu)} x^{\mu-\nu},$$

therefore

$$_0\mathsf{D}_x^{1-\alpha} x = \frac{1}{\Gamma(1+\alpha} x^{\alpha},$$

and

$$_0\mathsf{D}_x^{1-\alpha} 1 = \frac{1}{\Gamma(\alpha} x^{\alpha-1}.$$

As a result, we have

$$y(x) = \frac{A}{\Gamma(1+\alpha)} x^{\alpha} + \frac{B}{\Gamma(\alpha)} x^{\alpha-1} = (ax+b)x^{\alpha-1}$$

with $\alpha = A/\Gamma(1+\alpha)$ and $b = B/\Gamma(\alpha)$. On the other hand, any function representable in the form of (5.46) is a solution of Eq. (5.45). Really,

$$_0\mathsf{D}_x^{\alpha+1}\left[(ax+b)x^{\alpha-1}\right] = a\,{}_0\mathsf{D}_x^{\alpha+1} x^{\alpha} + b\,{}_0\mathsf{D}_x^{\alpha+1} x^{\alpha-1} = b\,{}_0\mathsf{D}_x^{\alpha+1} x^{\alpha-1} = b\frac{\partial^2 \Gamma(\alpha)}{\partial x^2} = 0.$$

Chapter 6

The scale correspondence principle

Concluding our book, we would like to discuss an important, in our mind, question: How do relate integer-order and fractional-order to each other? At first glance, it is not hard to answer it: when fractional order ν reaches an integer number n, we observe a conventional integer-order operator. Of course, we have to perform the limit transition $\nu \to n$, but this is not a difficult task.

Therefore, each equation of fractional order ν should take a classical form when ν becomes an integer, and this form must be related to a real physical process. This statement may be called the *exponent correspondence principle.* Below, we consider the second statement being more significant for understanding of a role of fractional processes. We shall call it the *scale correspondence principle.* However, we start with discussion of infinity.

6.1 Finity and infinity

Dealing with long-tailed distributions we admitted that the related physical variables can take any large values. The first thought which crosses one's mind after acquaintance with distributions having infinite variances or infinite mean values is how a physical variable can occur to be infinite. All physical variables must be measurable, but we can not measure infinite values! Thus, the distributions do not have a physical sense. Where is a mistake here? This reasoning is wrong because mean values and variances are determined through improper integrals of the type

$$\int_{-\infty}^{\infty} xp(x)dx, \qquad \int_{-\infty}^{\infty} x^2p(x)dx,$$

which are not observable variables. In reality, the measured X_j and estimations

$$\frac{1}{n}\sum_{j=1}^{n} X_j, \qquad \frac{1}{n}\sum_{j=1}^{n} X_j^2,$$

of the integrals are always finite. Experimentally, we can never prove that some random variable is unbounded and its pdf continues "up to infinity". But we always find arguments to claim principally boundedness of physical variables X_j and others connected to them. As noticed in [Mantegna & Stanley (1994)], "in real physical systems, an unavoidable cutoff is always present". For example, in the case of a single molecule embedded in solids, the authors write, due to the minimal length between the molecule and the nearest two-level systems, a cutoff is present in the distribution of the jumps of the resonance frequency. In the Boltzmann lattice gas model used for numerical simulation, the jumps of the particles are limited by finite size of the simulated system. There exist other examples of cutoff reasons.

6.2 Intermediate space-asymptotics

Note that using the Central Limit Theorem, we ignore the cut-off idea although the Gaussian distribution have an infinite support as well. This is because the tails of an original distributions fall very quickly in this case and from practical point of view may be considered as if they were truncated. Tracing the development of the sum Σ_n in the case $\langle X_j^2 \rangle < \infty$ one can observe that for small n the Σ_n pdf depends on individual pdf of X_j, while for large n it has a normal (Gaussian) shape independently of $p_{X_j}(x)$. Thus, the n-axis is divided into two domains: initial domain (small n) and asymptotic domain (large n) separated by transition strip.

Let us now imagine that $p(x)$ meets the demands of the generalized limit theorem (GLT) (they say that $p(x)$ or X_j belongs to the domain of attraction of a correspondent Lévy stable law). In the region of small n, the Σ_n distribution will depend on original distribution as in the previous case, but at large n it takes the Lévy stable form. But what happens when long tailed original pdf will be truncated by cutting a very small tip? The original pdf doesn't practically change in a wide region of the power type tail, but all moments become finite. It seems to be evident that if the power type tail is very long up to the cut off threshold and the cut probability is very small, the sum does not observe the defect and follows to GLT, at least the number n is not too large. Everyone can verify this by Monte Carlo simulations: the sums Σ_n follow the Lévy stable distributions although all random numbers simulated have bounded distributions.

The situation changes when n becomes so large that we observe significant number of random numbers in vicinity of the threshold. The sum starts to feel the original distribution boundedness and change its limit way according to GLT. As Fig. 6.1 and 6.2 show that this is not only a thought result or a visual impression from simulation data, but is confirmed by the χ^2-analysis. This phenomena has been revealed in [Uchaikin (2002c)] and named "intermediate asymptotics". Fig. 6.3 demonstrates the trajectory of truncated Lévy flights on small, middle and large scales. The form of the trajectory changes from the Brownian type to the Lévy one if we increase the resolution of observation.

It is clearly see that two domains of n exist now: final (Gaussian) asymptotics and intermediate (Lévy) asymptotics, separated by transition interval. This proves

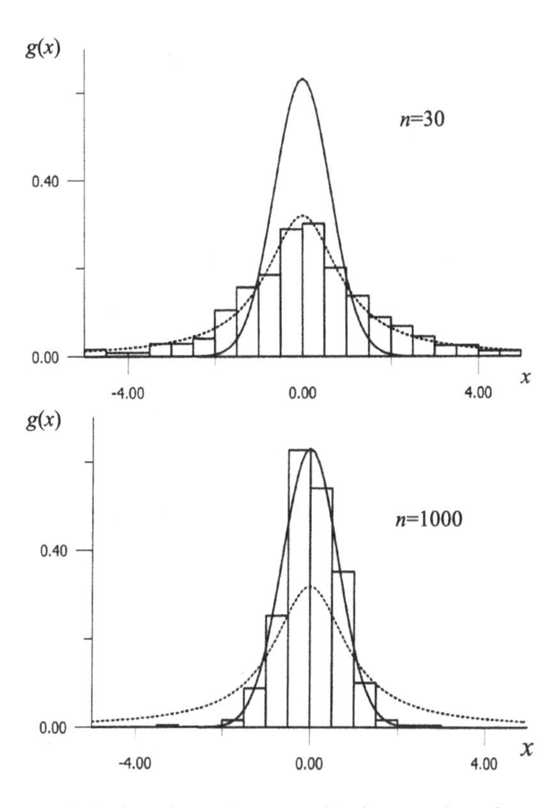

Figure 6.1 Comparison of the bar chart of renormalized sums of random variables having truncated power law tails with Cauchy (dotted line) and Gausian (solid line) distributions.

it would seem paradoxical statement: GLT can be applicable in the conditions of CLT if pdf has truncated power type tails.

6.3 Intermediate time-asymptotics

Before considering the similar situation with waiting time distribution, have a look at the Poisson process again. Let us take a fixed interval of unit length (0, 1), divide it into n bins, each of the equal length τ and consider a histogram showing the number of events in each bin (Fig. 6.4). Dealing with the standard Poisson process with rate μ, we can observe some bins to be empty if $\mu\tau$ is of order 1, and all bins to be filled if $\mu\tau \gg 1$. Relative fluctuations of the number of events are given by $1/\sqrt{\mu\tau}$ and tend to 0 as $\mu \to \infty$. For large values of μ, the distribution of events over bins looks almost uniform. If we introduce $R(\mu)$ as a proportion of empty bins, that is the ratio of mean number of bins to their total number, $R(\mu) = \langle n(\mu)\rangle/n$, we obtain a decreasing function of μ, rapidly vanishing with μ. It is very important

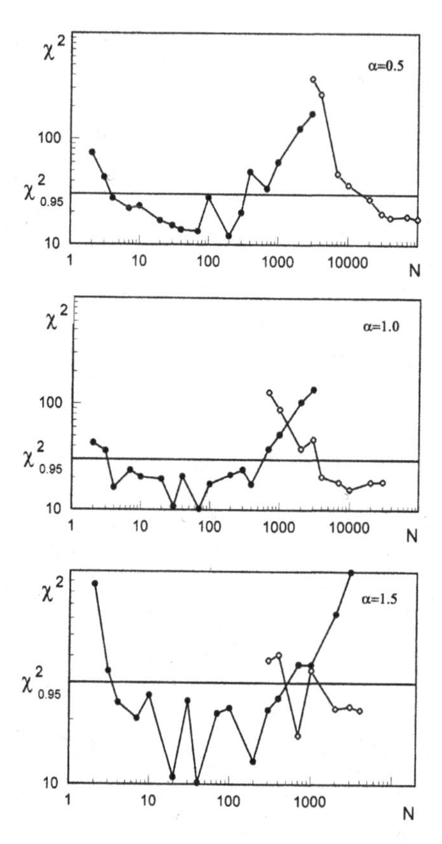

Figure 6.2 χ^2-test for the transition of sum distributions from the Lévy stable distribution (filled circles) to the Gaussian one (empty circles).

to stress that the property characterizes in asymptotic regime any renewal process with a finite mean waiting time.

Performing such simulation for fractional Poisson process, we can see perhaps the most specific property of fractional processes, called *intermittency*: the proportion of empty bins does not vanish with μ, but tends to a finite limit value depending on the order ν. In other words, in fractional processes empty bins are observed at all scales even if the total number of events on the interval under consideration becomes very large. These events form clusters on the time axis with voids between them.

The same behavior is observed in the case of truncated Lévy waiting times but up to some scale size only. When observation time exceed this limit, the distribution tends to be homogeneous.

Fig. 6.5 illustrates the transition from fractionally stable statistics to the Gaussian one in the case of a one-sided random walk with waiting time distribution having truncated power law tail. The upper graph is for exponentially distributed

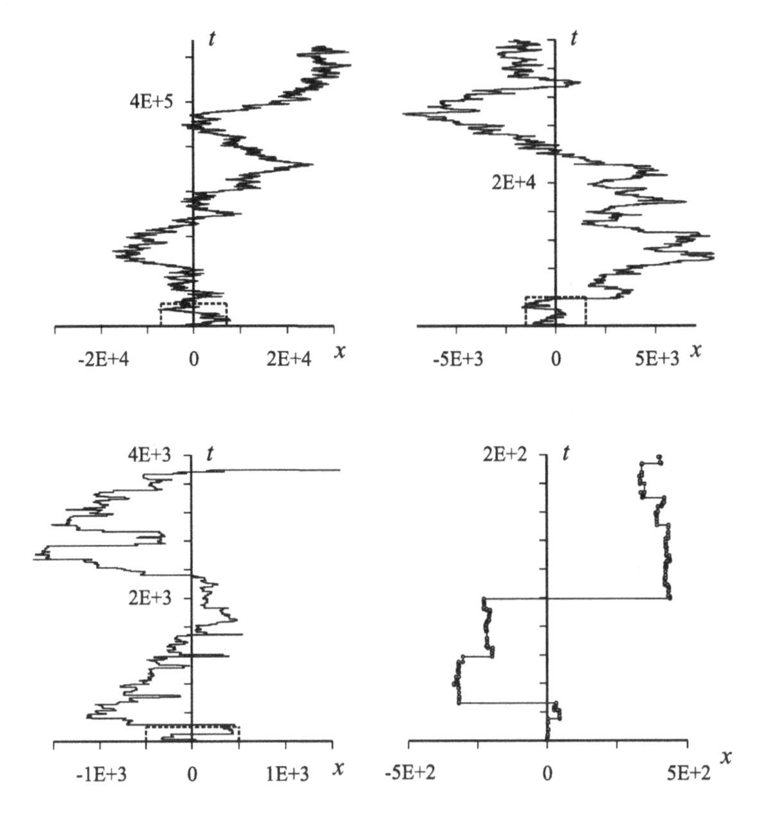

Figure 6.3 The trajectory of truncated Lévy flights (one realization) on different scales.

times between jumps. In the middle graph waiting times are not truncated, the exponent of the power law tail is equal to 0.5. In the lower graph waiting time distribution has the sharp truncation at time $\tau_{tr} = 1500$ (arbitrary units). The exponential distribution of waiting times has the same mathematical expectation as truncated power law in the lower graph. In the case of exponentially distributed waiting times, we see fast convergence to the Gaussian distribution, in the second one the distribution becomes the fractional stable one at some time t and remain at following times. In the case of truncated power law tails, the crossover between fractional stable and Gaussian statistics is observed.

6.4 Concluding remarks

Thus, fractional stable distributions play a role of intermediate asymptotics towards the Gaussian distribution for CTRF process with waiting time and path length characterizing by distributions with truncated power law tails. For example in the case of truncated Lévy flights, from physical point of view, the number n can be interpreted as a size of a system, and transition from large n's to small n's

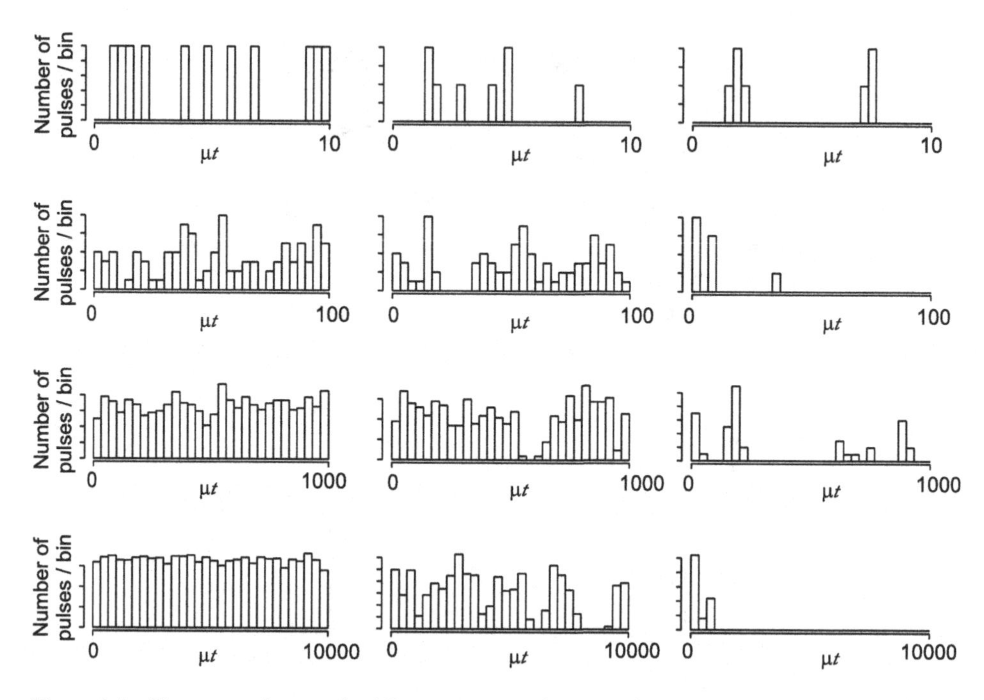

Figure 6.4 Histograms for standard Poisson process (left panel) and fractional Poisson processes with $\omega = 0.9$ and $\omega = 0.5$ (middle and right panels).

as transition from macroscale to mesoscale (nano) scale. In some sense, this is a demonstration of the scale correspondence principle: *the theoretical results should convert from the classical form to the fractional one without any special corrections "by hand"*.

Distributions with truncated power law tails are quite widespread in nanodynamics. For example, on-intervals distributed according to such law had been observed in a signal of blinking quantum dot fluorescence by Shimizu and coauthors [Shimizu et al. (2001)]. They consider this behavior as a temperature-dependent "saturation effect" that alters the long time tail of the distribution, the saturation arises due to a secondary mechanism that limits the maximum of on-state duration. In the subrecoil laser cooling process, pdf of recycling times may be truncated due to optical friction forces that limit large values of momentum [Bardou et al. (2002)]. Authors [Maruyama & Murakami (2003)] had demonstrated with the help of molecular dynamics simulation that the fast stick-slip diffusion of a nanocluster bound weakly to an atomically flat surface is a truncated Lévy walk.

Fractional stable distribution as asymptotic solution of the one-dimensional CTRF-model is a function of two variables, coordinate and time. Investigation of a crossover from non-Gaussian Lévy statistics to the Gaussian one with the help of fractional stable distributions allows to follow this phenomena as the scale (time or size) effect. For example, in Fig. 6.5 it has been demonstrated that fractionally

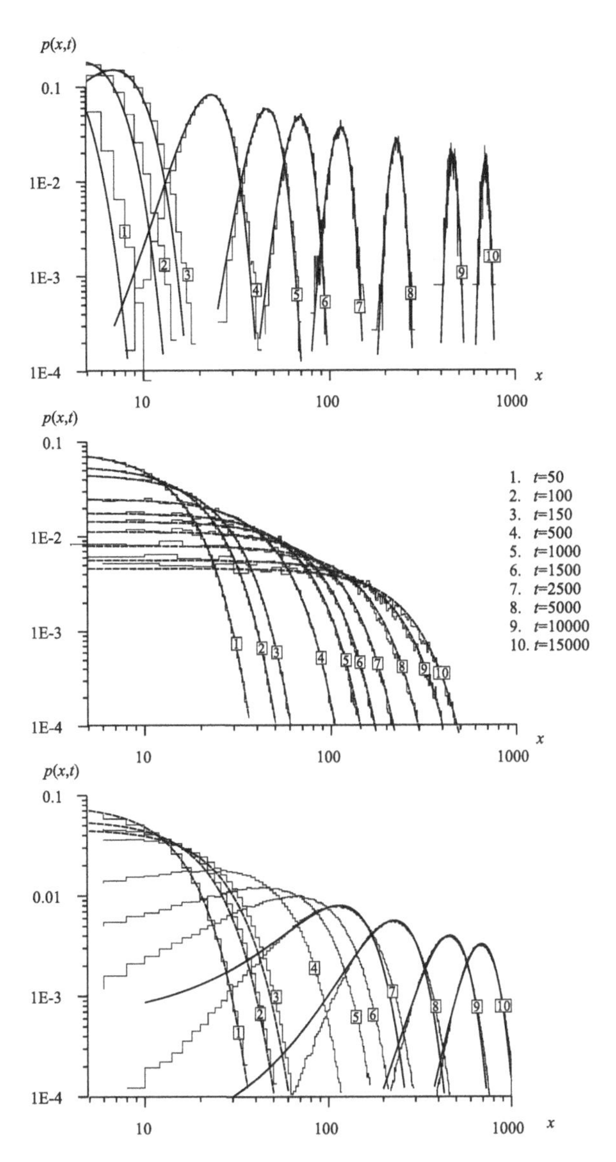

Figure 6.5 The transition from fractionally stable statistics to the Gaussian one in the case of a one-sided random walk with waiting times distributed according to law having truncated power law tail. Solid lines represent Gaussian pdfs, dotted lines correspond to fractionally stable densities.

stable statistics reduces to normal one at large times in the case of truncated power law tailed distribution of waiting times. The trajectory of truncated Lévy flights transforms its form from the Lévy type to the Brownian one with increasing of observable scale (Fig. 6.3).

Appendix A

One-sided stable laws

One-sided stable densities $g_+(t;\alpha)$, $\alpha < 1$, comprise a subset of the family of strictly stable laws defined in the following way: the density $g(t)$ is strictly stable if and only if for any two positive numbers b_1 and b_2 there exists a positive number b such that

$$\frac{1}{b_1 b_2} g(t/b_1) * g(t/b_2) = \frac{1}{b} g(t/b).$$

In other words, the shape of strictly stable distributions is invariant under convolutions (the best known representative of this class of distributions is the Gaussian distribution, corresponding to $\alpha = 2$, but it is not a member of the subset of one-sided distributions considered here).

The characteristic functions of one-sided stable distributions have the simple form (form C)

$$\varphi^{(\alpha)}(k) = \int_0^\infty e^{ikt} g_+(t;\alpha)dt = \exp\{-|k|^\alpha \exp[-(\alpha\pi/2)k/|k|]\}$$

According to Lemma 2.2.1 of Ref. [Schneider (1987)], the analytic continuation of the function $\varphi^{(\alpha)}(k)$ from the entire real k axis to the complex z plane with a cut along the ray $\arg z = -(3/4)\pi$ is given by the function

$$\varphi^{(\alpha)}(z) = \exp\{-(-iz)^\alpha\}, \quad \alpha < 1,$$

which implies that

$$g_+(\lambda;\alpha) = \int_0^\infty e^{-\lambda t} g_+(t;\alpha)dt,$$

the Laplace transform of the one-sided stable density $g_+(t;\alpha)$, has the form

$$g_+(\lambda;\alpha) = \varphi^{(\alpha)}(i\lambda) = \exp\{-\lambda^\alpha\} \tag{A.1}$$

The Mellin transform of the stable density can be expressed in terms of the ratio of two gamma functions [Uchaikin & Zolotarev (1999)]:

$$g_+(s;\alpha) \equiv \int_0^\infty t^s g_+(t;\alpha)dt = \Gamma(1-s/\alpha)/\Gamma(1-s). \tag{A.2}$$

We have the relationship

$$\frac{1}{\sqrt{4\pi}}\int_0^\infty e^{-r^2\tau^\alpha/4}g_+(\tau;\alpha)\tau^{\alpha/2}d\tau = \alpha^{-1}r^{-1-2/\alpha}g_+(r^{-2/\alpha};\alpha/2)\,, \tag{A.3}$$

whose validity can easily be proved by taking the Mellin transform and using (A.2).

The density $g_+(t;\alpha)$ can be expressed in terms of elementary functions only if $\alpha = 1/2$:

$$g_+(t;1/2) = \frac{1}{2\sqrt{\pi}}t^{-3/2}\exp\left(-\frac{1}{4t}\right).$$

This is the Lévy distribution, also known as the Smirnov distribution (named after N. V. Smirnov). At $\alpha = 1/3$ and $\alpha = 2/3$ the stable density (1.72) can be expressed in terms of the modified Bessel function and Whittaker function [Zolotarev (1983)]:

$$g_+(t;1/3) = (3\pi)^{-1}t^{-3/2}K_{1/3}\left(2/\sqrt{27t}\right), \tag{A.4}$$

$$g_+(t;2/3) = \sqrt{3/\pi}t^{-1}e^{-u/2}W_{1/2,1/6}(u),$$

where $u = (4/27)t^{-2}$. When α is rational, $g_+(t;\alpha)$ can be represented by a finite sum of generalized hypergeometric functions, e.g., $g_+(t;3/4)=$

$$=-t^{-1}\left(\frac{8}{3\pi}\right)\sum_{n=1}^{3}\sin\left(\frac{3n\pi}{4}\right)z^n_2F_2\left(\frac{1}{3}+\frac{n}{4},\frac{2}{3}+\frac{n}{4};\frac{1}{2}+\frac{n(n-1)}{8},\frac{n(7-n)}{8};-z^4\right),$$

where $z = -(3/t)^{3/4}/4$. For an arbitrary function of α, the functions $g_+(t;\alpha)$ are related to the Fox functions by

$$g_+(x;\alpha) = \alpha^{-1}x^{-2}H^{10}_{11}\left(x^{-1}\left|\begin{matrix}(\ -1\ ,\ 1\ )\\(-\alpha^{-1},\ \alpha^{-1})\end{matrix}\right.\right).$$

For numerical purposes, it is convenient to represent the density as the integral of a nonoscillating function. This representation was obtained by Zolotarev (1983) by deforming the integration contour in (1.72) in a special way:

$$g_+(t;\alpha)=\frac{\alpha t^{1/(\alpha-1)}}{\pi(1-\alpha)}\int_{-\pi/2}^{\pi/2}U_\alpha(\varphi)\exp\left\{-t^{\alpha/(\alpha-1)}U_\alpha(\varphi)\right\}d\varphi,$$

where

$$U_\alpha(\varphi)=\left[\frac{\sin[\alpha(\varphi+\pi/2)]}{\cos\varphi}\right]^{\alpha/(1-\alpha)}\frac{\cos[\pi\alpha/2-(1-\alpha)\varphi]}{\cos\varphi}.$$

It is also convenient to expand the density in a series that convergence for any positive t,

$$g_+(t;\alpha) = \frac{1}{\pi}\sum_{n=1}^{\infty}\frac{(-1)^{n-1}}{n!}\Gamma(1+n\alpha)\sin(\pi n\alpha)t^{-n\alpha-1}, \tag{A.5}$$

and to find the leasing term in the asymptotic expansion when $t \to 0$

$$g_+(t;\alpha) = At^{-\gamma}e^{-bt^\delta}, \tag{A.6}$$

where

$$A = [2\pi(1-\alpha)]^{-1/2}\alpha^{\frac{1}{2(1-\alpha)}}, \quad \gamma = \frac{1-\alpha/2}{1-\alpha}, \quad b = \frac{\alpha}{(1-\alpha)\alpha^\delta}, \quad \delta = \frac{\alpha}{(1-\alpha)}$$

Formula (A.6), which is exact at $\alpha = 1/2$, provides a fairly accurate approximation in the middle of the interval $(0, 1)$.

The expectation value and higher-order moments of these densities are infinite, but moments of order $\mu < \alpha$ (including negative orders) exist and are given by Eq. (A.2).

Tables of the distribution functions

$$G^{(\alpha)}(t) = \int_0^t g_+(\tau;\alpha)d\tau$$

and densities can be found in Refs. [Bol'shev (1970); Holt & Crow (1973)], respectively. Note that Holt and Crow (1973) use form (A) of the stable distribution, which is related to form (B) of Bolshov et al. (1970) used here,

$$g_A^{(\alpha)}(t) = c(\alpha)g_+(c(\alpha)t;\alpha),$$

where $c(\alpha) = [\cos(\pi\alpha/2)]^{1/\alpha}$. Moreover, the second parameter β of stable laws is defined differently by these researchers: according to Holt and Crow (1973), $\beta = -1$, for a one-sided distribution on the positive semiaxis, while according to Bolshev et al. (1970). $\beta = 1$ for the same case. It may have been these differences that prevented Holt and Crow (1973) from comparing their results with the earlier results of Bolshev et al. (1970) (at least they attempted to do so at the end of their paper; see p. 163 in Ref. [Holt & Crow (1973)]). In any case, our comparison of the results of the two groups, with full allowance for the foregoing, has shown that there is good agreement.

Stable laws play the same role in the summation of independent random quantities with infinite variances as the ordinary Gaussian law does in the case of finite variances. In particular, if independent random quantities $T_i \leq 0$ are distributed with a density $q_0(t)$ satisfying the condition (1.54) for $\alpha < 1$, for large values of n the normalized sum

$$S_n = \sum_{i=1}^n T_i / [nB\Gamma(1-\alpha)]^{1/\alpha} \tag{A.7}$$

is distributed with a density $g_+(t;\alpha)$. In other words, in the asymptotic region of large n, the distribution density $q_0^{(n)}(t)$ of the sum $\sum_{i=1}^n T_i$ has the form

$$q_0^{(n)}(t) \sim [nB^*]^{-1/\alpha} g_+\left((nB^*)^{-1/\alpha}t;\alpha\right), \tag{A.8}$$

where $B^* = B\Gamma(1-\alpha)$.

Appendix B

Fractional stable distributions

B.1. Let $S(\alpha_1, \theta_1)$ and $S(\alpha_2, \theta_2)$ be mutually independent, strictly stable random quantities represented in form C (see for details [Uchaikin & Zolotarev (1999)]). The random quantity

$$Y(\alpha_1, \alpha_2, \theta_1, \theta_2, \mu) \stackrel{d}{=} S(\alpha_1, \theta_1) / [S(\alpha_2, \theta_2)]^{\langle \mu \rangle},$$

where

$$S^{\langle \mu \rangle} = |S|^\mu \operatorname{sign} S, \qquad -\infty < \mu < \infty,$$

and the corresponding density

$$p_Y(x; \alpha_1, \alpha_2, \theta_1, \theta_2, \mu) = \int_{-\infty}^{\infty} g\left(xy^{\langle \mu \rangle}; \alpha_1, \theta_1\right) g(y; \alpha_2, \theta_2) |y|^\mu \, dy.$$

was called *fractional stable.*

Fractional stable random quantities are not rare in the literature. In particular, the random quantity

$$Y(2, \alpha, 0, 1, -1/2) = S(2, 0) \sqrt{S(\alpha, 1)}, \qquad \alpha < 1,$$

is called *sub-Gaussian*, in a more general case,

$$Y(\alpha_1, \alpha_2, 0, 1, -1/\alpha_1) = S(\alpha_1, 0) [S(\alpha_2, 1)]^{1/\alpha_1}, \qquad 0 < \alpha_1 \le 2, \qquad \alpha_2 < 1,$$

is referred to as a *substable* random quantity. The following relationships are known:

$$Y(2, 2, 0, 0, 1) \stackrel{d}{=} S(1, 0),$$

$$Y(\alpha_1, \alpha/\alpha_1, 0, 1, -1/\alpha_1) \stackrel{d}{=} S(\alpha, 0),$$

$$Y(1, 1, 1, 0, 1) \stackrel{d}{=} S(1, 0).$$

It is understood that

$$Y(\alpha, 1, \theta, 1, \mu) \stackrel{d}{=} S(\alpha, 0).$$

The distribution of the random quantity

$$Z(\alpha, \omega, \theta) = S(\alpha, \theta) / [S(\omega, 1)]^{\omega/\alpha}, \qquad \omega < 1$$

obtained by Kotulsky [29] by asymptotically solving the Montroll-Weiss problem

$$q(x;\alpha,\omega,\theta) \equiv p(x;\alpha,\omega,\theta,1,\omega/\alpha) = \int_0^\infty g(xy^{\omega/\alpha};\alpha,\theta)y^{\omega/\alpha}dy$$

also belongs too the class of fractional stable distributions.

We introduced the term "fractional stable distributions" in Ref. [?], and their properties were studied in Refs [Uchaikin (2000)]-[Uchaikin (2002b)]. Let us the most important ones.

B.2. If $0 < \alpha < 1$ and $\theta = 1$, the density $q\,(x;\alpha,\omega,1)$ is not zero only on the positive semiaxis. Otherwise, it differs from zero on the entire real axis.

B.3. The following inversion property takes place:

$$q\,(-x;\alpha,\omega,\theta) = q\,(x;\alpha,\omega,-\theta)\,.$$

If $\theta = 0$,

$$q\,(-x;\alpha,\omega,0) = q\,(x;\alpha,\omega,0)\,,$$

i.e., a fractional stable distribution with $\theta = 0$ is symmetric with respect to the origin.

B.4. According to the inversion property, it is sufficient to consider the class of fractional stable densities for all α, ω and θ only on the positive semiaxis. This allows the introduction of the one-sided Mellin transform

$$\bar{q}\,(s;\alpha,\omega,\theta) = \int_0^\infty x^s q\,(x;\alpha,\omega,\theta)\,dx, \qquad -1 < \text{Re } s < \alpha,$$

which, in view of specific features of fractional stable random quantities, is a much more convenient tool for analysis than the traditional characteristic function. We employ this transform and use the expression for the Mellin transform of the stable density,

$$\bar{g}\,(s;\alpha,\theta) = \rho\frac{\Gamma\,(1+s)\,\Gamma\,(1-s/\alpha)}{\Gamma\,(1+\rho s)\,\Gamma\,(1-\rho s)}, \quad \rho = (1+\theta)/2,$$

to obtain

$$\bar{q}\,(s;\alpha,\omega,\theta) = \rho\frac{\Gamma\,(1+s)\,\Gamma\,(1-s/\alpha)\,\Gamma\,(1+s/\alpha)}{\Gamma\,(1+\rho s)\,\Gamma\,(1-\rho s)\,\Gamma\,(1+\omega s/\alpha)}.$$

B.5. A fractional stable distribution has moments of all orders only if $\alpha = 2$. In this case, the region of θ values shrinks to a single value, $\theta = 0$, the distribution becomes symmetric, odd moments vanish, and even moments are

$$m^{(2n)}\,(2,\omega,0) \equiv \int_{-\infty}^\infty x^{2n} q\,(x;2,\omega,0)\,dx = \frac{4^n n!\Gamma\,(n+1/2)}{\sqrt{\pi}\Gamma\,(n\omega+1)}.$$

At $1 < \alpha < 2$, the second and higher moments are infinite; at $\alpha \leq 1$, even the mean value is infinite.

B.6. At the origin, we have

$$q\left(0;\alpha,\omega,\theta\right)=\frac{\Gamma\left(1+1/\alpha\right)\Gamma\left(1-1/\alpha\right)}{\pi\Gamma\left(1-\omega/\alpha\right)}\cos\left(\theta\pi/2\right)$$

and

$$Q\left(0;\alpha,\omega,\theta\right)\equiv\int\limits_{-\infty}^{0}q\left(0;\alpha,\omega,\theta\right)=\left(1-\theta\right)/2.$$

We note that $q\left(x;\alpha,\omega,\theta\right)$ has an integrable singularity at the origin if $\alpha\leq 1$ and $\omega<1$.

B.7. If $\omega\to 1$, the densities $q\left(x;\alpha,\omega,\theta\right)$ become stable densities $g\left(x;\alpha,\theta\right)$.

B.8. If $\omega<1$ and $\theta>0$, then

$$q\left(x;\alpha,\omega,\theta\right)\sim g\left(x;\alpha,\theta\right)/\Gamma\left(1+\omega\right),\qquad x\to\infty,$$

i. e., the tails of the fractional stable densities are as heavy (in terms of their power-law behavior) as those of the stable densities.

B.9. At $\alpha=2$, a fractional stable density can be represented in terms of a one-sided stable density via the relationship

$$q\left(x;2,\omega,0\right)=\frac{1}{\omega\left|x\right|^{1+2/\omega}}g_{\omega/2,1}\left(\left|x\right|^{-2/\omega}\right).$$

B.10. In Ref. [Uchaikin (2002a)], the characteristic functions of fractional stable distributions were derived,

$$\widetilde{q}(k;\alpha,\omega,\theta)=E_{\omega}(-\psi(k;\alpha,\theta)),$$

where

$$E_{\omega}(x)=\sum_{n=0}^{\infty}\frac{x^{n}}{\Gamma(1+n\omega)},$$

is the Mittag-Leffler function, and

$$\psi(k;\alpha,\theta)=-|k|^{\alpha}\exp\{-i\alpha\theta(\pi/2)\mathrm{sign}k\}.$$

In the same study, inverse-power series for fractional stable densities

$$q(x;\alpha,\omega,\theta)=\sum_{n=1}^{\infty}(-1)^{n-1}\frac{\Gamma(n\alpha+1)}{\Gamma(n\rho)\Gamma(1-n\rho)\Gamma(1+n\omega)}x^{-n\alpha-1}$$

were obtained.

B.11. One-dimensional fractional stable densities can in a natural way be extended to a multidimensional case [Uchaikin (2002b)],

$$q_{d}(\mathbf{x};\alpha,\omega,\Gamma)=\int\limits_{0}^{\infty}g_{d}(\mathbf{x}y^{\omega/\alpha};\alpha,\Gamma)g_{1}(y;\omega,1)y^{d\omega/\alpha}dy,$$

to describe the distribution of the random vector

$$\mathbf{Z}_{d}(\alpha,\omega,\Gamma)\equiv\mathbf{S}_{d}(\alpha,\Gamma)\Big/[S_{1}(\omega,1)]^{\omega/\alpha}.$$

Appendix C

Fractional operators: main properties

C.1 Axiomatics (Ross, 1975)

Involving the regularization procedure into foundation of fractional calculus foundation leads to development of an axiomatic approach. Slightly changing the Ross axiomatic (Ross, 1975), we reformulate it as follows:

For every function $f(z)$, $z = x + iy$, of a sufficiently wide class, and every number ν, irrational, fractional or complex, a function $\mathsf{D}_z^\nu f(z) = g(z)$ should be assigned subject to the following criteria:

1. If $f(z)$ is an analytic function of the complex variable z, the fractional derivative $\mathsf{D}_z^\nu f(z)$ is an analytic function of ν and z.

2. The fractional operator must be linear:

$$\mathsf{D}_z^\nu[af(z) + bg(z)] = a\mathsf{D}_z^\nu f(z) + b\mathsf{D}_x^\nu g(z).$$

3. The operation of order zero leaves the function unchanged:

$$\mathsf{D}_z^0 f(z) = f(z).$$

4. If ν is a negative integer, say $\nu = -n$, $n = 1, 2, 3, \ldots$, then $\mathsf{D}_z^{-n} f(z)$ must produce the same result as ordinary n-fold integration.

5. The law of exponents for operators of arbitrary negative orders holds:

$$\mathsf{D}_z^\mu\ \mathsf{D}_z^\nu f(z) = \mathsf{D}_z^{\mu+\nu} f(z), \quad \mu,\ \nu < 0.$$

6. Fractional operator of a positive order $\nu > 0$ is determined as follows:

$$\mathsf{D}_z^\nu f(z) = \text{p.f.}\mathsf{D}_z^{-\nu} f(z).$$

C.2 Interrelations between fractional operators

$${}_a\mathsf{D}_x^\nu = \mathsf{D}_x^n\ {}_a\mathsf{D}_x^{\nu-n}$$

$${}_x\mathsf{D}_b^\nu = (-\mathsf{D}_x)^n\ {}_x\mathsf{D}_b^{\nu-n}$$

$$
\begin{aligned}
{}_a^{\nu}\mathsf{D}_x &= {}_a\mathsf{D}_x^{\nu-n}\mathsf{D}_x^n \\
{}_x^{\nu}\mathsf{D}_b &= {}_x\mathsf{D}_b^{\nu-n}(-\mathsf{D}_x)^n
\end{aligned}
$$

$$
{}_a\mathsf{D}_x^{\nu} f(x) = {}_a^{\nu}\mathsf{D}_x f(x) + \sum_{j=0}^{[\nu]} \frac{(x-a)^{j-\nu}}{\Gamma(1+j-\nu)} f^{(j)}(a+), \ \nu > 0
$$

$$
{}_a^{\nu}\mathsf{D}_x f(x) = {}_a\mathsf{D}_x^{\nu}\left[f(x) - \sum_{j=0}^{[\nu]} \frac{(x-a)^j}{j!} f^{(j)}(a+) \right], \ \nu > 0
$$

$$
{}_a\mathsf{D}_x^{\nu} f(x) = {}_a\mathsf{D}_x^{\nu} g(x) \Leftrightarrow f(x) = g(x) + \sum_{j=1}^{n} c_j x^{\nu-j}, \ n-1 < \nu \leq n
$$

$$
{}_a^{\nu}\mathsf{D}_x f(x) = {}_a^{\nu}\mathsf{D}_x g(x) \Leftrightarrow f(x) = g(x) + \sum_{j=1}^{n} c_j x^{n-j}, \ n-1 < \nu \leq n
$$

$$
\begin{aligned}
{}_a\mathsf{D}_b^{\nu} &= \mathsf{D}_x^n \ {}_a\mathsf{D}_b^{\nu-n} \\
{}_a^{\nu}\mathsf{D}_b &= {}_a\mathsf{D}_b^{\nu-n} \ \mathsf{D}_x^n \\
{}_a\mathsf{D}_b^{\nu} &= \frac{1}{2\cos(\nu\pi/2)} \left({}_a\mathsf{D}_x^{\nu} + (-1)^n \ {}_x\mathsf{D}_b^{\nu} \right) \\
{}_a^{\nu}\mathsf{D}_b &= \frac{1}{2\cos(\nu\pi/2)} \left({}_a^{\nu}\mathsf{D}_x + (-1)^n \ {}_x^{\nu}\mathsf{D}_b \right)
\end{aligned}
$$

C.3 The law of exponents

For

$$
\mu > 0, \ \nu > 0, \ m, \ n \text{ integers such that}
$$

$$
m - 1 < \mu \leqslant m, \quad n - 1 < \nu \leqslant n, \quad \mu + \nu < n
$$

$$
{}_a\mathsf{D}_x^{-\mu} {}_a\mathsf{D}_x^{-\nu} f(x) = {}_a\mathsf{D}_x^{-\mu-\nu} f(x)
$$

$$
{}_a\mathsf{D}_x^{\mu} {}_a\mathsf{D}_x^{-\nu} f(x) = {}_a\mathsf{D}_x^{\mu-\nu} f(x), \quad \nu \geqslant \mu
$$

$$
{}_a\mathsf{D}_x^{-\mu} {}_a\mathsf{D}_x^{\nu} f(x) = {}_a\mathsf{D}_x^{-\mu+\nu} f(x) - \sum_{j=1}^{n} {}_a f^{(\nu-j)}(a+) \frac{(x-a)^{\mu-j}}{\Gamma(\mu-j+1)}
$$

$$
{}_a\mathsf{D}_x^{\mu} {}_a\mathsf{D}_x^{\nu} f(x) = {}_a\mathsf{D}_x^{\mu+\nu} f(x) - \sum_{j=1}^{n} {}_a f^{(\nu-j)}(a+) \frac{(x-a)^{-\mu-j}}{\Gamma(-\mu-j+1)}
$$

where

$$
{}_a f^{(\nu)}(a+) = \lim_{\varepsilon \downarrow 0} {}_a\mathsf{D}_{a+\varepsilon}^{\nu} f(a+\varepsilon).
$$

C.4 Differentiation of a product

$$ {}_a\mathsf{D}_x^\nu[f(x)g(x)] = \sum_{k=0}^{\infty} \frac{\Gamma(\nu+1)}{k!\Gamma(\nu-k+1)} \, {}_af^{(\nu-k)}(x)g^{(k)}(x), $$

$$ {}_a\mathsf{D}_x^\nu\,[f(x)g(x)] = \sum_{k=-\infty}^{\infty} c\binom{\nu}{ck+\mu} \, {}_af^{(\nu-ck-\mu)}(x){}_ag^{(ck+\mu)}(x),\ 0<c\le,-\infty<\mu<\infty $$

$$ {}_a\mathsf{D}_x^\nu\,[f(x)g(x)] = \int_{-\infty}^{\infty} \frac{\Gamma(\nu+1)}{\Gamma(\mu+1)\Gamma(\nu+1-\mu)} \, {}_af^{(\nu-\mu)}(x){}_ag^{(\mu)}(x)d\mu. $$

C.5 Integration by parts

$$ \int_a^b u(x)\ {}_av^{(-\nu)}(x)dx = \int_a^b v(x)v_b^{(-\nu)}(x)dx,\ \nu>0 $$

$$ \int_a^b u(x)\ {}_av^{\nu}(x)dx = \int_a^b v(x)\ {}^{\nu}u_b(x)dx - \sum_{j=0}^{n-1}(-1)^{n+j}\ {}_av^{(\nu+j-n)}(x)u^{(n-1-j)}(x)\Big|_a^b $$

$$ \int_a^b u(x)\ v_b^{\nu}(x)dx = \int_a^b v(x)\ {}_a^{\nu}u(x)dx - \sum_{j=0}^{n-1}\ u_b^{(\nu+j-n)}(x)v^{(n-1-j)}(x)\Big|_a^b $$

$$ \int_a^b u(x)\ {}_a^{\nu}v(x)dx = \int_a^b v(x)\ u_b^{\nu}(x)dx + \sum_{j=0}^{n-1}\ u_b^{(\nu+j-n)}(x)v^{(n-1-j)}(x)\Big|_a^b $$

$$ \int_a^b u(x)\ {}^{\nu}v_b(x)dx = \int_a^b v(x)\ {}_au^{\nu}(x)dx + \sum_{j=0}^{n-1}(-1)^{n+j}\ {}_au^{(\nu+j-n)}(x)v^{(n-1-j)}(x)\Big|_a^b $$

$$ \int_a^b u(x)\ {}_av_b^{(\nu)}(x)dx = (-1)^n\int_a^b v(x)\ {}_a^{\nu}u_b(x)dx + \sum_{j=0}^{n-1}(-1)^{n+j}\ {}_au_b^{(\nu+j-n)}v^{(n-1-j)}(x)\Big|_a^b $$

$$ \int_a^b u(x)\ {}^{(\nu)}{}_av_b(x)dx = (-1)^n\int_a^b v(x)\ {}_au_b^{(\nu)}(x)dx + \sum_{j=0}^{n-1}(-1)^{j}\ {}_au_b^{(\nu+j-n)}v^{(n-1-j)}(x)\Big|_a^b $$

C.6 Generalized Taylor series

$$f(x) = \sum_{j=0}^{\infty} \frac{(x-a)^{\nu+j}}{\Gamma(\nu+j+1)}\, {}_a f^{(\nu+j)}(a), \quad \nu > 0,$$

$$f(x) = \sum_{j=0}^{\infty} \frac{(x-a)^{\alpha j}}{\Gamma(1+\alpha j)}\, {}_a f^{(\alpha j)}(a), \; 0 < \alpha \leq 1.$$

$$f(x+\xi) = c \sum_{k=-\infty}^{\infty} \frac{1}{\Gamma(ck+\mu+1)}\, {}_a f^{(ck+\mu)}(x)\xi^{ck+\mu}, \quad 0 < c \leq 1.$$

$$f(x) = \frac{1}{2\pi i} \int_{\mu-i\infty}^{\mu+\infty} \Gamma(z)\, {}_0 f^{(-z)}(\infty) x^{-z} dz, \quad x > 0, \quad \mu > 0.$$

$$f(x) = \int_{-\infty}^{\infty} \frac{{}_a f^{\mu}(x_0)}{\Gamma(\mu+1)} (x-x_0)^{\mu} d\mu.$$

C.7 Expression of fractional derivatives through the integers

$${}_a\mathsf{D}_x^{\nu} f(x) = \sum_{k=0}^{\infty} \frac{(x-a)^{k-\nu}}{\Gamma(1+k-\nu)} f^{(k)}(a) + R_n(x), \quad \nu > 0.$$

C.8 Indirect differentiation: chain rule

For a monotonic function $g(x)$ having a continuous derivative

$${}_a\mathsf{D}_{g(x)}^{-\mu} f(x) = \frac{1}{\Gamma(\mu)} \int_a^x \frac{f(\xi)g'(\xi)d\xi}{[g(x)-g(\xi)]^{1-\mu}}, \quad \mu > 0, \quad a \geq -\infty$$

$${}_a\mathsf{D}_{g(x)}^{\nu} f(x) = \frac{1}{\Gamma(n-\nu)} \left(\frac{d}{g'(x)dx} \right)^n \int_a^x \frac{f(\xi)g^{(\nu)}(\xi)d\xi}{[g(x)-g(\xi)]^{\nu-n+1}}.$$

$${}_{g(x)}\mathsf{D}_b^{\nu} f(x) = \frac{1}{\Gamma(n-\nu)} \left(-\frac{d}{g'(x)dx} \right)^n \int_x^b \frac{f(\xi)g'(\xi)d\xi}{[g(\xi)-g(x)]^{\nu-n+1}}.$$

$$ {}_{x^\sigma}\mathsf{D}^{\nu}_{\infty} f(x) = \frac{\sigma}{\Gamma(-\nu)} \int\limits_x^\infty \frac{f(\xi)\xi^{\sigma-1}d\xi}{(\xi^\sigma - x^\sigma)^{1+\nu}}, \quad \nu < 0, $$

$$ {}_{x^\sigma}\mathsf{D}^{\nu}_{\infty} f(x) = f(x), \quad \nu = 0 $$

$$ {}_{x^\sigma}\mathsf{D}^{\nu}_{\infty} f(x) = \frac{\sigma^{-n}}{\Gamma(n-\nu)} \left(-\frac{d}{x^{\sigma-1}dx}\right)^{n+1} \int\limits_x^\infty \frac{f(\xi)\xi^{\sigma-1}d\xi}{(\xi^\sigma - x^\sigma)^{\nu-n+1}}, \quad \nu > 0. $$

C.9 Fractional powers of operators and Lévy stable variables

Fractional powers of closed linear operators were first constructed by Bochner (1949) and subsequently Feller (1952), for the Laplacian operator. These constructions depend in an essential way on the fact that the Laplacian generates a semigroup.

Let A_α, $\alpha \in (0,1]$, be the infinitesimal generating operator of the continuous semigroup $\mathsf{T}_\alpha(t)$,

$$ \mathsf{T}_\alpha(t) = \exp\{t\mathsf{A}_\alpha\}, \quad t \ge 0, $$

and

$$ \mathsf{A}_\alpha = -(-\mathsf{A}_1)^\alpha, $$

then the semigroups T_α and $\mathbf{T}_1$ are linked with each other via relation

$$ \mathsf{T}_\alpha(t) = \int_0^\infty \mathsf{T}_1(\tau) g_+\left(\frac{\tau}{t^{1/\alpha}};\alpha\right)\frac{d\tau}{t^{1/\alpha}}. $$

From the formal point of view these identities look almost obvious,

$$ \int_0^\infty \mathsf{T}_1(\tau) g_+\left(\frac{\tau}{t^{1/\alpha}};\alpha\right)\frac{d\tau}{t^{1/\alpha}} = \int_0^\infty \mathsf{T}_1(zt^{1/\alpha}) g_+(z;\alpha)dz $$

$$ = \int_0^\infty \exp\left\{zt^{1/\alpha}\mathsf{A}_1\right\} g_+(z;\alpha)dz = \int_0^\infty \exp\left\{-z\left[t^{1/\alpha}(-\mathsf{A}_1)\right]\right\} g_+(z;\alpha)dz $$

$$ = \exp\left\{-\left[t^{1/\alpha}(-\mathsf{A}_1)\right]^\alpha\right\} = \exp\left\{-t(-\mathsf{A}_1)^\alpha\right\} = \exp\{t\mathsf{A}_\alpha\} = \mathsf{T}_\alpha(t). $$

So, we arrive at:

$$ (-\mathsf{A}_1)^{-\alpha} = \int_0^\infty \exp\left\{-t(-\mathsf{A}_1)^\alpha\right\} dt = \int_0^\infty \mathsf{T}_\alpha(t)dt. $$

For more substantial proof see (Yosida, 1965).

Thus,

$$ \mathsf{T}_\alpha(t) = \left\langle \mathsf{T}_1(t^{1/\alpha} S_+(\alpha)) f \right\rangle $$

and

$$ (-\mathsf{A})^{-\alpha} f = \int_0^\infty \left[\int_0^\infty \exp\left\{-z[t^{1/\alpha}(-\mathsf{A})]\right\} dt\right] f g_+(z;\alpha)dz $$

$$ = \left\langle \left[\int_0^\infty \exp\left\{-S_+(\alpha)[t^{1/\alpha}(-\mathsf{A})]\right\} dt\right] f \right\rangle. $$

Bibliography

Abramowitz M., Stegun I. A., eds. Handbook of Mathematical Functions with Formulas, Graphs, and Mathematical Tables, New York: Dover Publications (1972).

Alisultanov Z. Z., Meilanov R. P. Some features of quantum statistical systems with an energy spectrum of the fractional-power type. *Theoretical and Mathematical Physics* (2012).

Alisultanov Z. Z., Meilanov R. P. Thermophysical propeties of quantum-statistical system with fractional power-law spectrum. *J. Sib. Fed. Univ. Math. Phys.* 5 (2012) 349358.

Amanatidis I., Kleftogiannis I., Falceto F. and Gopar V. A. Conductance of one-dimensional quantum wires with anomalous electron wave-function localization. *Phys. Rev. B* 85 (2012) 235450.

Anderson P. W., 1958, Absence of diffusion in certain random lattices, *Phys. Rev.* **109**, 1492-1505.

Argall, F., Jonscher, A. K. Dielectric properties of thin films of aluminium oxide and silicon oxide. *Thin Solid Films* 2(3) (1968) 185-210.

Arkhincheev V. E. Charge relaxation in fractal structures. *JETP Letters* 52, 1007-1009 (1990).

Arkhincheev V. E. Anomalous diffusion and drift in the comb model of percolation clusters. *JETP* 100, 292-300 (1991).

Arkhincheev V. E., Baskin E. M. Anomalous diffusion and drift in a comb of percolation clusters *JETP* 100 (1991) 292.

Arkhincheev V. E. On the relation between conduction and diffusion in a random walk along self-similar clusters. *JETP Letters* 67, 518-520 (1998).

Arkhincheev V. E. On the drift of a random walk on self-similar clusters. *JETP* 115, 1016-1023 (1999).

Arkhincheev V. E. Random walk on hierarchical comb structures. *JETP* 115, 1285-1296 (1999).

Arkhipov V. I., Popova Yu. A., Rudenko A. I. The effect of multiple trapping of carriers on the transient current in amorphous semiconductors. *Fizika i Tekhnika Poluprovodnikov* 17 (1983) 1817.

Arkhipov V. I., Kazakova L. P., Lebedev E. A., Rudenko A. I. *Fizika i Tekhnika Poluprovodnikov* 22 (1988) 723.

Arkhipov V. I., Nikitenko V. R. Dispersive transport in materials with non-monotonic energetic distribution of localized states. *Sov. Phys. Semiconductors* 23 (1989).

Arkhipov V. I., Perova I. A. Non-Langevin recombination in disordered dielectrics. *J. Phys. D: Appl. Phys.* 26 (1993) 1301.

Arkhipov V. I., Rudenko A. I. Drift and diffusion in materials with traps. II. Non-equilibrium transport regime. *Philos. Mag. B* 45 (1982) 189.

Arkhipov V. I., Rudenko A. I., Andriesh A. M., Iovu M. S., Shutov S. D. *Nestatsionarnye Inzhektsionnye Toki v Neupo-ryadochennykh Tverdykh Telakh* (*Nonstationary Injection Currents in Disordered Solids*). Exec. Ed. S. I. Radautsan, Kishinev: Shtiintsa, 1983 (in Russian).

Arkhipov V. I., Rudenko A. I., Sessler G. M. Radiation-induced conductivity and charge storage in irradiated dielectrics. *J. Phys. D: Appl. Phys.* 26 (1993) 1298.

Aroutiounian V. M., Ghoolinian M. Zh., Tributsch H. Fractal model of a porous semiconductor. *Applied Surface Science* 162-163 (2000) 122.

Arutyunyan N. H. Plane contact problem of the creep theory. *Journal of Applied Mathematics and Mechanics* 23 (1959) 901-924.

Averkiev N. S., Kazakova L. P., Lebedev E. A., Rud' Y. V., Smirnov A. N., Smirnova N. N. Optical and electrical properties of porous gallium arsenide. *Semiconductors* 34 (2000) 732.

Averkiev N. S., Kazakova L. P., Piryatinskiy Y. P., Smirnova N. N. Transient photocurrent and photoluminescense in porous silicon. *Semiconductors* 37 (2003) 1214.

Averkiev N. S., Kazakova L. P., Smirnova N. N. Charge carrier transport in porous silicon. *Semiconductors* 36 (2002) 355.

Aydiner Ekrem. Anomalous rotational relaxation: A fractional Fokker-Planck equation approach. *Phys. Rev. E* 71 (2005) 046103.

Babenko Yu. I. Teplo- i massoperenos (Heat and Mass Transfer). Leningrad: Chimiya, 1986 (in Russian).

Bakunin O. G. The correlation and percolation properties of turbulent diffusion. *Physics-Uspekhi* 173 (2003) 757-768.

Balakrishnan V. Anomalous diffusion in one dimension. *Physica A* 132 (1985) 569.

Baleanu D., Guvenc Z. B., Tenreiro Machado J. A. *New Trends in Nanotechnology and Fractional Calculus Applications.* Springer (2010).

Baleanu D., Tenreiro Machado J.A. and Luo A.C J. (editors). *Fractional Dynamics and Control.* Springer, New York (2012).

Balescu R. Anomalous transport in turbulent plasmas and continuous time random walks. *Phys. Rev. E* 51 (1995) 4807-4822.

Bardou F. Rare events in quantum tunneling. *Europhys. Letters* 39 (1997) 239-244.

Bardou F., Bouchaud J.-P., Aspect A. and Cohen-Tannoudji C. *Lévy Statistics and Laser Cooling: How Rare Events Bring Atoms to Rest.* Cambridge (2002).

Baranovskii S. D., Zvyagin I. P., Cordes H., Yamasaki S., Thomas P. Electronic transport in disordered organic and inorganic semiconductors. *Journal of Non-Crystalline Solids* 299-302 (2002) 416-41

Bardou F., Bouchaud J.-P., Aspect A., Cohen-Tannoudji C. *Lévy Statistics and Laser Cooling.* Cambridge Univ. Press (2002).

Barkai E., Fleurov V., Klafter J. One-dimensional stochastic Lévy-Lorentz gas. *Phys. Rev. E* 61 (2000) 1164-1169.

Barkai E., Metzler R., Klafter J. From continuous time random walks to the fractional Fokker-Planck equation. *Phys. Rev. E* 61 (2000) 132.

Barkai E. Fractional Fokker-Planck equation, solution, and application. *Phys. Rev. E* 63 (2001) 046118-1.

Barkai E., Jung Y., Silbey R. Time-dependent fluctuations in single molecule spectroscopy: a generalized Wiener-Khintchine approach. *Phys. Rev. Lett.* 87 (2001) 207403.

Barthelemy P., Bertolotti J., Wiersma D. S. A Lévy flight for light. *Nature* 453 (2008) 495.

Bässler H. Charge transport in disordered organic photoconductors. A Monte Carlo simulation study. *Phys. Status Solidi B* 175 (1993) 15.

Beenakker C. W. J., Rejaei B. Nonlogarithmic repulsion of transmission eigenvalues in a disordered wire. *Phys. Rev. Letters* 71 (1993) 3689-3692.

Beenakker C. W. J., Rejaei B. Exact solution for the distribution of transmission eigenvalues in a disordered wire and comparison with random-matrix theory. *Phys. Rev. B* 49 (1994) 7499-7510.

Beenakker C. W. J. Random-matrix theory of quantum transport. *Rev. Mod. Phys.* 69 (1997) 731-808.

Beenakker C. W. J., Groth C. W., Akhmerov A. R. Nonalgebraic length dependence of transmission through a chain of barriers with a Lévy spacing distribution. *Phys. Rev. B* 79 (2009) 024204.

Bel G., Barkai E. Weak ergodicity breaking in the continuous-time random walk. *Phys. Rev. Letters* 94 (2005) 240602.

Bening V. E., Korolev V. Yu., Kolokol'tsov V. N., Saenko V. V., Uchaikin and Zolotarev V. M. Estimation of Parameters of Fractional Stable Distributions. *Journal of Mathematical Sciences* 123 (2004) 3722-3732.

Biberman L. M. To theory of resonance radiation diffusion. *J. of Experimental and Theoretical Physics* 17 (1947) No 5.

Bisi O., Ossicini S., Pavesi L. Porous silicon: a quantum sponge structure for silicon based optoelectronics. *Surface Science Reports* 38 (2000) 1-126.

Bisquert J. Fractional diffusion in the multiple-trapping regime and revision of the equivalence with the continuous-time random walk. *Phys. Rev. Lett.* 91 (2003) 010602-1.

Bisquert J. Interpretation of a fractional diffusion equation with nonconserved probability density in terms of experimental systems with trapping or recombination. *Phys. Rev. E* 72 (2005) 011109.

Blom P. W. M., Vissenberg M. C. J. M. Dispersive Hole Transport in Poly(p-Phenylene Vinylene). Phys. Rev. Letters, 80 (1998) 3819.

Blumen A. Excitation transfer from a donor to acceptors in condensed media: a unified approach *Nuovo Cimento Soc. Ital. Fis. B* 63 (1981) 50-58.

Blumen A., Klafter J., and Zumofen G., 1986, Optical Spectroscopy of Glasses, ed. I.Zschokke, Reidel, Dodrecht.

Blumen A., Gurtovenko A., Jespersen, S. Anomalous diffusion and relaxation in macromolecular systems. *Journal of Non-Crystalline Solids* 305 (2002) 71-80.

Boden N., Bushby R. J., Clements J. Mechanism of quasionedimensional electronic conductivity in discotic liquid crystals. *J. Chem. Phys.* 98 (1993) 5920.

Bodurka J., Seitter R. O., Kimmich R., Gutsze, A. Field-cycling nuclear magnetic resonance relaxometry of molecular dynamics at biological interfaces in eye lenses: The Le'vy walk mechanism *J. Chem. Phys.* 107 (1997) 5621.

Bolshev L. N., Zolotarev V. M., Kedrova E. S., Rybinskaya M. A. Tables of the cumulative functions of one-side stable distributions. *Teor. Veroyatnost. i Primenen.* 15(2) (1970) 309319.

Bonch-Bruevich V. L., Kakashnikov S. G. *Physics of semiconductors.* M.: Science, 1977.

Von Borgnis F. Feste Dielectrika im electrishen Wechselfeld. *The European Physical Journal* 108 (1938) 107-127.

Bouchaud J. P., Georges A. Anomalous diffusion in disordered media: statistical mechanisms, models and physical applications. *Physics Reports* 195 (1990) 127–293.

Brouers F. Sotolongo-Costa O., Gonzalez A., Pirard J. Entropic origin of dielectric relaxation universalities in heterogeneous materials (polymers, glasses, aerogel catalysts). *Physica Status Solidi* (c) 2 (2005) 3529-3531.

Brouwer P. W. Quantum transport in disordered wires: Equivalence of the one-dimensional σ model and the Dorokhov-Mello-Pereyra-Kumar equation. *Phys. Rev. B* 53 (1996) 1490-1501.

Brown W. F. *Dielectrics.* Berlin: Springer-Verlag, (1956).

Bulyarsky S. B., Grushko N. S. *Generatzionno-rekombinatzionnye protzessy v aktivnykh elementakh.* (*Generation-recombination processes in the active elements.*) Moscow: MSU, 1995, 402 pp.

Bulyarsky S. V., Rud Yu. V., Vostretsova L. N., Kagarmanov A. S., Trifonov O. A. Tunnel recombination in semiconductor structures with nanodisorder. *Semiconductors* 43 (2009) 460.

Burioni R., Caniparoli L., Vezzani A. Lévy walks and scaling in quenched disordered media. *Phys. Rev. E* 81 (2010) 060101(R).

Calderwood J. H. The skewed arc representation for the frequency dependence of complex permittivity. *Journal of Molecular Liquids* 114 (2004) 59-62.

Caponetto R., Dongola G., Fortuna L., Petras I. *Fractional Order Systems: Modeling and Control Applications.* World Scientific Publishing Company, 2010.

Cartea Á., del-Castillo-Negrete D. On the fluid limit of the continuous-time random walk with general Lévy jump distribution functions. *Physical Review E* 76 (2007) 041105.

Cassele M. Distribution of transmission eigenvalues in disordered wires. *Phys. Rev. Lett.* 74 (1995) 2776-2779.

Chakraborty T. Quantum Dots: *A survey of the properties of artificial atoms.* Elsevier 1999.

Chandre C., Leonchini X., Zaslavsky G. (editors), *Chaos, Complexity and Transport, Theory and Applications.* World Scientific, Singapore (2008).

Chen W. Time-space fabric underlying anomalous diffusion. *Chaos, Solitons and Fractals* **28** (2006) 923-929.

Chen W., Sun H., Zhang X., and Korousak D. Anomalous diffusion modeling by fractal and fractional derivatives. *Computers and Mathematics with Applications* **59** (2010) 1754-1758.

Choudhury K. R., Winiarz J. G., Samoc M., Prasad P. N. Charge carrier mobility in an organic-inorganic hybrid nanocomposite. *Applied Physics Letters* 82 (2003) 406.

Chukbar K. V. Stochastic transport and fractional derivatives. *Journal of Experimental and Theoretical Physics* 81 (1995) 1025-1029.

Chung I., Bawendi M. G. Relationship between single quantum-dot intermittency and fluorescence intensity decays from collections of dots. *Phys. Rev. B* 70 (2004) 165304.

Coffey W. T., Kalmykov Yu. P., Titov S. V. Anomalous dielectric relaxation in the context of the Debye model of noninertial rotational diffusion. *Journal of Chemical Physics* 116 (2002) 6422-6426.

Cole K. S., Cole R. H. Dispersion and absorption in dielectrics. *Journal of Chemical Physics* 9 (1941) 341-350.

Compte A. Stochastic foundations of fractional dynamics. *Physical Review E* 53 (1996) 4191-4193.

Cox D. R., Smith V. L. *Renewal theory.* Sowiet radio, Moscow (1967).

Curie J. Recherches sur le pouvoir inducteur specifique et la conductibilite des corps cristallises. *Ann. Chim. Phys.* 17 (1889) 385-434.

Das S. *Functional Fractional Calculus for System Identification and Controls.* Springer-Verlag Berlin Heidelberg (2008).

Davidson D. W., Cole R. H. Dielectric relaxation in glycerol, propylene glycol, and n-propanol. *Journal of Chemical Physics* 19 (1951) 1484-1490.

Debye P. *Polare Moleceln.* S. Hirzel, Leipzig, (1929).

Debye P. *Polar Molecules.* Dover, New York (1954).
Degiorgio V., Piazza R., Mantegazza F., Bellini T. Stretched-exponential relaxation of electric birefringence in complex liquids. *Journal of Physics: Condensed Matter*, 2 (1990) SA69-SA78.
Déjardin J.-L. Fractional dynamics and nonlinear harmonic responses in dielectric relaxation of disordered liquids. *Phys. Rev. E* 68 (2003) 031108.
Dorokhov O. N. Transmission coefficient and the localization length of an electron in N bound disordered chains. *JETP Letters* 36 (1982) 318-321.
Drndic M., Jarosz M. V., Morgan N. Y., Kastner M. A., Bawendi M. G. Transport properties of annealed CdSe colloidal nanocrystal solids. *Journal of Applied Physics* 92 (2002) 7498.
Dybiec B., Gudowska-Nowak E. and Hänggi P. Lévy-Brownian motion on finite intervals: Mean first passage time analysis. *Phys. Rev. E* **73** (2006) 046104.
Dykhne A. M., Kondratenko P. S., Matveev L. V. Impurity transport in percolation media. *JETP Letters* 80 (2004) 464.
Efetov K. B. Supersymmetry and theory of disordered metals *Adv. Phys.* 32 (1983) 53.
Efetov K. B., Larkin A. I. Kinetics of a quantum particle in long metallic wires. *Sov. Phys. JETP* 58 (1983) 444.
Efros A. L., Rosen M. Random telegraph signal in the photoluminescence intensity of a single quantum dot. *Phys. Rev. Letters* 78 (1997) 1110.
Emelyanova E. V., Arkhipov V. I. A model of photoinduced anisotropy in amorphous semiconductors. *Fizika i Tekhnika Poluprovodnikov* 32 (1998) 995.
Empedocles S. A., Bawendi M. G. Influence of spectral diffusion on the line shapes of single CdSe nanocrystallite quantum dots. *J. Phys. Chem. B* 103 (1999) 1826.
Enck R. G., Pfister G. In: *Photoconductivity and Related Phenomena*, Elseiver, New York (1976).
Ermakov S. M., Mikhailov G. A. *Statistical Modeling.* Nauka, Moscow, 1982 (in Russian).
Falceto F., Gopar V. A. Conductance through quantum wires with Lévy-type disorder: universal statistics in anomalous quantum transport. *Europhysics Letters* 92 (2010) 57014.
Feder E. *Fractals.* Moscow (1991).
Feldman Yu., Puzenko A., Ryabov Ya. Non-Debye dielectric relaxation in complex materials. *Chemical Physics* 284 (2002) 139-168.
Feller W. *An Introduction to Probability Theory and its Application*, Vol 1, John Wiley & Sons, Inc., New York (1967).
Feller W. *An Introduction to Probability Theory and its Application*, Vol 2, John Wiley & Sons, Inc., New York (1971).
Fernández-Marin A. A., Méndez-Bermúdez J. A., Gopar V. A. Photonic heterostructures with Lévy-type disorder: Statistics of coherent transmission. *Phys. Rev. A* 85 (2012) 035803.
Fischbein M. D., Drndic M. CdSe nanocrystal quantum-dot memory. *Applied Phys. Letters* (2005) 86 193106.
Forster T. Z. *Naturforsch. Teil A* 4 (1949) 321-342.
Fox C. *The G and H functions as symmetrical Fourier kernels.* Trans. Am. Math. Soc. 98 (1961) 395-429.
Frahm K. Equivalence of the Fokker-Planck approach and the nonlinear σ model for disordered wires in the unitary symmetry class. *Phys. Rev. Letters* 74 (1995) 4706-4709.
Frantsuzov P. A., Marcus R. A. Explanation of quantum dot blinking without the long-lived trap hypothesis. *Phys. Rev. B* 72 (2005) 155321.
Fröhlich H. *Theory of Dielectrics. Dielectric Constant and Dielectric Loss.* Oxford, (1949).

Fukunaga M. On initial value problems in fractional differential equations. *International Journal of Applied Mathematics* 9 (2002) 219-236.

Fus V., Yaan K. Recombination in a-Si: H. Temperature and field quenching of photoluminescence. In: *Amorphous silicon and related materials.*, Fritzsche (editor) H. (1991).

Gafvert U., Adeen L., Tapper M., Ghasemi P. and Jonsson B., "Dielectric spectroscopy in time and frequency domain applied to diagnostics of power transformers", Proceedings of the *6th International Conference on Properties and Applications of Dielectric Materials*, IEEE, Piscataway, NJ, USA, pp.825-830, 2000.

Gaman V. I. *Physics of Semiconductor Devices.* Tomsk: NTL (2000).

Gemant A. A method of analyzing experimental results obtained from elastoviscous bodies. *Physics* 7 (1936) 311-317.

De Gennes, P. G. *Scaling Concepts in Polymer Physics.* Cornell University Press, Ithaca (1979).

Ginger D. S., Greenham N. C. Charge injection and transport in films of CdSe nanocrystals. *Journal of Applied Physics* 87 (2000) 1361.

Ghosh P., Sarkar A., Meikap A. K., Chattopadhyay S. K., Chatterjee S. K., Ghosh M. Electron transport properties of cobalt doped polyaniline. *J. Phys. D: Appl. Phys.* 39 (2006) 3047.

Glarum S. H. Dielectric relaxation of isoamyl bromide. *J. Chern. Phys.* 33 (1960) 639-643.

Glöckle W. G., Nonnenmacher T. F. Fox function representation of non-Debye relaxation processes. *Journal of Statistical Physics* 71 (1993) 741-757.

Glöckle W. G., Nonnenmacher T. F. Fractional relaxation and the time-temperature superposition principle. *Rheologica acta* 33 (1994) 337-343.

Gnedenko B. V., Kolmogorov A. N. *Limit distributions for sums of independent random variables.* Addison-Wesley (1954).

Gorenflo R., Mainardi F. Fractional oscillations and Mittag-Leffler functions." In: Recent Advances in Applied Mathematics (RAAM '96; Proc. Int. Workshop, Kuwait, May 1996), Kuwait, 193-208.

Gorenflo R., Mainardi F. *Fractals and Fractional Calculus in Continuum Mechanics.* Editors: Carpinteri A. and Mainardi F. (Vienna: Springer) (1997) p. 223.

Gorenflo R., Mainardi F., Srivastava H. M. Special functions in fractional relaxation-oscillation and fractional diffusion-wave phenomena. In: D. Bainov (editor). *Proceedings VIII International Colloquium on Differential Equations.* Plovdiv, VSP, Utrecht (1998) 195202.

Gorenflo R., Luchko Yu., and Mainardi F. Wright functions as scale-invariant solutions of the diffusion-wave equation. *J. Comput. Appl. Math.* 118 (2000) 175191.

Grigolini P., Rocco A., and West B. J. Fractional calculus as a macroscopic manifestation of randomness, *Phys. Rev. E* **59** (1999) 2603-2613.

Grushko N. S., Loginova E. A., Potanakhina L. N. A tunnel recombination process in spatially heterogeneous structures. *Semiconductors*, (2005) 40, 5, 584-588.

Grushko N. S., Potanakhina L. N. The temperature dependence of the average length of a jump in structures based on InGaN/GaN. *Applied Physics* (2008), 1, 17-19.

Grünewald M., Thomas P. *Phys. Status Solidi B* 94, 125 (1979).

Gusarov G. G., Korobko D. A., Orlov V. A., Uchaikin V. V. Calculation of transient current in amorphous semiconductors. *Proceedings of the Ulyanovsk State University* (Physical series) 6 (1999) 26.

Guyomar D., Ducharne B., Sébald G. Dynamical hysteresis model of ferroelectric ceramics under electric field using fractional derivatives. *J. Phys. D: Appl. Phys.* 40 (2007) 6048-6054.

Guyomar D., Ducharne B., Sébald G. Time fractional derivatives for voltage creep in ferroelectric materials; theory and experiment. *J. Phys. D: Appl. Phys.* 41 (2008) 125410.

Hack M. N. Long time tails in decay theory. *Phys. Lett.* 90A (1982) 220-221.

Hartenstein B., Bässler H., Jakobs A., Kehr K. W. Comparison between multiple trapping and multiple hopping transport in a random medium. *Phys. Rev. B* 54 (1996) 8574-8579.

Havriliak S., Negami S. A complex plane analysis of α-dispersions in some polymer systems. *Journal of Polymer Science* 14 (1966) 99-117.

Heisenberg W., Die Rolle dor phänomenologischen Theorien im System der theoretischen Physik. In: *Preludes in Theoretical Physics* (in Honor of V. F. Weisskopf). Amsterdam (1966).

Heiss W. Dieter (editor). *Quantum Dots: a Doorway to Nanoscale Physics, Lect. Notes Phys.* Springer, Berlin Heidelberg (2005).

Hegger H., Huckestein B., Hecker Klaus, Janssen M., Freimuth A., Reckziegel G., Tuzinski R. Fractal conductance fluctuations in gold nanowires. *Phys. Rev. Lett.* 77 (1996) 3885-3888.

Herrmann R. *Fractional calculus. An introduction for Physicists.* World Scientific (2011).

Hilfer R. Foundations of fractional dynamics. *Fractals* 3 (1995) 211.

Hilfer R. *Applications of Fractional Calculus in Phisics.* World Scientific (2000).

Hilfer R. Fractional diffusion based on Riemann-Liouville fractional derivatives. *J. Phys. Chem B* 104 (2000) 3914.

Hippel A. R. *Dielectrics and Waves.* New York, 1954.

Holt D. R., Crow E. L. Tables and graphs of the stable probability density functions. *Journal of Research of the National Bureau of Standards* 77B (1973) 143.

Hvam J. M., Brodsky M. H. Dispersive transport and recombination lifetime in phosphorus-doped hydrogenated amorphous silicon. *Phys. Rev. Letters* 46 (1981) 371-374.

Inokuti M., Hirayama F. Influence of Energy Transfer by the Exchange Mechanism on Donor Luminescence. *Journal of Chemical Physics* 43 (1965) 1978-1989.

Isichenko M. B. Percolation, statistical topography, and transport in random media *Rev. Mod. Phys.* 64(4) (1992) 961.

Jackson W., Kakalios J. In Amorphous silicon and related materials – H. Fritzsche (editor). 1991.

Jonscher A. K. A new model of dielectric loss in polymers. *Colloid and Polymer Science* 253 (1975) 1231.

Jonscher A. K. The "universal" dielectric response. *Nature* 267 (1977) 673.

Jonscher A. K. A new understanding of the dielectric relaxation of solids. *Journal of Materials Science* 16 (1981) 2037-2060

Jonscher A. K. *Dielectric Relaxation in Solids.* Chelsea Dielectric Press, London (1983).

Jonscher A. K. *Universal Relaxation Law.* Chelsea-Dielectrics Press, London (1996).

Jung Y., Barkai E., Silbey R. J. Lineshape theory and photon counting statistics for blinking quantum dots: a Lévy walk process. *Chemical Physics* 284 (2002) 181.

Jurlewicz A., Weron K. Relaxation dynamics of the fastest channel in multichannel parallel relaxation mechanism. *Chaos, Solitons and Fractals* 11 (2000) 303.

Jurlewicz A., Weron K. Continuous-time random walk approach to modeling of relaxation: the role of compound counting processes. *Acta Physica Polonica B* 39 (2008) 1055-1066.

Kaczer B., Arkhipov V., Degraeve R., Collaert N., Groeseneken G., Goodwin M. Temperature dependence of the negative bias temperature instability in the framework of dispersive transport. *Appl. Phys. Lett.* 86 (2005) 143506-1.

Kakalios J., Jackson W. In: *Amorphous silicon and related materials.* H. Fritzsche (editor). Mir, Moscow (1991).

Kazakova L. P., Lebedev E. A. Transient current in the structures amorphous, porous semiconductor – crystalline semiconductor. *Semiconductors* 32 (1998) 187-191.

Kazakova L. P., Minbaeva M. G., Minbaev K. D. Charge Carrier Drift Mobility in Porous Silicon Carbide. *Semiconductors* 38 (2004) 1118.

Kilbas A. A., Srivastava H. M., Trujillo J. J. *Theory and Applications of Fractional Differential Equations.* Elsevier (2006).

Kimmich R. *NMR: Tomography, Diffusometry, Relaxometry.* Springer, Berlin (1997).

Khintchin A. Ya. *Mathematical Foundations of Statistical Mechanics.* Moscow-Leningrad, Gostechizdat (1943).

Klafter J., Blumen A. Models for dynamically controlled relaxation *Chem. Phys. Lett.* 119 (1985) 377- 382.

Klafter J., Shlesinger M. On the relationship among three theories of relaxation in disordered systems. *Proc. Nat. Acad. Sci. USA* Vol. 83, February (1986).

Klages R., Radons G., Sokolov I. M. (editors). *Anomalous Transport: Foundations and Applications*, Wiley-VCH, 2008.

Kliem H., Arlt G. A relation between dielectric distribution findings and structural properties of amorphous matter. *CEIDP Annual Report*, Vol. 56 (1987) 325331.

Klimontovich Yu. L. *Statistical Theory of Open Systems.* Kluwer, Dordrecht (1995).

Kohlrausch R. Theorie des Elektrischen Ru"ckstandes in der Leidener Flasche *Pogg. Ann. Phys. Chem.* 91 (1854) 179-214.

Kohlrausch R. Theorie des elektrischen Ru"ckstandes in der Leidener Flasche *Annalen der Physik und der Physikalischen Chemie* 91 (1854) 179.

Kohno H., Yoshida H. Multiscaling in semiconductor nanowire growth. *Phys. Rev.* E 70 (2004) 062601.

Kohno H. Self-Organized Nanowire Formation of Si-Based Materials. In: *One-Dimensional Nanostructures Lecture Notes in Nanoscale Science and Technology* Vol. 3 (2008) 61-78.

Kolokoltsov V., Korolev V. Yu., Uchaikin V. V. Fractional stable distributions. *J. Math. Sci* 105 (2001) 2569-2576.

Kolomietz B. T, Lebedev E. A., Kazakova L. P. Features of charge carrier transport in glassy As_2Se_3. *Semiconductors* 12 (1978) 1771.

Komatsu H. Fractional powers of operators, IV Potential operators, *J. Math. Soc. Japan* Vol. 21, 2 (1969) 221-228.

Koponen I. Analytic approach to the problem of convergence of truncated Lvy flights towards the Gaussian stochastic process. *Phys. Rev. E* 52 (1995) 1197-1199.

Korolev V. Yu., Uchaikin V. V. Some limit theorems for generalized renewal processes with heavy tails, *Proab. Theory and Applications* 45 (2000) 809-811.

Kotulski M. Asimptotic distributions of continuous-time random walks: a probabilistic approach. *J. Stat. Phys.* 81 (1995) 777-792.

Krylov N. S. and Fock V. A. On the two main interpretations of the uncertainty relation for energy and time, *Zh. Exper. Teoret. Fiz.* 17 (1947) 93-107 (in Russian).

Kubo R. *Statistical Mechanics. An Advanced Course with Problem and Solutions.* North-Holland Publishing Company, Amsterdam (1965).

Kuchler A., Bedel T. Dielectric Diagnosis of Water Content in Transformer Insulation Systems. *European Trans. Electr. Power* Vol. 11 (2001) pp. 65-68.

Kuno M., Fromm D. P., Hamann H. F., Gallagher A., Nesbitt D. J. "On"/"off" fluorescence intermittency of single semiconductor quantum dots. *J. Chem. Phys.* 115 (2001) 1028.

Kusnezov D., Bulgac A., and Do Dang G. Quantum Lèvy processes and fractional kinetics, *Phys. Rev. Lett.* 82 (1999) 1136-1139.

Lagutin A. A., Uchaikin V. V. Anomalous diffusion equation: Application to cosmic ray transport. *Bulletin of the Russian Academy of Sciences: Physics* 73 (2009) 561563.

Lakshmikantham V., Leela S., Vasundhara Devi J. *Theory of Fractional Dynamic Systems.* Cambridge Scientific Publishers (2009).

Lamperti J. An occupation time theorem for a class of stochastic processes. *Trans. Amer. Math. Soc.* 88 (1958) 380.

Laskin N. Fractional Poisson process. *Communications in Nonlinear Science and Numerical Simulation* 8 (2003) 201.

Lavenda B. H. *Statistical Physics. A Probabilistic Approach.* John Wiley and Sons, New York (1991).

Lax M. Fluctuations from the nonequilibrium steady state. *Rev. Mod. Phys.* 32 (1960) Sec. I, 25.

Leadbeater M., Falko V. I., Lambert C. J. Lévy flights in quantum transport in quasiballistic wires. *Phys. Rev. Lett.* 81 (1998) 1274-1277.

Lévy P. *Processes stochastiques et mouvement Brownien.* 2nd ed., Gauthier-Villars, Paris (1965).

Logvinova K., Neel M.-C. A fractional equation for anomalous diffusion in a randomly heterogeneous porous medium, *Chaos* 14 (2004) 982-987.

Logvinova K. and Neel M.-C. *In Solute spreading in homogeneous aggregated porous media, in Advances in Fractional Calculus,* Eds: J.Sabatier, O.P.Agraval and Machado, Dordrecht, the Netherlands, Springer (2007) 185-198.

Lorenzo A., Hartley T. Initialized fractional calculus *Int. J. Appl. Math.* 3, (2000) 249.

Luedtke W. D., Landmann U. Slip Diffusion and Lévy Flights of an Adsorbed Gold Nanocluster *Phys. Rev. Lett.* 82 (1999) 3836-3838.

Luo A. C. J., Afraimovich V. (editros) *Long-range Interactions, Stochasticity and Fractional Dynamics.* Higher Education Press, Beijing, Springer Heidelberg (2010).

Madan A., Shaw M. P. *The Physics and Applications of Amorphous Semiconductors.* Boston: Academic Press, 1988.

Main C., Bruggemann R., Webb D. P., Reynolds S. Determination of gap-state distributions in amorphous semiconductors from transient photocurrents using a fourier transform technique. *Solid State Commun.* 83 (1992) 401.

Mainardi F. and Gorenflo R. On Mittag- Leffler-type functions in fractional evolution processes. *J. Comput. Appl. Math.*, 118 (2000) 283299.

Mainardi F., Pagnini G. The Wright functions as solutions of the time fractional diffusion equations. *Appl. Math. Comput.* 141 (2003) 5162.

Mainardi F., Gorenflo R., and Vivoli A. Renewal processes of Mittag-Leffler and Wright type. *Fractional Calculus Appl. Anal.*, 8 (2005) 738.

Mainardi F. *Fractional Calculus and Waves in Linear Viscoelasticity. An Introduction to Mathematical Models.* Imperial College Press (2010).

Mandelbrot B. B. *The Fractional Geometry of Nature.* Freeman, New York (1983).

Mandelstam L. I., Leontovich M. A. To the theory of the absorption of sound in liquids, Zh. Eksp. Teor. Fiz. [Sov. Phys. JETP] 7 (1937) 438-453.

Mantegna R. N., Stanley H. E. Stochastic Process with Ultraslow Convergence to a Gaussian: The Truncated Lévy Flight. *Phys. Rev. Lett.* 73 (1994) 2946-2949.

Margolin G., Berkowitz B. Spatial behavior of anomalous transport. Phys. Rev. E 65 (2002) 031101.

Margolin G., Berkowitz B. Application of continuous time random walks to transport in porous media, *J. Phys. Chem. B* 104 (2000) 39423947 [with a minor correction, published in *J. Phys. Chem. B* 104 (2000) 8762].

Margolin G., Berkowitz B. Continuous time random walks revisited: First passage time and spatial distributions. Physica A 334 (2004) 4666.

Margolin G., Barkai E. Aging correlation functions for blinking nanocrystals, and other on-off stochastic processes. *J. Chem. Phys.* 121 (2004) 1566.

Margolin G., Barkai E. Nonergodicity of blinking nanocrystals and other L'evy-walk processes. *Phys. Rev. Lett.* 94 (2005) 080601.

Martin J. E., Adolf D., Wilcoxon J. P. Viscoelasticity near the sol-gel transition *Phys. Rev. A* 39 (1989) 1325.

Maruyama Y., Murakami J. Truncated Lévy walk of a nanocluster bound weakly to an atomically flat surface: Crossover from superdiffusion to normal diffusion. *Phys. Rev. B* 67 (2003) 085406-1-5.

Mathai A. M., Saxena R. K. *The H-function with Applications in Statistics and Other Disciplines.* Weley, New Delhi (1978).

Maxwell J.C. On the dynamical theory of gases *Philos. Trans. Roy. Soc.* 157 (1867) 49.

Meerschaert M. M., Scheffler H.-P. *Limit Distributions for Sums of Independent Random Vectors*, Wiley (2001).

Meerschaert M. M. & Scheffler H.-P. Limit theorems for continuous-time random walks with infinite mean waiting times. *J. Appl. Probab.* 41 (2004) 623-638.

Meilanov R. P. Generalized quantum kinetic equation for conducting disordered media *Russian Physics Journal* 32 (1989) 331-336.

Meilanov R. P., Sadykov S. A. Fractal model for polarization switching kinetics in ferroelectric crystals. *Technical Physics* 44 (1999) 595-596.

Mello P. A., Pereyra P., Kumar N. Macroscopic approach to multichannel disordered conductors. *Annals of Physics* 181 (1988) 290-317.

Metzler R., Klafter J., Sokolov I. Anomalous transport in external fields: Continuous time random walks and fractional diffusion equations extended. *Phys. Rev. E* 58 (1998) 1621.

Metzler R., Barkai E., Klafter J. Anomalous diffusion and relaxation close to thermal equilibrium: A fractional fokker-planck equation approach. *Phys. Rev. Lett.* 82 (1999) 3563.

Metzler R., Klafter J. Subdiffusive transport close to thermal equilibrium: From the Langevin equation to fractional diffusion *Phys. Rev. E* 61 (2000) 6308-6311.

Metzler R., Klafter J. The random walk's guide to anomalous diffusion: a fractional dynamics approach. *Physics Reports* 339 (2000) 1–77.

Metzler R., Compte A. Generalized Diffusion-Advection Schemes and Dispersive Sedimentation: A Fractional Approach. *J. Phys. Chem. B* 104 (2000) 3858-3865.

Metzler R. and Klafter J. The restaurant at the end of the random walk: recent developments in the description of anomalous transport by fractional dynamics, Physics A: Math Gen. 37 (2004) 161-208.

Mikhailov G. A., Voytischek A. V. *Numerical Statistical Simulation. Monte-Carlo Methods.* Academia, Moscow, 2006.

Miller K., Ross B. *An Introduction to the Fractional Calculus and Fractional Differential Equations.*– Jon Wiley inc., N.Y. (1993). p. 336.

Monin A. S. Equation of turbulent diffusion *Doklady AN SSSR* 105 (1955) 256-259.

Monroe D. Hopping in exponential band tails. *Phys. Rev. Lett.* 54 (1985) 146.

Montroll E. W., Weiss G. H. Random walks on lattices. II. *J. Math. Phys.* 6 (1965) 167.

Morgan N. Y., Leatherdale C. A., Drndic M., Vitasovic M., Kastner M. A., Bawendi M. G. Electronic transport in films of colloidal CdSe nanocrystals. *Phys. Rev. B* 66 (2002) 075339.

Mott N. F. Electrons in disordered structures. *Advances in Physics* 50 (2001) .

Mott N. F., Davis E. A. *Electronic Processes in Non-crystaUine Materials.* Oxford University Press, Oxford (1971).

Muttaliib K. A., Gopar V. A. Generalization of the DMPK equation beyond quasi one dimension. *Phys. Rev. B* 66 (2002) 115318.

Muttalib K. A., Klauder J. R. Generalized Fokker-Planck Equation for Multichannel Disordered Quantum Conductors. *Phys. Rev. Lett.* 82 (1999) 4272.

Nagase T., Naito H. Determination of free carrier recombination lifetime in amorphous semiconductors: application to the study of iodine doping effect in arsenic triselenide. *J. Non-Cryst. Sol.* 227-230 (1998) 824.

Nagase T., Kishimoto K., Naito H. High resolution measurement of localized-state distributions from transient photoconductivity in amorphous and polymeric semiconductors. *J. Appl. Phys.* 86 (1999) 5026.

Nakhushev A. M. Inverse problems of degenerating equations and the Volterra integral equations of the third kind. *Differentsial'nye Uravneniia* 10 (1974) 100-111.

Nakhushev A. M. *Fractional calculus and its applications.* Moscow: Fizmatlit, 2003.

Nakhusheva V. A. *Differential Equations of Mathematical Models of Non-Local Processes.* Nauka, Moscow (2006).

Naito H., Ding J., Okuda M. Determination of localized-state distributions in amorphous semiconductors from transient photoconductivity. *Appl. Phys. Lett.* 64 (1994) 1830.

Ngai K. L., White C. T. Frequency dependence of dielectric loss in condensed matter *Phys. Rev. B.* 20 (1979) 2475.

Ngai K. L., 1986, In *Non-Debye Relaxation in Condensed Matter*, eds. Ramakrishnan T. V. and Raj Lakschmi L., World Scientific, Singapore.

Nigmatullin R. R. On the theory of relaxation for systems with "remnant" memory. *Phys. Status Solidi B* 124 (1984) 389.

Nigmatullin R. R. To the theoretical explanation of the "universal response". *Phys. Stat. Sol.(b)* 123 (1984) 739-745.

Nigmatullin R. R. The realization of the generalized transfer equation in a medium with fractal geometry *Phys. Stat. Sol.* 133 425 (1986).

Nigmatullin R. R. Non-integer integral and its physical interpretation. *Theoretical and Mathematical Physics* 90, 242-251 (1992).

Nigmattulin R. R., Goncharov V. A., Ryabov Ya. E. Extended abstract of the *XXVII congress Ampere.* Kazan (1994) p. 251.

Nigmatullin R. R., Ryabov Ya. E. Cole-Davidson dielectric relaxation as a self-similar relaxation process. *Physics of Solid State* 39 (1997) 87.

Nigmatullin R. R., Osokin S. I., Smith G. The justified data-curve fitting approach: recognition of the new type of kinetic eqations in fractional derivatives from analysis of raw dielectric data. *J. of Physics D: Applied Physics* 36 (2003) 2281-2294.

Nigmatullin R. R. Theory of dielectric relaxation in non-cristalline solids: from a set of micromotions to the averaged collective motion in the mesoscale region. *Physica B* 358 (2005) 201-215.

Nigmatullin R. Sh., Vyaselev M. R., Oscillographic polarography with application of the stepwise voltage *Journ. Analytic Chemistry* 19 (1964) 545-552 (in Russian).

Nigmatullin R. Sh., Belavin V. A., 1964, An electrical fractionally differentiating and and integrating two-pole network, Proc. Kazan Aviation Institute, Issue 82, 58-65 (in Russian).

Nikitenko V. R., Bässler H. Tunneling controlled electroluminescence switching in bilayer light emitting diodes. *J. Appl. Phys.* 185 (1999) 6515-6519.

Nikitenko V. R., Nikolaenkov D. V., Kundina Yu. F., Tyutnev A. P., Saenko V. S., Pozhidaev E. D.Features of transient current in the dielectric in the case of pulse generation of excess carriers in the surface region. *Zhurnal nauchnoy i prikladnoy fotographii,* 2001, .46, 4, 44-51.

Nikitenko V. R., Tyutnev A. P., Saenko V. S., Pozhidaev E. D. The electric polarization of geminate pairs in polymer model with complete suppression of the initial recombination. *Russian Journal of Physical Chemistry B,* 2001, 20, 1, 98-103.

Nikitenko V. R., Salata O. V., Bässler H. Comparison of models of electroluminescence in organic double-layer light-emitting diodes. *J. Appl. Phys.*, 2002, v. 92, p. 2359-2367.

Nikitenko V. R., Jupton J. M. Recombination kinetics in wide gap electroluminescent conjugated polymers with on-chain emissive defects. *J. Appl. Phys.*, 2003, v. 93, p. 5973- 5977.

Nikitenko V. R. Thesis for Doctor's degree, Moscow, 2006.

Nikitenko V. R., Tyutnev A. P. Transient current in thin layers of disordered organic materials under conditions of non-equilibrium transport of charge carriers. *Semiconductors* 41 (2007) 1118.

Nikitenko V. R. *Non-stationary processes of transport and recombination of charge carriers in thin layers of organic materials.* MEPhI (2011).

Niss K., Jakobsen B. Dielectric and Shear Mechanical Relaxation in Glass Forming Liquids. A thorough analysis and experimental test of the DiMarzio-Bishop model. Master thesis (supervisor: Niels Boye Olsen), 2003, 140 pp.

Nonnenmacher T. F., Nonnenmacher D. J. F. Towards the formulation of a nonlinear fractional extended irreversible thermodynamics. *Acta Physica Hungarica* 66 (1989) 145–154.

Nonnenmacher T. F. and Metzler R. Applications of Fractional Calculus Techniques to Problems in Biophysics. In: *Applications of Fractional Calculus in Physics.* R. Hilfer (editor), 377-427, World Scientific, Singapore (2000).

Noolandi J. Multiple-trapping model of anomalous transit-time dispersion in *a*-Se. *Phys. Rev. B* 16 (1977) 4466.

Novikov V. V., Privalko V. P. Temporal fractal model for the anomalous dielectric relaxation of inhomogeneous media with chaotic structure. Phys. Rev. E 64 (2002) 031504.

Novikov D. S. *Transport in Nanoscale Systems.* (PhD-thesis, MIT) (2003) 225 p.

Novikov D. S., Drndic M., Levitov L. S., Kastner M. A., Jarosz M. V., Bawendi M. G. Lévy statistics and anomalous transport in quantum dot arrays. *Phys. Rev. B* 72 (2005) 075309.

Novikov V. V., Wojciechowski K. W., Komkova O. A., Thiel T. Anomalous relaxation in dielectrics. Equations with fractional derivatives. *Material Science – Poland* 23 (2005) 977.

Ochmann M., Makarov S. Representation of the absorption of nonlinear waves by fractional derivatives. *J. Am. Acoust. Soc.* 94 (1993) 392.

Oldham K., Spanier J. *The Fractional Calculus.*– Academic Press, N.Y.–London (1974) p. 243.

Orenstein J., Kastner M. A., Vaninov V. Transient photoconductivity and photo-induced optical absorption in amorphous semiconductors. *Philos. Mag. B* 46 (1982) 23.

Osadko I. S. Power-law statistics of intermittent photoluminescence in single semiconductor nanocrystals. *JETP Letters* 79 (2004) 522-526.

Osadko I. S. Blinking fluorescence of single molecules and semiconductor nanocrystals. *Physics Uspekhi* 49 (2006) 19.

O'Shaughnessy B., Procaccia I. Analytical solutions for diffusion on fractal objects. *Phys. Rev. Lett.* 54 (1985) 455-458.

Ovchinnikov A. A., Timashev S. F., Beliy A. A. *The kinetics of diffusion-controlled chemical processes.* – .: 1986.

Palmer R. G., Stein D., Abrahams E. S., Anderson P. W. Models of hierarchically constrained dynamics for glassy relaxation. *Phys. Rev. Lett.* 53 (1984) 958-961.

Paradisi P., Cesari R., Mainardi F., Tampieri F.. The fractional Fick's law for non-local transport processes. *Physica A* 293 (2001) 130.

Pfister G. Dispersive low-temperature transport in a-selenium. *Phys. Rev. Letters* 86 (1976) 271.

Pfister G., Scher H. Time-dependent electrical transport in amorphous solids: As_2Se_3. *Phys. Rev. B* 15 (1977) 2062.

Phillips J. C. Stretched exponential relaxation in molecular and electronic glasses. *Rep. Prog. Phys.* 59 (1996) 1133-1207.

Plonka A. Phenomenological Approach to Thermally Assisted Tunneling. *J. Phys. Chem. B* 104 (2000) 3804-3807.

Podlubny I. *Fractional Differential Equations.* Academic Press (1999).

Pollak M., Geballe T. H. Low-frequency conductivity due to hopping processes in silicon. *Phys. Rev.* 122 (1961) 1742-1753.

Porto M., Bunde A., Havlin S., Roman H. E. Structural and dynamical properties of the percolation backbone in two and three dimensions. *Phys. Rev. E* 56 (1997) 1667.

Potapov A. A. *Fractals in Radiophysics and Radiolocation: Topology of Sampling*, University Book Press (2005).

Prabhakar T. R. A singular integral equation with a generalized Mittag-Leffler function in the kernel. *Yokohama Math. J.* Vol. 19 (1971) pp. 7-15.

Pskhu A. V. *Partial differential equations of fractional order.* Nauka, Moscow (2005).

Raicu V. Dielectric dispersion of biological matter: Model combining Debye-type and universal responses. *Phys. Rev. E* 60 (1999) 4677.

Raju G. G. *Dielectrics in Electric Fields.* Dekker (2003).

Ramakrishnan T. V., Raj Lakshmi M. *Non-Debye Relaxation in Condensed Matter.* World Scientific, Singapore (1987).

Ramirez-Bon R., Sanchez-Sinencio F., Gonzalez de la Cruz G. and Zelaya O. Dispersive electron transport in polycrystalline films of CdTe. *Phys. Rev. B* 48 (1993) 2200.

Rao P. N., Schiff E. A., Tsybeskov L., Fauchet P. Photocarrier drift-mobility measurements and electron localization in nanoporous silicon *Chem. Physics* 284 (2002) 129.

Redner S., Kang K. Kinetics of the "scavenger" reaction. *J. Phys. A* 17 (1984) 451-455.

Rekhviashvili S. Sh. Transient electrical conductivity of polymers in the model with fractional integro-differentiation. *Physics of the Solid State* 49 (2007) 1522.

Repin O. N., Saichev A. I. Fractional Poisson law. *Radiophysics and Quantum Electronics* 43 (2000) 738.

Richardson L. F. Atmospheric diffusion shown on a distance-neighbour graph. *Proc. Roy. Soc.* 110 (1926) 709.

Van Roosebroeck W. Current-Carrier Transport and Photoconductivity in semiconductors with trapping. *Phys. Rev.* 119 (1960) 636.

Rudenko O. V. *Theoretical Foundations of Non-linear Acoustics.* Studies in Soviet Science Ser., Consultants Bureau, New York (1977).

Rudenko A. I., Arkhipov V. I. Trap-controlled transient current injection in amorphous materials. *J. Non-Cryst. Solids* 30 (1978) 163-189.

Rudenko A. I., Arkhipov V. I. III. Analysis of transient current and transit time characteristics. *Philos. Mag. B* 45 (1982) 209.

Rybicki J., Chybicki M. Multiple-trapping transient currents in thin insulating layers with spatially nonhomogeneous trap distribution *J. Phys.: Condensed Matter* 1 (1989) 4623.

Saha T. K. Review of modern diagnostic techniques for assessing insulation condition in aged transformers. *IEEE Transactions on Dielectrics and Electrical Insulation* Vol. 10 No. 5 (2003) pp. 903-917.

Saha T. K., Purkait P. Investigations of temperature effects on the dielectric response measurements of transformer oil-paper insulation system. *IEEE Transactions on Power Delivery* Vol. 23 No. 1 (2008) pp. 252-260.

Sahimi M. Non-linear and non-local transport processes in heterogeneous media: from long-range correlated percolation to fracture and materials breakdown. *Phys. Rep.* 306 (1998) 214.

Saichev A. I., Zaslavsky M. Fractional kinetic equations: solutions and applications Chaos 7 (1997) 753.

Samarsky A. A. *The theory of difference schemes.* – .: Nauka, 1989, 616 .

Samko S. G., Kilbas A. A., Marichev O. I. *Integrals and derivatives of fractional order and some applications.* Minsk: Nauka i Technika, (1987) [*Fractional Integrals and Derivatives – Theory and Application.* Gordon and Breach, New York (1993)].

Schaufler S., Schleich W. P., Yakovlev V. P. Keyhole look at Lévy flights in subrecoil laser cooling. *Phys. Rev. Lett.* 83 (1999) 3162.

Scher H., Lax M. Continuous time random walk model of hopping transport: Application to impurity conduction. *J. Non-Cryst. Solids* 8/10 (1972) 497.

Scher H., Lax M. Stochastic transport in a disordered solid. I. Theory. *Phys. Rev. B* 7 (1973) 4491–4502.

Scher H. and Montroll E. W. Anomalous transit-time dispersion in amorphous solids. *Phys. Rev. B* 12 (1975) 2455-2477.

Schmidlin F. W. Theory of multiple trapping. *Solid State Commun.* 22 (1977) 451.

Schneider W. R., Wyss W. *Fractional diffusion and wave equations. J. Math. Phys.* 30 (1989) 134.

Schneider W. R. In: *Lecture Notes in Physics*, Vol. 1250, Springer, Berlin (1987).

von Schweidler E. Studien ber anomalien im verhalten der dielektrika (Studies on the anomalous behaviour of dielectrics). *Ann. Phys., Lpz.* 24 (1907) 711-70 (in German).

Seki K., Wojcik M., Tachiya M. Fractional reaction-diffusion equation. *J. Chem. Phys.* 119 (2003) 7525.

Seki K., Murayama K. and Tachiya M. Dispersive photoluminescence decay by geminate recombination in amorphous semiconductors. *Phys. Rev. B* 71 (2005) 235212.

Seki K., Wojcik M., Tachiya M. Dispersive-diffusion-controlled distance-dependent recombination in amorphous semiconductors. *Phys. Rev. B* 71 (2005) 235212.

Shah K. S., Lund J. C., Olschner F., Bennett P., Zhang J., Moy L. P., and Squillante M. R. Electronic noise in lead iodide X-ray detectors. *Nucl. Instrum. Methods Phys. Res. A* 353 (1994) 85.

Shah K. S., Bennett P., Klugerman M., Moy L. P., Entine G., Ouimette D., and Aikens R. Lead iodide films for x-ray imaging. *Proc. SPIE* 3032 (1997) 395.

Shapiro F. R., Adler D. Equilibrium transport in amorphous semiconductors. *J. Non-Cryst. Solids* 74 (1985) 189.

Shimizu K. T., Neuhauser R. G., Leatherdale C. A., Empedocles S. A., Woo W. K., Bawendi M. G. Blinking statistics in single semiconductor nanocrystal quantum dots. *Phys. Rev. B* 63 (2001) 205316.

Scaife B. K. P. *Principles of Dielectrics.* Oxford University Press, London (1998).

Schlesinger M. Asymptotic solutions of continuous-time random walks. *J. Stat. Phys.* 10 (1974) 421.

Shlesinger M. F., Montroll E. W. On the WilliamsWatts function of dielectric relaxation. *Proc. Natl. Acad. Sci. USA* 81 (1984) 1280-1283.

Shlesinger M., Klafter J. Hierarchical and fractal properties of disordered systems. In: *Fractal in Physics.* L. Pietronero, E. Tosatti (editors). North Holland, Amsterdam (1986).

Shlesinger M. F., Zaslavsky G. M., Klafter J. Strange kinetics. *J. Nature* 363 (1993) 31.

Shklovsky B. I., Efros A. L. *Electronic properties of doped semiconductors.* Nauka, Moscow (1979).

Shutov S. D., Iovu M. A., Iovu M. S. The drift mobility of holes in thin films of glassy arsenic trisulfide. *Semiconductors* 13 (1979) 956.

Si-Ammour A., Djennoune S., Bettayeb M. A sliding mode control for linear fractional systems with input and state delays. *Communications in Nonlinear Science and Numerical Simulatation* 14 (2009) 2310.

Sibatov R. T., Uchaikin V. V. Fractional diffusion p-n-junction with dispersive transport. *Journal of the Voronezh State Technical University. Series of Physical-Mathematical Modeling* 8 (2006) 136.

Sibatov R. T., Uchaikin V. V. Fractional differential kinetics of charge transport in desordered semiconductors. *Semiconductors* 41 (2007) 346.

Sibatov R. T., Uchaikin V. V. Fractional differential approach to dispersive transport in semiconductors. *Physics Uspekhi* 52 (2009) 1019–1043.

Sibatov R. T., Uchaikin V. V. Fractional differential approach to dispersive transport in semiconductors. *Uspekhi Fizicheskikh Nauk*, 2009, 179, 10 1079-1104.

Sibatov R. T., Uchaikin V. V. Statistics counts photon fluorescence blinking of quantum dots. *Optics and Spectroscopy*, 2010, 108, 5, 761-767.

Sibatov R. T., Uchaikin V. V., Uchaikin D. V., Shulezhko V. V. Fractional differential equation for the dielectric medium with a frequency response Gavrilyak-Negami, *Nelineyniy Mir*, 9, 5. 294-300 (2011)

Sibatov R. T. Conductance distribution of linear chain of tunnel barriers with fractal disirder. *JETP Letters* 93 (2011).

Sibatov R. T. Statistical interpretation of transient current power-law decay in colloidal quantum dot arrays. *Physica Scripta* 84 (2011) 025701.

Sibatov R. T. Fractional theory of anomalous kinetics in disordered semiconductor and dielectric systems (in Russian), Thesis for a Docrot's degree (2012)

Silver M., Cohen L. Monte Carlo simulation of anomalous transit-time dispersion of amorphous solids. *Phys. Rev. B* 15 (1977) 3276.

Smyth C. P. *Dielectric Behavior and Structure.* New York (1955).

Sokolov I. M. Thermodynamics and fractional Fokker-Planck equations. *Phys. Rev. E* 63 (2001) 056111.

Sokolov I. M., Blumen A., Klafter J. Linear response in complex systems: CTRW and the fractional Fokker-Planck equations. *Physica A* 302 (2001) 268.

Sokolov I. M., Metzler R. Towards deterministic equations for Lévy walks: The fractional material derivative. *Phys. Rev. E* 67 (2003) 010101 (R).

Sokolov I. M., Klafter J., and Blumen A. Fractional kinetics. *Physics Today* 55 (2002) 48-54.

Sokolov I. M., Chechkin A. V., and Klafter J. Fractional diffusion equation for a power-law-truncated Lévy process. *Physica A* 336 (2004) 245-251.

Spear V. Transport with the participation of tails states zones in amorphous silicon. In: *Amorphous Silicon and Related Materials.* H. Fritzsche (editor) (1991).

Stanislavsky A. A. Subordinated Random Walk Approach to Anomalous Relaxation in Disordered Systems. *Acta Physica Polonica B* 34 (2003) 3649.

Stanislavsky A. A. Probabilistic interpretation of the integral of the fractional order. *Theoretical and Mathematical Physics* 138 (2004) 418-431.

Stanislavsky A., Weron K. Exact solution of averaging procedure over the Cantor set. *Physica* A 303 (2002) 57-66.

Stanislavsky A., Weron K. Two-time scale subordination in physical processes with long-term memory, *Annals of Physics* 323 (2007) 643-653.

Stefani F. D., Zhong X., Knoll W., Han M., Kreiter M. Memory in quantum-dot photoluminescence blinking. *New Journal of Physics* 7 (2005) 197.

Street R. A., Ready S. E., Lemmi F., Shah K. S., Bennett P., and Dmitriyev Y. Electronic transport in polycrystalline PbI_2 films. *Journal of Applied Physics* 86(5) (1999).

Street R. A., Song K. W., and Northrup J. E., Cowan S. Photoconductivity measurements of the electronic structure of organic solar cells. *Phys. Rev. B* 83 (2011) 165207.

Suvakov M., Tadić B. Modeling collective charge transport in nanoparticle assemblies *J. Phys.: Condens. Matter* 22 (2010) 163201.

Takeda M., Kimura K., Murayama K. Transient Photocurrent Studies on Amorphous and β-Rhombohedral Boron*J. of Solid State Chemistry* 133 (1997) 201.

Tang J., Marcus R. A. Mechanisms of fluorescence blinking in semiconductor nanocrystal quantum dots. *J. Chem. Phys.* 123 (2005) 054704.

Tang Yang, Wang Zidong, Fang Jian-an Pinning control of fractional-order weighted complex networks. *Chaos* 19 (2009) 013112.

Tarasov V. Fractional Generalization of Liouville Equations, arXiv:nlin/0312044v1 (2003).

Tarasov V. E. Fractional equations of Curie-von Schweidler and Gauss laws. *J. Phys.: Condens. Matter* 20 (2008) 145212.

Tarasov V. E. Universal Electromagnetic Waves in Dielectric. *Journal of Physics: Condensed Matter*, 20 (2008) 175223.

Tarasov V. E. *Fractional Dynamics: Applications of Fractional Calculus to Dynamics of Particles, Fields and Media.* Springer, 2010.

Tenreiro Machado J. A. A probabilistic interpretation of the fractional-order differentiation, *Fractional Calculus and Applied Analysis* 6 (2003) 73-80.

Tenreiro Machado J. A. and Galhano A. Fractional Dynamics: A Statistical Perspectives, Journal of Computational and Nonlinear Dynamics 3 (2008) 021201(1-5)

Tiedje T. In.: *The Physics of Hydrogenated Amorphous Silicon II. Electronic and Vibrational Properties.* – Edited by Joannoulos J. D. and Lucovsky G. – Springer-Verlag, 1984, 448 p.

Tiedje T., Cebulka J. M., Morel D. L., Abeles B. Evidence for exponential band tails in amorphous silicon hydride. *Phys. Rev. Lett.* 46 (1981) 1425.

Tiedje T., Rose A. W. A physical interpretation of dispersive transport in disordered semiconductors. *Solid State Commun.* 37 (1981) 49.

Tsallis C. Possible generalization of Boltzmann-Gibbs statistics. *J. Stat. Phys.* 52 (1988) 479-487.

Tsallis C. Introduction to Nonextensive Statistical Mechanics, Springer, 2009.

Tunaley J. K. E. A physical process for $1/f$ noise in thin metallic films. *Journal of Applied Physics* 43 (1972) 4783.

Tunaley J. K. E. Some stochastic processes yielding a $f^{-\nu}$ type of spectral density. *J. Appl. Phys.* 43 (1972) 4777.

Tunaley J. K. E. Conduction in a random lattice under a potential gradient. *J. Appl. Phys.* 43 (1972) 4783.

Tyutnev A. P. et al. Dielektricheskie Svoistva Polimerov v Polyakh Ioniziruyushchikh Izluchenii (Dielektriki i Radiatsiya, Kn. 5) [Di- electric Properties of Polymers in Ionizing Radiation Fields (Di- electrics and Radiation, Book 5)]. Moscow: Nauka (2005) (in Russian).

Tyutnev A. P., Nikitenko V. R., Kundina Yu. F., Saenko V. S., Pozhidaev E. D., Vannikov A. V. Theoretical issues of transport photo-and radiation-generated charge carriers in polymers. *Zhurnal nauchnoy i prikladnoy fotographii* 46 (6) (2001) 21-27 (in Russian).

Tyutnev A. P., Saenko V. S., Nikitenko V. R., Kundina Yu. F., Pozhidaev E. D., Vannikov A. V. The bipolar nature of the radiation conductivity of polystyrene. *Zhurnal nauchnoy i prikladnoy fotographii* 46 (6) (2001) 28-35 (in Russian).

Tyutnev A. P., Saenko V. S., Pozhidaev E. D., Kolesnikov V. A. Charge carrier transport in polyvinylcarbazole. *J. Phys.: Condens. Matter* 18 (2006) 6365.

Uchaikin V. V. Anomalous diffusion of particles with a finite free-motion velocity. *Theoretical and Mathematical Physics* 115 (1998) 496-501.

Uchaikin V. V. Subdiffusion and stable laws. *J. of Exper. and Theor. Phys.* 115 (1999) 2113–2132.

Uchaikin V. Montroll-Weiss problem, fractional equations, and stable distributions. *International Journal of Theoretical Physics* 39 (2000) 2087.

Uchaikin V. Anomalous transport equations and their application to fractal walking. *Physica A* 305 (2002) 205.

Uchaikin V. Fractional Stable Distributions 2. – *Nottingham Trent University Preprint No.12/02* (2002).

Uchaikin V. V. Non-Gaussian stable laws as the intrmediate asymptotics on the way to Gaussian one, *Surveys in Applied and Industrial Mathematics* 9 (2002) 477.

Uchaikin V. V. Multidimensional symmetric anomalous diffusion. *Chemical Physics* 284 (2002) 507-520.

Uchaikin V. V. Relaxation processes and fractional differential equations. *International Journal of Theoretical Physics* 42 (2003) 121-134.

Uchaikin V. V. Self-similar anomalous diffusion and stable laws. *Physics Uspekhi* 173 (2003).

Uchaikin V. V. Anomalous diffusion and fractional stable distributions. Journal of Experimental and Theoretical Physics, 97 (2003) 810-825.

Uchaikin V. V. Self-similar anomalous diffusion and stable laws. *Uspekhi Fizicheskikh Nauk* 173 (2003) 847.

Uchaikin V. V. (2004). Fractal walks and walks on fractals. *Technical Physics* 74 7, 123126.

Uchaikin V. V. The method of fractional derivatives. - Ulyanovsk: Artichoke, 2008, 510.

Uchaikin V. V., Korobko D. A. Fractal model of transport: small-angle approximation. *Technical Physics* 74 (2004) 12.

Uchaikin V. V. and Saenko V. V. Stochastic solution of partial differential equations of fractional orders. *Siberian Journal of Computational Mathemaics* 6 (2003) 197-203.

Uchaikin V. V., Sibatov R. T. (2004) Lévy walks on a one-dimensional fractal Lorentz gas with trapping atoms. // *Mathematics and Statistics Research Report*, 4/04, NG1 4BU, Nottingham, 1-14.

Uchaikin V. V., Sibatov R. T. Fractional derivatives in the theory of semiconductors. *Obozrenie prikladnoy i promishlennoy matematiki* 12 (2005) 540-542.

Uchaikin V. V., Sibatov R. T. Fractional differential model of recombination in disordered semiconductors. *Obozrenie prikladnoy i promishlennoy matematiki* 14 (2007) 938 (in Russian).

Uchaikin V. V., Sibatov R. T. In "Nonlinear Science and Complexity" , Edited by A. C. J. Luo, L. Dai, H. R. Hamidzadeh (World Scientific, Singapore) (2007).

Uchaikin V. V., Sibatov R. T. Fractional differential kinetics of dispersive transport as the consequence of its self-similarity. *JETP Lett.* 86 (2007) 512516.

Uchaikin V. V., Uchaikin D. V. About memory regeneration effect in dielectrics. *Scientist Notes of Ulyanovsk State University. Physical Series* 1(17) (2005) 14 (in Russian).

Uchaikin V. V., Uchaikin D. V. Proc. Int. Conf. on "Chaos, Complexity and Transport" (France, Marseille, 2007) p. 337.

Uchaikin V. V., Zolotarev V. M. Chance and Stability: Stable Distributions and their Applications – VSP, Ultrecht, The Netherlands, 1999.

Uchaikin V. V., Ambrozevich S. A., Sibatov R. T. *Proc. XI Int. Conf. on "Physics of Dielectrics"* (Sankt-Peterburg, 2008) p. 129.

Uchaikin V. V., Cahoy D. O., Sibatov R. T. Fractional processes: from Poisson to branching one. *International Journal of Bifurcation and Chaos* 18 (2008) 2717.

Uchaikin V. V., Sibatov R. T. Fractional theory for transport in disordered semiconductors. *Communications in Nonlinear Science and Numerical Simulation* 13 (2008) 715727.

Uchaikin V. V., Sibatov R. T. Fractionally stable laws: a new kind of statistics in nanodynamics. – Symposium on Fractional Signals and Systems, Lisbon, Portugal, November 4-6, 2009, pp. 1-46.

Uchaikin V. V., Sibatov R. T. Statistical model of fluorescence blinking. *Journal of Experimental and Theoretical Physics* 109 (2009) 537–546.

Uchaikin V. V., Sibatov R. T., Uchaikin D. V. Memory regeneration phenomenon in dielectrics: the fractional derivative approach. *Physica Scripta* 136 (2009) 014002.

Uchaikin V. V., Sibatov R. T. Fractionally stable laws: a new kind of statistics in nanodynamics. Proceedings of the Symposium on Fractional Signals and Systems, M. Ortigueira et.al. (eds.) Lisbon, Portugal, November 4-6, 2009.

Uchaikin V., Sibatov R. Fractional Boltzmann equation for multiple scattering of resonance radiation in low-temperature plasma. *J. Phys. A: Math. Theor.* 44 (2011).

Uchaikin V. V. Fractional derivatives for physicists and engineers. Vol. II. 2012.

Valdes-Parada F. J., Alberto Ochoa-Tapia J., Alvarez-Ramirez J. Effective medium equation for fractional Cattaneos diffusion and heterogeneous reaction in disordered porous media, *Physica A* 369 (2006) 318-328.

Valdes-Parada F. J., Alberto Ochoa-Tapia J., Alvarez-Ramirez J. Effective medium equation for fractional Ficks law in porous media. *Physica A* 373 (2007) 339-353.

Vannikov A. V., Matveev V. K., Sichkar V. P., Tyutnev A. P. *Radiation effects in polymers. Electrical properties.* Moscow, "Nauka", 1982.

Vázquez L. Fractional diffusions with internal degrees of freedom, *J. Comp. Math.* 21 (2003) 491-494.

Vázquez L. A fruitful interplay: from non-locality to fractional calculus, in: *Nonlinear Waves: Classical and Quantum Aspects*, F. Kh. Abdullaev, V. V. Konotop (editors). Kluwer Acad.Publ., the Netherlands (2004) 129-133.

Vlad M. O. An inverse scaling approach to a multi-state random activation energy model *Physica A* 184 (1992) 303.

Vlad M. O., Schönfisch B., Mackey M. C. Fluctuationdissipation relations and universal behavior for relaxation processes in systems with static disorder and in the theory of mortality. *Physical Review* E 53 (1996) 4703-4710.

Vlad M. O., Metzler R., Ross J. Generalized Huber kinetics for nonlinear rate processes in disordered systems: nonlinear analogs of stretched exponential, *Physical Review* E 57 (1998) 6497-6505.

Volterra V. *Lec'ons sur la Théorie Mathématique de la Lutte pour la Vie.* Paris, Gauthier-Villars, 1931.

Watt A., Eichmann T., Rubinsztein-Dunlop H., Meredith P. Carrier transport in PbS nanocrystal conducting polymer composites. *Applied Physics Letters* 87 (2005) 253109.

Weron A. and Weron K. Stable measures and processes in statistical physics, *Lecture Notes Math.* 1153, Springer, Berlin (1985) 440-452.

Weron K. How to obtain the universal response law in the Jonscher screened hopping model for dielectric relaxation. *Phys.: Condens. Matter* 3 (1991) 221.

Weron K., Jurlewicz A. Two forms of self-similarity as a fundamental feature of the power-law dielectric response. *J. Phys. A Math. Gen.* 26 (1993) 395-410.

Weron K., Kotulski M. On the equivalence of the parallel channel and the correlated cluster relaxation models. *Journal of Statistical Physics* 88 (1996) 1241.

Weron K., Jurlewicz A., Magdziarz M. HavriliakNegami response in the framework of the continuous-time random walk. *Acta Physica Polonica B* 36 (2005) 1855-1868.

West B., Deering W. Fractal physiology for physicists: Lévy statistics. *Physics Reports* 246 (1994) 1-100.

West B. J., Grigolini P., Metzler R., Nonnenmacher T. F. Fractional diffusion and Lévy stable processes. *Phys. Rev. E* 55 (1997) 99-106.

West B. J., Bologna M., Grigolini P. *Physics of Fractal Operators.* Springer-Verlag, New York (2002).

West B. In: *Fractals, Diffusion, and Relaxation in Disordered Complex Systems*, 2006.

Westerlund S. Dead matter has memory. *Physica Scripta* 43 (1991) 174.

Williams G., Watts D. C. Non-symmetrical dielectric relaxation behaviour arising from a simple empirical decay function *Trans. Faraday Soc.* 66 (1970) 80.

Wyss W. The fractional diffusion equation. *J. Math. Physics* 27 (1986) 2782-2785.

Xia Yuan, Wu Jichun, Zhou Luying. Numerical solutions of time-space fractional advectiondispersion equations. ICCES, vol.9, no.2, (2009)117-126.

Yosida K. *Functional Analysis*, Springer-Verlag (1980).

Young W., Pumir A., Pomeau Y. Anomalous diffusion of tracer in convection rolls *Phys. Fluids A* 1 (1989) 462.

Zaslavsky G. M. Fractional kinetic equation for Hamiltonian chaos. *Physica D* 76 (1994) 110-122.

Zaslavsky G. M. Chaos, fractionalkinetics, and anomalous transport. *Physics Reports* 371 (2002) 461580.

Zaslavsky G. M. *Chaos and Fractional Dynamics.* Oxford Univ. Press (2005).

Zhang W., Shimizu N. Damping Properties of the Viscoelastic Material Described by Fractional Kelvin-Voigt Model. *JSME Int. J.* C 42 (1999) 825.

Zolotarev V. M. *One-dimensional stable distributions.* .: Fizmatlit, 1983, 304.

Zolotarev V. M., Uchaikin V. V., Saenko V. V. Superdiffusion and stable laws. *JETP* 88 (1999) 780-787.

Zubov V. V., Zisin Yu. A., Tuturov Yu. F. et al. Polymer Science, Series A. Polymer Physics, 14, 12 (1972) 2634.

Zumofen G., Blumen A., Klafter J. Concentration fluctuations in reaction kinetics. *J. Chem. Phys.* 82 (1985) 3198-3206.

Zumofen G., Klafter J. Scale-invariant motion in intermittent chaotic systems. *Phys. Rev. E* 47 (1993) 851-863.

Zvyagin I. P. Kineticheskie Yavleniya v Neuporyadochennykh Polu- provodnikakh (Kinetic Phenomena in Disordered Semiconductors) Moscow: MSU (1984).

Index